Experiencing Animal Minds

Critical Perspectives on Animals

THEORY, CULTURE, SCIENCE, AND LAW

Critical Perspectives on Animals
THEORY, CULTURE, SCIENCE, AND LAW

Series editors: Gary L. Francione and Gary Steiner

The emerging interdisciplinary field of animal studies seeks to shed light on the nature of animal experience and the moral status of animals in ways that overcome the limitations of traditional approaches to animals. Recent work on animals has been characterized by an increasing recognition of the importance of crossing disciplinary boundaries and exploring the affinities as well as the differences among the approaches of fields such as philosophy, law, sociology, political theory, ethology, and literary studies to questions pertaining to animals. This recognition has brought with it an openness to a rethinking of the very terms of critical inquiry and of traditional assumptions about human being and its relationship to the animal world. The books published in this series seek to contribute to contemporary reflections on the basic terms and methods of critical inquiry, to do so by focusing on fundamental questions arising out of the relationships and confrontations between humans and nonhuman animals, and ultimately to enrich our appreciation of the nature and ethical significance of nonhuman animals by providing a forum for the interdisciplinary exploration of questions and problems that have traditionally been confined within narrowly circumscribed disciplinary boundaries.

The Animal Rights Debate: Abolition or Regulation? Gary L. Francione and Robert Garner
Animal Rights Without Liberation: Applied Ethics and Human Obligations, Alasdair Cochrane

Experiencing Animal Minds

An Anthology of Animal-Human Encounters

EDITED BY

Julie A. Smith
and Robert W. Mitchell

COLUMBIA UNIVERSITY PRESS

New York

Columbia University Press
Publishers Since 1893
New York Chichester, West Sussex
cup.columbia.edu

Library of Congress Cataloging-in-Publication Data
Experiencing animal minds : an anthology of animal-human encounters /
edited by Julie A. Smith and Robert W. Mitchell
p. cm. — (Critical perspectives on animals : theory, culture, science, and law)
Includes bibliographical references and index.
ISBN 978-0-231-16150-3 (cloth : acid-free paper) — ISBN 978-0-231-16151-0 (pbk. : acid-free paper — ISBN 978-0-231-53076-7 (e-book)
1. Animal psychology. 2. Human-animal relationships. 3. Cognition in animals. 4. Consciousness in animals. 5. Philosophy of mind. 6. Animal behavior. 7. Animal communication. Smith, Julie A. (Julie Ann), 1944 Oct. 27– II. Mitchell, Robert W., 1958–

QL785.E975 2013
591.5—dc23

2012012481

Columbia University Press books are printed on permanent and durable acid-free paper.
This book is printed on paper with recycled content.
Printed in the United States of America

c 10 9 8 7 6 5 4 3 2 1
p 10 9 8 7 6 5 4 3 2 1

Cover design: Martin Hinze
Photo image: Sarolta Bán.

Contents

Experiencing Animal Minds

Engaging Animal Minds

Matters of Perspective

ROBERT W. MITCHELL AND JULIE A. SMITH

One of the most well-known stories about perspective taking concerns an animal: the story about the six blind men and the elephant.[1] The story has various incarnations, but the gist is that each blind man makes contact with a different part of an elephant and decrees that the entire elephant is understandable on the basis of the part he touches. One point of the story is that we should not make judgments based on limited information; another is that human knowledge is perspectival—we approach things from our own limited perspectives. Importantly, the blind men in the parable do not communicate with each other or the elephant. Indeed, the elephant seems a passive object of their individualized conjecture. An objective of the current volume is to bring together many humans to touch one critical part of the elephant (and many other animals)—its own point of view. Thus our focus is the minds of animals, that is, their ways of apprehending themselves, others, and the world. We also propose to exemplify the variety of methods humans might use to try to do this.

The authors in the current volume strive to explore, analyze, query, and generally poke at the perspectives of animals. Some approach from a scientific or social-scientific perspective, others from an orientation derived from the humanities: literature, art, philosophy, history, or personal experience.

We present all methods as equally valid and invite readers to consider how the approaches might relate to each other. Whereas some readers may want to create an ultimate view of an animal by combining the rich set of diverse accounts offered herein, we suspect that perspectives are always to some degree independent, much as viewing the elephant from the back results in a different bit of knowing than viewing the elephant from the front. While we can combine perceptions to create knowledge of the whole elephant, we can never literally see the elephant clearly from all sides at once. In some cases we can only with difficulty hold one way of framing the subject while entertaining another. Still the attempt can be fun *and* informative.

Part 1: Living with Animals

Living with animals provides experiences with and knowledge of animals that are less likely to arise in other contexts where our relationship with animals may be less salient to us. Not only do we look at animals we live with, and they at us, but we each touch and smell and otherwise sense each other and also develop expectations about each other. Living and interacting with animals toward whom we feel affection heightens awareness of their fears, passions, needs, and interest in social involvement. We experience these animals as intentional agents with minds of their own; we collect anecdotes that point to the animals' capacities for companionability. Some of us come from long and rich experience with a single species. Davis has devoted her life to providing a sanctuary for chickens, those much maligned creatures viewed by most humans as so mentally vacuous that they are entitled to exploit them on a staggering scale. In Davis's presentation, one is clearly aware of her belief in a fundamental nature of chickens that calls the ethics of current human practices into question: their capacity for reciprocal relations. Davis lives with chickens as mutual agents who initiate contact and response. Less interested in precise calculations in an economy of exchange, Davis rather insists on a general idea of give and take as the definition of meaningful relationships.

Acutely attentive to the influence of representations of pets on our experiences of them, Schlosser is also interested in human-animal reciprocity. She describes how selected artists involve pets in the processes of making art in order to understand better their subjectivity and agency. Narrowing the field of exploration to artworks that examine touch-contact, Schlosser shows how these works offer complex understandings of the meaning of touch as a medium of exchange between animal and human participants as well as between them and the viewers of artworks. She locates her own work within this paradigm. Thus, while Davis explores reciprocity laterally, through one human life extended through time with one species, communicating her views through descriptions of episodes of mutuality, Schlosser proceeds vertically. She focuses on one kind of event within the domestic partnership and explores its rich variations as captured by a group of artists all intent on referencing its meanings. For both Davis and Schlosser, however, the domestic arrangement provides a space for interactions that will lead to insights about animal minds less likely to occur in other contexts.

As Braz's essay demonstrates, however, it is not necessary that animals be domesticated for humans to develop a rich appreciation of their lived experiences. People living with wild animals can also become attuned to nonhuman nuances even before they achieve a sensitizing affection. As Braz illustrates in his history of Grey Owl's writings about beavers, Grey Owl came to know beavers while he was a trapper. But, once attentive to them, he believed he saw regularities in their behavior that he took to be meaningful. He began to see them as communicating with him via something comparable to human language. Through this process, beavers transformed Grey Owl into an advocate on their behalf. As with the accounts of Davis and Schlosser, the Grey Owl story suggests that human engagement with animals allows people to discover an expressiveness on the part of animals, one that they can recognize and understand.

Part 2: Anthropomorphisms

The essays in this section trace the sources, manners, and effects of anthropomorphism, each focusing on a different context. Nineteenth-century British writers purposely sought to distinguish animal intelligence from human intelligence by employing the idea of sagacity. Boddice examines how authors used the concept, comparing the ways that dogs, foxes, and rats were represented as sagacious. All three had that mixture of intelligence, pluck, and savvy attention required for sagacity; but foxes and rats were seen to use sagacity to thwart humans and thus to deserve no pity.

Discussing primates, the essays of Waller and Desmond also address anthropomorphic representation. Waller presents the scientific study of primate psychology as derived more from popular cultural ideas about primates as humanlike than from scientific ideas about animal psychology emerging from studies of evolution and ecology. In Waller's analysis, cultural forms influence scientific representations, which influence what scientists look for in primates. In essence, the study of primate psychology is at the mercy of popular expectations. Desmond also sees ideas about primates in popular culture and science as mutually influential. Specifically, she sees scientific and popular discourses about primate art issuing from a shared desire to see the paintings as an animal protoaesthetic underlying the human capacity for expressive practice. She posits that such desire may be linked to the value of primate art within a transnational art market.

All the essays in part 2 attempt to identify a source and manner of anthropomorphism that has not been fully explored elsewhere. And each makes clear that if anthropomorphism means representing animals as "us," then the "us" can be as divergent as the animals themselves.

Part 3: Embodiments and Interembodiments

Authors in this section are divided in their attention to whole body and part body implications of embodiment. Argent, Warkentin, Smith, and Mitchell each focus on embodiment as traditionally conceived: the whole body is minded, inseparable from mental properties and a source for social understandings derived through encounters. All four present some kind of matching between bodies of individuals of the same or different species as a basis for psychological understanding between them. By contrast, Hart and Hart and Dillard-Wright assess the importance of individual systems within the body: respectively, the brain and the heart. Rohman explicitly acknowledges the minded and perceiving body in her view of poetry as "rooted in bodily experience" and art as functioning to "produce sensations."

The first four authors provide a response of sorts to claims of unknowability concerning nonhuman internal states and experiences. Argent presents horse-horse, human-horse, and human-human interactions as guided by nonverbal communication, directing particular attention to synchronous movement that develops between two species (as well as between horses) as evidence of attunement to others, a form of empathic intelligence apparently more pronounced in horses than in humans. Consistent with Argent's ethological directive of attending to interactive bodily activity as a source for understanding animal mind, Warkentin reads such activity using models provided by ecological psychology, cultural anthropology, and phenomenology. Applying these models to her observations of human-whale interactions in aquariums (whales inside, humans outside), she attempts to tease out the ways that interspecies bodily gestures in this context promote animal-human intersubjectivity or, conversely, lead to "toxic intercorporeity." Smith takes as her subject dog-human communication as conveyed by Cesar Millan's idea of energy in his books and television show *The Dog Whisperer*. She hypothesizes that Millan communicates with dogs and they with him by means of the feeling of movement each experiences when watching movement enacted by the other.

Mitchell examines the question of how we come to expect that others have inner experiences, and what we can know about these experiences, by presenting Strawson's response to Wittgenstein's argument for the unknowability of others' experiences. Strawson believed that our knowledge of other minds derives from our ability to match our internal experience to perceptions that are intersubjectively available. Following Strawson's argument, Mitchell posits that inner experiences are like perceptions, that our knowledge that others have inner experiences derives from our capacity for kinesthetic-visual matching, and that our understanding of others' inner experiences as directly experienced by them is always inherently incomplete whether we speak the same language or no language at all.

All the authors in this section assume that bodily organization clearly determines the expression of the mental. A dolphin cannot express joy in a wild canter as a horse does, and an ape cannot know an object through sound as does an echolocating bat. Hart and Hart posit that the structure and organization of the brain's neurons in elephants directly explain these animals' psychological skills and limitations. Dillard-Wright views the respiratory and circulatory subsystems (but particularly the latter, which includes the heart) as salient components of "thought," viewed liberally. Once again, as with other authors in this volume, meaning is derived from interaction, but now the interacting agencies are the heart and the mind, representing a physiological pas de deux.

Bodily pleasure is the indirect focus of Rohman's essay. Rohman, following Grosz, presents art as emerging from an experience of intense feeling, whether derived from perceptual stimulation or innate bodily energy. This excess or intensity of feeling leads to art in the form of revelry, joy, play. This idea of art makes it inherent in animal life. It is a decentering of art as essentially human and even of art as intentionally created. In Rohman's view, animal play is an exuberant exhibition of pleasure arising from the expenditure of excess bodily energy; and this suggests that, for animals, existence itself leads to art. In other words, the life spark leads to play's intense energy, which is expended in some form that should be called art. Rohman theorizes a protoaesthetic that Desmond examines more skeptically as a basis for the commodification of primate art.

Part 4: Animal Versus Human Consciousness

The authors in this section focus on limitations of animal mind and reevaluate its implications. Steiner posits that most nonhuman animals are, unlike humans, without abstract thought and rationality. Carefully defining "conceptual abstraction" and "predicative intentionality" as linguistically based, he argues that nonhuman animals do not have the same cognitive apparatus that humans do. In particular, they lack the mental capacities needed to form representational content separate from any particular perceptual experience. However, he believes that any human superiority derived from the lack of abstract thought in animals is irrelevant to ethical concerns about treatment of animals. Morin focuses on particular forms of abstract thought—self-reflection and self-rumination—he believes are absent in nonhuman animals and argues against the idea that these self-focused forms of thought are present in beings that lack inner speech.

Focusing on studies of caching behavior in scrub jays, Droege argues that, although the jays have sophisticated mental abilities, including the ability to reason, they are not conscious. Droege notes that the jays are able to discriminate when an event occurred, what happened during the event, and where the event was located; however the jays are not able to mentally reexperience a past event. Without this capacity, they lack a sense of time, which Droege maintains is the sine qua non of consciousness.

Part 5: Tailoring Representations to Audiences

Even if, as in the previous section, animals are said to lack abstract thought, language, self-reflection, or a sense of time, almost no one believes that they are nonmental entities (as Droege's essay attests). Authors in this penultimate section each attend to a particular method for presenting some quality of animal mind to some audience. Forms of presentation include experimental design, visual displays, and narrative. All the authors suggest that the form of presentation cannot be separated from the characterization of animal mind.

Lurz's essay addresses a central problem in comparative psychology—discerning whether or not apes have a theory of mind, that is, whether they interpret the activities of others as based on mental states. Lurz recounts that most of the research on apes' behaviors is uninterpretable: One cannot tell whether they have a theory of behavior and its consequences or a theory of mind. He proposes an experimental protocol that actually allows researchers to differentiate between these two mindsets, as it requires apes to recognize that what another sees from one vantage point differs from what it sees from another—evidence of having a theory of mind.

Ullrich discusses the methods artists use to understand and convey animal experience to human viewers. These methods include artistic observation and imitation, but Ullrich's primary focus is on identification and empathy as a goal of artistic production. She discusses artists who attempt to recreate artistically for the viewer a context for animal experience and to present it in such a way as the animal itself might encounter it. This includes eliciting in the viewer the feelings of a particular situation that an animal might have in that context. She views these attempts to understand animals' points of view as valiant but inadequate: valiant because they acknowledge animals as experiencing beings and try to capture their world anti-anthropomorphically, inadequate because of their restriction to vision and visual-based forms, animals being governed by other perceptual modes as well. She pushes for a heterophenomenological approach that may or may not be achievable by visual artists. In contrast to Ullrich, Lowe is less sanguine about the value of attempting to create identification with animals through visual representation. His essay focuses on the ways "images and spectacles" are used to influence public opinion about the treatment of animals. Showing how theories of "the spectacular" apply to strategies of the animal protection movement, Lowe evaluates whether the deployment of visual images usefully shapes the public's moral imagination.

Sickler, Fraser, and Reiss describe a two-year public research development project at the Wildlife Conservation Society's New York Aquarium. They began the project by gathering narratives derived from lay people's ideas about dolphin cognition. They then studied the diverse preexisting beliefs people had about dolphins in order to create methods to provide information about the animals that was understandable within these beliefs. In this way, the authors report, they allowed their audience to co-construct a narrative of dolphins.

Part 6: Synthesis

The story of the elephant is intended to raise doubts about how well we can know anything. Applied to the topic of animals' minds, it suggests that human concerns and purposes drive—indeed invade, permeate, corrupt—each and every attempt to understand and represent animals' mental lives. In the parable of the elephant, no method can escape the limitations of the perceiver: his agendas, her mistaking of the part for the whole. Each viewer is isolated from the others; all viewers remain alienated from the entity they attempt to know. Indeed, many of the essays herein address the problem of creating a human perspective on animal perspective. But one might reimagine this story. One might insist that knowing even one small part of the elephant is no mean feat, particularly when one is exploring it along with the usefulness of a particular methodology. Also, one might actually revise the story so that the blind men talk to each other. The result might be a composite understanding, wherein each examiner contributes knowledge of his part to a collaborative whole. While one may not be able to see all sides at once, he can walk around the subject, visiting the parts, being influenced by what he has already encountered as he approaches from a new angle. One might even imagine a result that is not a patchwork of single views, but a kind of depth composite—each blind man constructing a whole elephant from an idea of one part with the result of a layered view, somewhat like drawings on transparencies arranged in a stack. No one view is perfectly aligned with any other, but from the stack will come interesting, complex, provocative contours with indistinct but richly intriguing perimeters that are as instructive in their own ways as lines of sharp clarity.

An even more radical understanding of how the individual perspectives on animal minds herein might relate to each other is to think of each essay as an intervention into the human collective cultural consciousness (and unconsciousness) about animals. In this reading each author brings to life an elephant, creates her own elephant that could not have existed otherwise. In this account there is no one objective elephant, nor ever can be, one that is understood to stand free and clear of perspective. There is only a panorama of elephants. This is not to say that each is only a fiction of its author. Rather, it is to hypothesize that each elephant is really about a relationship between the author and the animal; each author chooses a particular way of seeing the elephant, a way of interacting with it that produces a very particular interdependence. It is to look at each essay as a document about the possibilities of one animal-human connection—be it personal, intellectual, emotional, or some combination of these. Giving up the idea that there is one objective elephant we must struggle to know, we might expect that each essay gives a snapshot in time of an always shifting and multivalent relationship that continually brings new elephants into being.

In any case, whether the reader wishes to relate the essays objectively or subjectively, relate to them individually, or see them as pointing to some whole, we very much intend that the book will be enjoyable as well as instructive to the degree that the reader entertains interconnections between its essays. We have intentionally grouped essays with very different disciplinary methodologies and controlling themes. We have worked to bring not only multiple academic fields into the discussion but the views of other kinds of specialists. The parts are organized by a sometimes unusual common thread between essays that will suggest one way of reading them as a subgroup. But the parts are also intended to create a bit of dissonance. We want readers to think about how one essay ought to be paired with another in a different part and whether the essays in each group really have a shared comprehensive idea of animal minds. We intend the gaps to be productive of new apprehensions and of meaning making.

Note

1. This centuries-old anecdote equates visual blindness with restricted thinking, an attitude we don't share. We intend here for visual blindness to signify the common human failure to see beyond one's own perspective.

PART I

Living with Animals

1

The Mental Life of Chickens as Observed Through Their Social Relationships

KAREN DAVIS

> I looked at the Chicken endlessly, and I wondered. What lay behind the veil of animal secrecy?
>
> (GRIMES 2002:59)

In this essay I discuss the social life of chickens and the mental states I believe they have and need in order to participate in the social relationships I have observed in them. What follows is a personalized, candid discussion of what I know, what I think I know, and what I am unsure of but have observed relevant to the minds of chickens in their relationships with each other, with other species, and with me.

Chickens evolved in the foothills of the Himalayan mountains and the tropical forests of Southeast Asia where they have lived and raised their families for thousands of years. Most people I talk to had no idea that chickens are natives of a rugged, forested habitat filled with vibrant tropical colors and sounds. Similarly surprising to many is the fact that chickens are endowed with memory and emotions, and that they have a keenly developed consciousness of one another and of their surroundings.

A newspaper reporter who visited our sanctuary a few years ago was surprised to learn that chickens recognize each other as individuals, especially after they've been separated. A friend and I had recently rescued a hen and a rooster in a patch of woods alongside a road in rural Virginia on the Eastern Shore. The first night we managed to get the hen out of the tree, but the rooster got away. The following night, after hours of playing hide and seek with him in the rain, we succeeded in netting the rooster, and the two were reunited at our sanctuary. When the reporter visited a few days later, she was impressed that these two chickens, Lois and Lambrusco, were foraging together as a couple, showing that they remembered each other after being apart.

Chickens form memories that influence their social behavior from the time they are embryos, and they update their memories over the course of their lives. I've observed their memories in action at our sanctuary. For instance, if I have to remove a hen from the flock for two or three weeks in order to treat an infection, when I put her outside again, she moves easily back into the flock, which accepts her as if she had never been away. There may be a little showdown, a tiff instigated by another hen, but the challenge is quickly resolved. Best of all, I've watched many a returning hen be greeted by her own flock members led by the rooster walking over and gathering around her conversably, as if they were saying to her, "Where have you been?" and "How are you?" and "We're glad you're back."

My Experience with Mother Hens and Their Families

> We have before our eyes every day the manner in which hens care for their brood, drooping their wings for some to creep under, and receiving with joyous and affectionate clucks others that mount upon their backs or run up to them from every direction; and though they flee from dogs and snakes if they are frightened only for themselves, if their fright is for their children, they stand their ground and fight it out beyond their strength.
>
> (PLUTARCH 1939 [AD 70]:341)

The purpose of our sanctuary on the Virginia Eastern Shore is to provide a home for chickens who already exist, rather than adding to the population and thus diminishing our capacity to adopt more birds. For this reason we do not allow our hens to hatch their eggs in the spring and early summer as they would otherwise do, given their association with the roosters in our yard. All of our birds have been adopted from situations of abandonment or abuse or else they were no longer wanted or able to be cared for by their previous owners. Our two-acre sanctuary is a fenced open yard that shades into tangled wooded areas filled with trees, bushes, vines, undergrowth, and the soil chickens love to scratch in all year round. It also includes several smaller fenced enclosures with chicken wire roofs, each with its own predator-proof house, for chickens who are inclined to fly over fences during chick-hatching season and thus be vulnerable to the raccoons, foxes, owls, possums, and others inhabiting the woods and fields around us.

I learned the hard way about the vulnerability of chickens to predators. Once, a hen named Eva, who had jumped the fence and been missing for several weeks, reappeared in early June with a brood of eight fluffy chicks. This gave me a chance to observe directly some of the maternal behavior I had read so much about. We had adopted Eva into our sanctuary along with several other hens and a rooster confiscated during a cockfighting raid in Alabama. Watching Eva travel around the yard, outside the sanctuary fence with her tiny brood close behind her, was like watching a family of wild birds whose dark and golden feathers blended perfectly with the woods and foliage they melted in and out of during the day. Periodically, at the edge of the woods, Eva would squat down with her feathers puffed out, and her peeping chicks would all run under her wings for comfort and warmth. A few minutes later the family was on the move again.

Throughout history, hens have been praised for their ability to defend their young from an attacker. I watched Eva do exactly this one day when a large dog wandered in front of the magnolia tree where she and her chicks were foraging. With her wings outspread and curved menacingly toward the dog, she rushed at him over and over, cackling loudly, all the while continuing to push her chicks behind herself with her wings. The dog stood stock still before the excited mother hen and soon ambled away, but Eva maintained her aggressive posture of self-defense, her sharp, repetitive cackle and attentive lookout for several minutes after he was gone.

Eva's behavior toward the dog differed radically from her behavior toward me, demonstrating her ability to distinguish between a likely predator and someone she perceived as presenting no dire threat to her and her chicks. She already knew me from the sanctuary yard, and, though I had never handled her apart from lifting her out of the crate she'd arrived in from Alabama several months earlier, when I started discreetly stalking her and her family, to get the closest possible view of them, the most she did when she saw me coming was dissolve with her brood into the woods or disappear under the magnolia tree. While she didn't see me as particularly dangerous, she nevertheless maintained a wary distance that, over time, diminished to where she increasingly brought her brood right up to the sanctuary fence, approaching the front steps of our house and ever closer to me—but not too close just yet. When she and her chicks were out and about, and I called to her, "Hey, Eva," she'd quickly look up at me, poised and alert for several seconds, before resuming her occupation.

One morning, I looked outside expecting to see the little group in the dewy grass, but they were not there. Knowing that mother raccoons prowled nightly looking for food for their own youngsters in the summer, I sadly surmised they were the likely reason that I never saw my dear Eva and her chicks again.

Inside the sanctuary I broke the no chick-hatching rule just once. Upon returning from a trip of several days, I discovered that Daffodil, a soft white hen with a sweet face and quiet manner, was nestled deep in the corner of her house in a nest she'd pulled together from the straw bedding on the dirt floor. Seeing there were only two eggs under her, and fearing they might contain embryos mature enough to have well-developed nervous systems by then, I left her alone. A few weeks later, on a warm day in June, I was scattering fresh straw in the house next to hers, when all of a sudden I heard the tiniest peeps. Thinking a sparrow was caught inside, I ran to guide the bird out. But those peeps were not from a sparrow; they arose from Daffodil's corner. Adjusting my eyes, I peered down into the dark place where Daffodil was, and there I beheld the source of the tiny voice—a little yellow face with dark bright eyes was peeking out of her feathers.

I kneeled down and stared into the face of the chick who looked intently back at me, before it hid itself, then peeked out again. I looked closely into Daffodil's face as well, knowing from experience that making direct eye contact with chickens is crucial to forming a trusting, friendly relationship with them. If chickens see people only from the standpoint of boots and shoes, and people don't look them in the eye and talk to them, no bond of friendship will be formed between human and bird.

I've seen this difference expressed between hens we've adopted into our sanctuary from an egg production facility, for example, and chickens brought to us as young birds or as someone's former pet. Former egg-industry hens tend to look back at me, not with that sharp, bright, direct focus of a fully confident chicken, but with a watchful opacity that no doubt in part reflects their having spent their entire previous lives in cages or on crowded floors in dark, polluted buildings that permanently affected their eyes before coming to our sanctuary. Psychologically, it's as if they've pulled down a little curtain between themselves and human beings that does not prevent friendship but infuses their recovery with a settled strain of fear. I'll say more about these hens presently.

From the very first, a large red rooster named Francis regularly visited Daffodil and her chick in their nesting place, and Daffodil acted happy and content to have him there. Frequently, I found him quietly sitting with her and the little chick, who scrambled around both of them, in and out of their feathers. Though roosters will mate with more than one hen in the flock, a rooster and a hen will also form bonds so strong that they will refuse to mate with anyone else. Could it be that Francis was the father of this chick and that he and Daffodil knew it? He certainly was uniquely and intimately involved with the pair, and it wasn't as though he was the head of the flock, the one who oversaw all of the hens and the other roosters and was thus fulfilling his duty in that role. Rather, Francis seemed simply to be a member of this particular family. For the rest of the summer, Daffodil and her chick formed a kind of enchanted circle with an inviolable space all around themselves, as they roamed together in the yard, undisturbed by the other chickens. Not once did I see Francis or any of the other roosters try to mate with Daffodil during the time she was raising her frisky chick—the little one I named Daisy who grew up to be Sir Daisy, a large, handsome rooster with white and golden-brown feathers.

My Relationship with the Hens in Our Sanctuary

> The industry must convey the message that hens are distinct from companion species to defuse the misperceptions.
>
> (SHANE 2008:3)

The poultry industry represents chickens bred for food as mentally vacuous, eviscerated organisms. Hens bred for commercial egg production are said to be suited to a caged environment, with no need for personal space or normal foraging and social activity. They are characterized as aggressive cannibals who, notwithstanding their otherwise mindless passivity and affinity for cages, cannot live together in a cage without first having a portion of their sensitive beaks burned off—otherwise, it is said, they will tear each other up. Similarly, the instinct to tend and fuss over her eggs and be a mother has been rooted out of these hens (so it is claimed), and the idea of one's having a social relationship with such hens is dismissed as silly sentimentalism. I confess I have yet to meet a single example of these so-called cannibalistic cage-loving birds.

Over the years, we have adopted hundreds of "egg-type" hens into our sanctuary straight from the cage environment, which is all they ever knew until they were rescued and placed gently on the ground where they felt the earth next to their bodies for the first time in their lives. To watch a little group of nearly featherless hens with naked necks and mutilated beaks respond to this experience is deeply moving. Because their bones have never been properly exercised and their toenails are long and spindly from never having scratched vigorously in the ground, some hens take a few days or longer learning to walk normally and to fly up to a perch and settle on it securely, but their desire to do these things is evident from the time they arrive.

Chickens released from a long siege in a cage and placed on the ground almost invariably start making the tentative, increasingly vigorous gestures of taking a dust bath. They paddle and fling the dirt with their claws, rake in particles of earth with their beaks, fluff up their feathers, roll on their sides, pause from time to time with their eyes closed, and stretch out their legs in obvious relish at being able to bask luxuriously and satisfy their urge to clean themselves and to be clean.

Carefully lifting a battered hen, who has never known anything before but brutal handling, out of a transport carrier and placing her on the ground to begin taking her first real dust bath (as opposed to the "vacuum" dust baths hens try to perform in a cage) is a gesture from which a trusting relationship between human and bird grows. If hens were flowers, it would be like watching a flower unfold or, in the case of a little flock of hens set carefully on the ground together, a little field of flowers transforming themselves from withered stalks into blossoms. For chickens, dust bathing is not only a cleansing activity; it is also a social gathering. Typically, one hen begins the process and is quickly joined by other hens and maybe one or two roosters. Soon the birds are buried so deep in their dustbowls that only the moving tail of a rooster or an outspread wing can be seen a few feet away. Eventually, one by one, the little flock emerges from their ritual entrancement all refreshed. Each bird stands up, vigorously shakes the dirt particles out of his or her feathers, creating a fierce little dust storm before running off to the next engaging activity.

Early on as I began forming our sanctuary and organization in the 1980s, I drove one day from Maryland to New York to pick up seven former battery-caged hens. Instead of crating them in the car, I allowed them to sit together in the back seat on towels, so they wouldn't be cramped yet again in a dark enclosure, unable to see out the windows or to see me. Also, I wanted to watch them through my rearview mirror and talk to them.

Once their flutter of anxiety and fear had subsided, the hens sat quietly in the car, occasionally standing up to stretch a leg or a wing, all the while peering out from under their pale and pendulous combs (the bright red crest on top of chickens' heads grows abnormally long, flaccid and yellowish-white in the cage environment) as I drove and spoke to them of the life awaiting. Then an astonishing thing happened. The most naked and pitiful looking hen began making her way slowly from the back seat, across the passenger seat separator, toward me. She crawled onto my knee and settled herself in my lap for the remainder of the trip.

The question has been asked whether chickens can form intentions. Do they have "intentionality"? Do they consciously formulate purposes and carry them out? In the rearview mirror I watched Bonnie, that ravaged little hen, make a difficult yet beeline trip from the backseat of the car into my lap. Reliving the scene in my mind, I see her journey as her intention to reach me. Once she obtained her objective, she rested without further incident.

Intentionality in chickens is shown in many ways. An example is a hen's desire not only to lay an egg, but to lay her egg in a particular place with a particular group of hens or in a secluded spot she has chosen—and she has definitely *chosen* it. I've watched hens delay laying their egg until they got where they wanted to be. Conscious or not at the outset, once the intention has been formed, the hen is consciously and emotionally committed to accomplishing it. No other interpretation of her behavior makes sense by comparison. Sarah, for example, a white leghorn hen from a battery-cage egg-laying operation, came to our sanctuary with osteoporosis and a broken leg. During her recovery, she lived in the house with me. As she grew stronger, she became determined to climb the front stairs of our house, one laborious step at a time, just so that she could lay her egg behind the toilet in the bathroom next to the second floor landing. This was a hen, remember, who had never known anything before in her life but a crowded metal cage among thousands of cages in a windowless building. I was Sarah's friendly facilitator. I cheered her on, and the interest I showed in her and her wishes and successes was a critical part of her recovery, both physical and mental.

These days in the morning when I unhook the door of the little house in which eight hens and Sir Valery Valentine the rooster spend the night, brown Josephine runs alongside me and dashes ahead down to the Big House where she waits in a state of eager anticipation while I unlatch the door to let the birds who are eagerly assembled on the other side of that door out into the yard. Out they rush, and in goes Josephine, straight to the favorite spot shaped by herself and her friends into a comfy nest atop three stacked bales of straw that, envisioned in her mind's eye, she was determined to get to. Why else, unless she remembered the place and her experience in it with anticipatory pleasure, would she be determined day after day to repeat the episode?

In her mind's eye as well is my own role in her morning ritual. I hold the keys to the little straw kingdom Josephine is eager to reenter, and she accompanies me trustingly and expectantly as we make our way toward it. Likewise, our hen Charity knew that I held the keys to the cellar where she laid her eggs for years in a pile of books in a cabinet beside a table I worked at. Unlike Josephine, Charity wanted to lay her egg in a private place, free of the fussing of hens gathered together and sharing their nest, often accompanied by a rooster boisterously crowing the egg-laying news amid the cacophony of cackles. Charity didn't mind my presence in the cellar. She seemed to like me sitting there, each of us intent on our silent endeavor. If the cellar door was closed, blocking her way to the basement when she was ready to lay her egg, she would pace back and forth in front of the window on the opposite side of the house where I sat at my desk facing the window. If I didn't respond quickly enough, she'd start pecking at the window with an increasing bang to get me to move. By the time I ran up the steps and opened the cellar door, she'd already be standing there, having raced around the house as soon as she saw me get up. Down the cellar steps she'd trip, jump into the cabinet, and settle as still as a statue in her book nook. After she had laid her egg and spent a little time with it, she let me know she was ready to go back outside, running up the steps to the landing where she waited until I opened the door and out she went.

Do events like these suggest that the chickens regard me as a chicken like themselves? I don't really think so, other than perhaps when they are motherless chicks and I am their sole provider and protector, similar to the way children raised by wolves imprint on and behave like wolves. I see the ability of chickens to bond with me and be endearingly companionable as an extension of their ability to adapt their native instincts to habitats and human-created environments that stimulate their natural ability to perceive analogies and fit what they find where they happen to be to the fulfillment of their own needs and desires.

The inherently social nature of chickens enables them to socialize successfully with a variety of other species and to form bonds of interspecies affection. Having adopted into our sanctuary many incapacitated young chickens from the "broiler" chicken (meat) industry, I know how quickly they learn to recognize me and my voice and their own names. They twitter and chirp when I talk to them and they turn their heads to watch me moving about or away from them. Living in the house until they are well enough to go outside if they ever can, they quickly learn the cues I provide that signify their comfort and care.

This is not to suggest that chickens are unlimitedly malleable. Mother hens and their embryos have a genetic repertoire of communications that are too subtle for humans to decipher entirely, let alone imitate. Chickens have ancestral memories that predispose the development of their behavior. Even chickens incubated in mechanical hatcheries and deprived of parental influence—virtually all of the birds at our sanctuary—behave like chickens in essential ways. For instance, they all follow the sun around the yard. They all sunbathe, dropping to the ground and lying on their sides with one wing outspread, then turning over and spreading out the other wing while raising their neck feathers to allow the warm sunlight and vitamin D to penetrate their skin. Similar to dust bathing, sunbathing is a social as well as a healthful activity for chickens, where you see one bird drop to the ground where the sun is shining, followed by another and then another. If you don't know what they are doing, you would think they had died the way they lie still with their eyes closed, flopped like mops under the sun.

I'm aware when I am in the yard with them that the chickens are constantly sending, receiving, and responding to many signals that elude me. They also exhibit a clear sense of distinction between themselves, as chickens, and the three ducks, two turkeys, and peacock Frankincense who share their sanctuary space. And they definitely know the difference between themselves and their predators, such as foxes and hawks, whose proximity raises a sustained alarm through the entire flock. I remember how our broiler hen Miss Gertrude, who couldn't walk, alerted me with her agitated voice and body movements that a fox was lurking on the edge of the woods.

While all of our sanctuary birds mingle together amiably, typically the ducks putter about as a trio, and Frankincense the peacock displays his plumage before the hens, who view him for the most part impassively. The closest interspecies relationship I've observed among our birds is between the chickens and the turkeys.

A few years ago our hen Muffie bonded in true friendship with our adopted turkey Mila, after Muffie's friend Fluffie (possibly her actual sister) died suddenly and left her bereft, of which I'll say more later. Right from the start, Muffie and Mila shared a quiet affection, foraging together and sometimes preening each other very delicately. One of their favorite rituals was in the evenings when I changed their water and ran the hose in their bowls. Together Muffie and Mila would follow the tiny rivulets along the ground, drinking as they went, Muffie darting and drinking like a brisk brown fairy, Mila dreamily swaying and sipping, piping her intermittent flute notes.

Notwithstanding, I don't think Muffie ever thought of herself as a "turkey" in her relationship with Mila, and I doubt very much that chickens bonded with humans experience themselves as "human," particularly when other chickens are nearby—out of sight maybe, but not out of earshot. (Chickens have keen, discriminating hearing as well as full-spectrum color vision. Chick embryos have been shown to distinguish the crow of a rooster from other sounds from inside their shells.)

Chickens in my experience have a core identity and sense of themselves as chickens. An example is a chick I named Fred, sole survivor of a classroom hatching project in which embryos were mechanically incubated. Fred was so large, loud, and demanding, from the moment he set foot in our kitchen, I assumed he'd grow up to be a rooster. He raced up and down the hallway, hopped up on my shoulder, leapt to the top of my head, ran across my back, down my arm, and onto the floor when I was at the computer, and was generally what you'd call "pushy," but adorably so. I remember one day putting Fred outdoors in an enclosure with a few adult hens on the ground, and he flew straight up the tree to a branch, peeping loudly, apparently wanting no part of them.

"Fred" grew into a lustrously beautiful black hen whom I renamed Freddaflower. Often we'd sit on the sofa together at night while I watched television or read. Even by herself, Freddaflower liked to perch on the arm of the sofa in front of the TV when it was on, suggesting she liked to be there because it was our special place. She ran up and down the stairs to the second floor as she pleased, and often I would find her in the guestroom standing prettily in front of the full-length mirror preening her feathers and observing herself. She appeared to be fully aware that it was she herself she was looking at in the mirror. I'd say to her, "Look, Freddaflower—that's you! Look how pretty you are!" And she seemed already to know that.

Freddaflower loved for me to hold her and pet her. She demanded to be picked up. She would close her eyes and purr while I stroked her feathers and kissed her face. From time to time, I placed her outside in the chicken yard, and sometimes she ventured out on her own, but she always came back. Eventually I noticed she was returning to me less and less and for shorter periods. One night she elected to remain in the chicken house with the flock. From then on until she died of ovarian cancer in my arms two years later, Freddaflower expressed her ambivalence of wanting to be with me but also wanting to be with the other hens, to socialize and nest with them and participate in their world, reliving the ancestral experiences she carried within herself.

My Relationship with the Roosters in Our Sanctuary

A less happy ambivalence appeared in a soft-colored gray and white rooster I named Ruby when he was brought to our sanctuary as a young bird by a girl who swore he was a hen. Following me about the house on his brisk little legs, even sleeping beside me on my pillow at night, Ruby grew up to be a rooster. In spite of our close relationship during his first months of life, once he became sexually mature, Ruby's attitude toward me changed. In the yard with the other chickens, he showed no disposition to fight. He didn't attack other birds or provoke antagonisms. He fit in with the existing flock of hens and roosters, but toward me and other people he became compulsively aggressive. As soon as I (or anyone) appeared in the yard, Ruby ran from wherever he was and physically attacked us. Having to work in the yard under his vigilant eye, I took to carrying a bottomless birdcage and placing it over him while I worked. When finished I would lift it off him and walk backward toward the gate with the birdcage in front of me as a shield.

What I saw taking place in Ruby was a conflict he couldn't control, and from which he suffered emotionally, between an autonomous genetic impulse, on the one hand, and his personal desire, on the other, to be friendly with me. He got to where, when he saw me coming with the birdcage, he would walk right up and let me place it over him as if grateful for my protection against a behavior he didn't want to carry out. Even more tellingly, he developed a syndrome of coughs and sneezes whenever I approached, symptomatic, I believed, of his inner turmoil. He didn't have a respiratory infection, and, despite his antagonism toward me, I never felt that he hated me, but rather that he suffered from his dilemma, including his inability to manage it.

My personal experience with our sanctuary roosters confirms the literature I've read about wild and feral chickens documenting that the majority of roosters do not physically and compulsively attack one another. Chickens maintain a social order in which every member of the flock has a place and finds a place. During the day our roosters and hens break up into small, fluctuating groups that are somewhat, but by no means rigidly, territorial. Antagonisms between roosters are resolved with bloodless showdowns and face-offs. The most notable exception is when a new rooster is introduced into an existing flock, which may provoke a temporary flare-up, but even then there is no predicting.

Last year I placed newcomer Benjamin in a yard already occupied by two other roosters, Rhubarb and Oliver, and their twenty or so hens, and he fit in right away. Ruby won immediate acceptance when I put him outside in the chicken yard after living in the house with me for almost six months. In dealing with Ruby I found an unexpected ally in our large red rooster Pola, who was so attentive to me all I had to do was call him and he bolted over from his hens and let me pick him up and hold him. I have a greeting card photograph of Pola and me "crowing" together, my one hand clasped over his swelled-out chest, my other hand holding his claw, in a duet I captioned "With Heart and Voice."

Playfully, I got into the habit of yelling "Pola, help!" whenever Ruby acted like he was ready to come after me, which worked as well as the birdcage. Hearing my call, Pola would perk up, race over to where Ruby was about to charge, and run him off with such cheerful alacrity it was as if he knew this was our little game together. I'd always say, "Thank you, Pola, thank you!" and he acted very pleased with his performance and the praise I lavished on him for "saving" me. He stuck out his chest, stretched up his neck, flapped his wings vigorously, and crowed triumphantly a few times.

Roosters crow to announce their accomplishments. Even after losing a skirmish, a rooster will often crow as if to compensate for his loss or deny its importance or call it a draw. Last summer, as I sat reading outside with the chickens, I was diverted by our two head roosters, Rhubarb and Sir Valery Valentine, crowing back and forth at each other in their respective yards just a few feet apart. It looked like Sir Valery was intentionally crossing a little too far into Rhubarb's territory, and Rhubarb kept dashing at him to reinforce the boundary. There was not a hint of hostility between them; rather the contest, I decided, as I watched them go at it, was being carried out as a kind of spirited mock ritual in which each rooster rushed at the other, only to halt abruptly on his own side of the invisible buffer zone they apparently had agreed upon. At that point, each rooster paced up and down on his own side, steadily eyeing the other bird and crowing at him across the divide. After ten minutes or so, they each backed off and were soon engrossed in other activities.

Roosters are so energetic and solicitous toward their hens, so intensely focused on every aspect of their social life together, that one of the saddest things to see is a rooster in a state of decline due to age, illness, or both. An aging or ailing rooster who can no longer hold his own in the flock suffers severely. He droops, and I have even heard a rooster cry over his loss of place and prestige within his flock. This is what happened to our rooster Jules—"Gentleman Jules," as my husband fondly named him—who came to our sanctuary in the following way.

One day I received a phone call from the resident of an apartment building outside Washington, DC, saying that a rooster was loose in the complex and was being chased by children who were throwing stones at him. After two weeks of trying, she managed to lure the rooster into the laundry room and called me to come get him. Expecting to find a cowering and emaciated creature needing to be carefully lifted out of a corner, I discovered instead a bright-eyed, perky, chatty little fellow with glossy black feathers like Freddaflower. I drove him to our sanctuary and set him outside with the flock, which at the time included our large white broiler rooster Henry and our feisty bantam rooster Bantu, who loved nothing better than sitting in the breeze under the trees with his two favorite large brown hens, Nadia and Nadine.

Jules was a sweet-natured rooster, warm and affectionate to the core. He was a natural leader, and the hens loved him. Our dusky brown hen Petal, whom we'd adopted from another sanctuary, was especially devoted to Jules. Petal had curled gnarly toes, which didn't stop her from whisking away from anyone she didn't want to come near her; otherwise she sat still watching everything, especially Jules. Petal never made a sound; she didn't cluck like most hens—except when Jules left her side a little too long. Then, all of a sudden, the silent and immobile hen with the watchful eye let out a raucous *squawk, squawk, squawk* that didn't stop until Jules had lifted his head up from whatever he was doing and, muttering to himself, run over to comfort his friend.

Two years after coming to live with us, Jules developed a respiratory infection that, with treatment, seemed to go away but left him weak and vulnerable. He returned to the chicken yard, only to find himself supplanted by Glippie, with whom he had used to be cordial but was now dueling, and he didn't have the heart or strength for it. His exuberance ebbed out of him, and he became sad; there is no other word for the total condition of mournfulness he showed. His voice, which had always been cheerful, changed to moaning tones of woe. He banished himself to the outer edges of the chicken yard where he paced up and down, bawling so loudly I could hear him crying from inside the house. I brought him in with me and sought to comfort my beloved bird, who showed by his whole demeanor that he knew he was dying and was hurt through and through by what he had become. Jules developed an abdominal tumor. One morning our veterinarian placed him gently on the floor of his office after a final and futile overnight stay. Jules looked up at me from the floor and let out a low groan of "ooooohh" so broken that it pierced me through. I am pierced by it now, remembering the sorrow expressed by this dear sweet creature, Gentleman Jules, who had loved his life and his hens and was leaving it all behind.

My Experience of Empathy and Affection in Chickens

> I perceive in your literature the proposal that chickens be treated as pets. I have been involved with many thousands of chickens and turkeys and I don't think they are good pets, although it is evident that almost any vertebrate may be trained to come for food.
>
> (THOMAS JUKES, PERSONAL COMMUNICATION, SEPTEMBER 4, 1992)

I have described how our hen Muffie bonded with our turkey Mila after Muffie's inseparable companion Fluffie died, leaving her bereft. Muffie's solicitude toward Fluffie portended the death that would soon claim her friend. Like Jules, Fluffie developed an infection that treatment seemed to heal, but she never fully recovered. One day I looked out the kitchen window and saw Muffie straddled on top of Fluffie with her wings extended over her. I called my husband to come take a look at this moving, yet disturbing scene. We saw it repeated several times over the next few days.

On a late afternoon I went outside to put Muffie and Fluffie in for the night, but found them already in their house in the straw. Fluffie stood drooping, with her head and tail curved toward the ground, and Muffie stood motionless beside her. I rushed Fluffie to the veterinarian and brought her home with medicine, but she died that same night in the small bedroom where she and Muffie had liked to perch on top of the bookcase in front of the big window overlooking the yard.

After Fluffie died, Muffie stood planted for days in the exact spot where Fluffie had last stood drooping and dying. Now Muffie drooped in her place. She no longer scampered into the woods or came bursting into the kitchen to jump up on the sink and peck holes in the sponge floating on top of the dishwater. She was not interested in me or the other chickens. Two weeks of this dejection and I said, "We must get Muffie a new sister." That is how Petal, who had loved Jules, came to live in our sanctuary. The minute

Petal appeared, Muffie lost her torpor and became a bustling "police miss," picking on Petal and patrolling everything Petal did until finally the two hens became amiable, but they were never pals.

Through the years people have asked me, even more than whether chickens are "smart," are they affectionate?—toward people, they particularly want to know. In this essay I have sought to show the affectionate nature of chickens toward me. Because I don't just feed them but I also talk to them and look them in the eye and express my feelings for them, the birds at our sanctuary gather around me and stand there serenely preening themselves or sit quietly on the ground next to my chair while I read and chat with them.

Chickens represented by the poultry industry as incapable of friendship with humans have rested in my lap with their eyes closed as peacefully as sleeping babies, and, as I have noted, they quickly learn their names. A little white hen from the egg industry named Karla became so friendly all I had to do was call out "Karla!" and she would break through the other hens and head straight toward me, knowing she'd be scooped off the ground and kissed on her sweet face and over her closed eyes. And I can still see Vicky, our large white hen from a "broiler breeder" operation, whose right eye had been knocked out, peeking around the corner of her house each time I shouted, "Vicky, what are you doing in there?" And there was Henry, likewise from a broiler breeder operation, who came to our sanctuary dirty and angry after falling out of a truck on the way to a slaughter plant. Lavished with my attention, Henry, who at first couldn't bear to be touched, became as pliant and lovable as a big shaggy dog. I couldn't resist wrestling him to the ground with bearish hugs, and his joy at being placed in a garden where he could eat all the tomatoes he wanted was expressed in groans of ecstasy. He was like, "Are all these riches of food and affection really for me?"

One of my most poignant memories is of a large black, beautiful hen I named Mavis. Mavis had been dropped off at a shelter by a man who'd exhibited her at agricultural fairs. She must have spent her whole life immobilized on the floor of a cage with a keeper who treated her like an object. During her first two weeks at our sanctuary, Mavis could not even stand up without crumbling to the ground, and she was deeply shy and inexpressive. In the chicken yard she sat alone by the fence and poked around a little by herself without showing or attracting interest. I saw no sign that she was ever going to recover from the emotional and sensory deprivation of her previous life.

During this time, we had three adult broiler hens: Bella Mae, Alice, and Florence. They were the opposite of Mavis. All I had to do was crouch down in the yard, and here came one of my three graces, as I called them—Bella Mae, for example—bumping up against me with her ample breast for an embrace. Immediately, Alice and Florence would hastily plod over on their heavy feet to participate in the embracement ceremony. Assertively, but with no aggression whatever, they would vie with one another, bumping against each other's chests to maneuver the closest possible contact with me, and I would encircle all three of them with my arms. One day, as we were doing this, I looked up and saw Mavis just a few feet away, staring at us. The next time, the same thing happened. There was Mavis with her melancholy eyes watching me hugging the three white hens. And then it struck me—Mavis wants to be hugged. I withdrew from the hens, walked over and knelt beside Mavis and pulled her gently toward me. It didn't take much. She rested against me in a completeness of comfort that seemed to include her gratitude that her shy desire had been understood.

In my first years of keeping chickens, there were no predators, until a fox found us, and we built our fences—but only after eleven chickens disappeared rapidly under our nose. The fox would sneak up in broad daylight, raising a clamor among the birds. Running out of the house, I'd see no stalker, just sometimes a soul-stabbing bunch of feathers on the ground in the midst of panic. When our bantam rooster Josie was taken, his companion Alexandra ran shrieking through the kitchen, jumped up on a table, could not stop shrieking, and was never the same afterward. The fox killed Pola, our big red rooster who had so gallantly responded to my calls begging him to "save" me from Ruby. I am sure he was attacked while trying to protect his hens, while I sat obliviously at the computer. It was too much. I sat on the kitchen floor crying and screaming. At the time I was caring for Sonja, a big white warm-natured, bouncy hen I was treating for wounds she'd received before I rescued her. As I sat on the floor exploding with grief and guilt, Sonja walked over to where I sat weeping. She nestled her face next to mine and began purring with the ineffable soft purr that is also a trill in chickens. She comforted me even as her gesture deepened the heartache I was feeling in that moment about the painful mystery of Pola and the mystery of all chickens. Did Sonja know why I was crying? I doubt it, but maybe she did. Did she know that I was terribly sad and distressed? There is no question in my mind about that. She responded to my grief with an expression of empathy that I have carried emotionally in my life ever since.

It is experiences such as this one, and others I've described in this essay, that have made me a passionate advocate for chickens. I do not seek to sentimentalize chickens but to characterize them as best I can within the purview of my own observations and relationships with them. In the 1980s I wrote an essay about an abandoned crippled broiler hen named Viva, who, more than any other single cause, led me to found United Poultry Concerns in 1990. It is hard for me to evoke in words how expressive she was in spite of her handicap and despite the miserable life she had had before I lifted her out of her misery and brought her home with me.

My experience with chickens for more than twenty years has shown me that chickens are conscious and emotional beings with adaptable sociability and a range of intentions and personalities. If there is one trait above all that leaps to my mind in thinking about chickens when they are enjoying their lives and pursuing their own interests, it is cheerfulness. Chickens are cheerful birds—quite vocally so—and when they are dispirited and oppressed their entire being expresses this state of affairs as well. The fact that chickens become lethargic in continuously barren environments, instead of proving that they are stupid or impassive by nature, shows how sensitive these birds are to their surroundings, deprivations, and prospects. Likewise, when chickens are happy, their sense of well-being resonates unmistakably.

References

Grimes, W. 2002. *My Fine Feathered Friend.* New York: North Point.

Plutarch. 1939 [AD 70]. "On Affection for Offspring." Trans. W. C. Helmbold. In *Moralia,* 6:328–357. Cambridge: Loeb Classical Library.

Shane, S. M. 2008. "Where Do We Go from Here?" *Egg Industry* 113, no. 12 (December): 1–3.

2

Tangible Affiliations

Photographic Representations of Touch Between Human and Animal Companions

JULIA SCHLOSSER

Photographic images of pet or companion animals, especially in a modernist fine art context, are often seen as nostalgic or sentimental. The animal's presence is usually highly scripted: formally arranged in the image, the animal frequently stands as a metaphor for human behavior, as an indicator of class or economic status, or as a symbol of its role as commodity item (see Baker 2001; Schlosser 2007). For many postmodern artists, pets are somehow "less than" wild animals, having given up their independence to rely on humans for their care (Baker 2000:168–169). This disparaging view within the fine arts emerges in part from ambivalence within the pet-keeping population itself. Even though pet keeping in many Western cultures is at an all-time high, and people claim that they value their pets highly and consider them as friends (Olmert 2009:242), they still often maintain an inherent prejudice, which intimates that living with and loving a pet animal is a shameful substitute for "real" or more legitimate human relationships (Levinson 1972:168; Serpell 1986:vix).

This chapter looks at the work of four women artists, all working with photographic media, whose art focuses on relationships with pet or companion animals: Martha Casanave, Barbara Dover, Carolee Schneemann, and myself. All four have chosen to include touch as an aspect of their exploration of their relationships with companion animals. Made in a deliberate and sometimes provocative manner, this choice of subject matter places them in danger of being disenfranchised in both the art and academic worlds. The choice of the domestic setting as the backdrop for their artworks, the societal association of touch with nurturance and play, and an open acknowledgment of their relationship with a pet animal have exposed these artists to the criticism of unacceptable subject matter for "serious" artists.[1] Casanave, Schneemann, and myself add another complexity to the picture: We are both subject and object of our work. A photographer who chooses to photograph herself along with her pet cannot retreat safely behind the lens of her camera to manipulate the image of her subject. Instead she must grapple with the logistics of taking the picture from that awkward vantage point, giving up a measure of control in the art-making process. She also appears in front of the camera as a coequal with the animal.

Touch is an often overlooked but nonetheless crucial aspect of human-animal relationships as well as a way to express that relationship in a visual art context. Circumventing both the one-sided activity of conveying information via language, which the animal is unable to reciprocate, and the act of looking from a privileged distance, touch is instead potentially more reciprocal. It brings human and animal into physical proximity and can compel further contact. Touch by a human hand often allows both to move into a more equal eye-level view, one with the other. An unconstrained animal is free to touch back or not by her choice, and a touch provides both participants with a reciprocal exchange of visceral or embodied information about the other (Weber 1990:24–25; see also Merleau-Ponty 1962 [1945]), a felt-sense that often functions outside the patterns of language. Animals can perceive information in many different ways than we do, so the type of information exchanged through touch may be particular to each party. Touch allows for a further articulation of the physical and psychological presence of each member of the partnership. Recently, when I cupped my hand momentarily around the stomach of a pregnant cat, my entire perception of her experience was instantly altered through the information conveyed from the sensation of the hot, strangely prickly, curiously alive movement of her belly.

The human touching an animal is privy to many types of information, including data about temperament and health. I frequently touch my animals' ears for a quick check of their temperature, feel for a scratchy coat, which would indicate a loss of health, or pick them up to get an approximate sense of their weight gain or loss. However, this type of touch "functions more like sight (i.e. cognitively) . . . and need not entail closeness or reciprocity" (Weber 1990:25; see also Wyschograd 1981). This cognitive touch can even have a distancing function, a way of "mastering experience and appropriating it" (Weber 1990:25). The type of touch that allows familiarity and a growing sense of the other's mind must be more reciprocal. Polly Ullrich (1996–1997) observes a trend toward the importance of such touch in contemporary artwork. Applying the theories of phenomenologists such as Maurice Merleau-Ponty to the work of a group of postmodern artists, she foregrounds the reciprocality of touch, noting that "the most singular characteristic of touch is that in touching, one is touched in return" (p. 13). Ullrich further emphasizes the importance of these tangible affiliations by saying, "Touch implies mutuality, not dominance" (p. 13). Touch—mutual and consensual in particular—provides a unique perceptual opportunity for both participants to experience each other.

Touch has been disparaged within philosophical and artistic discourses (see Weber 1990; Candlin 2006; Perricone 2007). Philosopher Renée Weber (1990) remarks, "Touch as an interactional modality has been neglected by philosophy. Especially in the last three hundred years, since the time of Descartes, only mental and verbal exchanges have been considered important" (p. 13). The tactile is somehow "less than" the visual and the auditory. Artist and museum studies lecturer Fiona Candlin (2006) argues that we need to reassess the pejorative role that touch has been accorded in the recent historiography of art in light of contemporary museum practice that emphasizes touch. She writes "We do not pass through touch to reach vision. . . . Once we start thinking about vision and touch as being intimately related then correlative equations of touch with body and not mind, nature and not culture, the past and not present, become equally untenable" (p. 150). Candlin urges readers to "avoid thinking of vision as being more conceptual than touch," and reminds us to "ask how touch and the tactile qualities can lead an audience into a discussion of content" and how they can "generate meaning" (p. 152). Although not intended as such, Candlin's points are quite helpful for a discussion of the ways that photographs of animals and humans touching can generate meaning for the viewer about animal subjectivity.

Traditional photographic representations of pet animals often convey no sense of the vibrant self-directedness and agency of the animal subjects their human companions witness on a daily basis. Traditional photographic images of pets reflect the view that animals, because they lack language, have no "voices." The act of touch provides a unique way to dislocate the assumption that lack of language means an incapacity for agency. By introducing an avenue through which human and animal interact, that is, the sensory experience of touch, the photographer can enable the animal to exhibit a similar sense of subjectivity to that of the other subject in the frame, the human.

Photographs also pull viewers in via their own viscerally imaginal involvement. Meg Daly Olmert (2009) researches the social bonding mechanisms between humans and animals. "From day one, all social mammals learn that the warmth and touch of another is good" (p. 31). She comments later, "We and other mammals have a common social biology that allows us to approach, interact, and relax in each other's company" (p. 220). The release of the hormone oxytocin accompanies and facilitates the bonding both between a mother and child who are touching, and between humans and companion animals. Simply viewing photographs of a subject interacting with another activates the same release of oxytocin that would be triggered by an actual interaction. Olmert notes, "Brain researchers . . . used fMRI imaging to see what goes on in a mother's head when she's looking at her baby. In their study, they don't even have the mother look at her actual child but merely a photo of the baby. 'Just watching' the image of her child was enough to trigger a powerful response in brain cells loaded with oxytocin" (p. 30).

Intimate touching involves an exchange between the parties, and it cements and enriches relationships in reciprocal but difficult to fully articulate ways. Olmert writes, "Mutual touch still raises and lowers our body temperature, speeds up the healing of wounds, takes away physical and emotional pain, and often is the only way we can communicate deep feelings for which there are no words" (p. 227). Weber explores the history of the philosophical analysis of the sense of touch and in the process develops three models of touch. For her, "the most comprehensive model," the one that not only incorporates touch as tactile stimulation or direct contact but also serves as a method of "'reaching' the other at some level deeper than the visible and behavioral one," is the "field model" (1990:14). Weber begins her discussion by announcing, "My focus is . . . holistic. It deals with the touch of a person and person, or of person and animal—but not the touch of hand and skin, or hand and fur, nor the idea of tactile stimulation" (p. 11). For Weber, a more thoughtful analysis leads away from the idea that touch functions simply as a physiological conduit streaming information in one direction from an exterior entity. Instead it becomes a mutual, participatory action that offers two beings nuanced perceptions of the other, affirming their intrinsic relatedness and providing solace and comfort. Weber continues by explaining that the holistic field model is based on the "underlying assumption of Eastern thought. At some deep level in virtually all the Eastern metaphysics, we are all interdependent and interconnected. We're actually not separate entities" (p. 40).

Philosopher Christopher Perricone (2007) uses art historian Bernard Berenson's ideas about the connection between touch and the aesthetic value of art-historical masterpieces—what Berenson calls "tactile values"—as a stepping-off point for his further articulation of the deeply embodied, intrinsic, but often overlooked place that touch holds for viewers who are integrating information about artwork. For Berenson, as viewers our "tactile imagination" is stimulated when an image maker is able to activate our sense of touch or tactile sense, and this provides a "quality of life-likeness" or "material significance" to our "retinal impressions" (Perricone 2007:91). For Perricone, the ways that we know artwork and the ways that we know each other are both intimate and associated with touch and with our "animal nature" (p. 96). Like Olmert, Weber, and others who are profoundly concerned with articulating this sensation, Perricone understands touch as communicating something deeper than language or sight. "There is a 'knowledge' communicated through the body that produces a person's healthy development as well as builds friendship and trust between one person and another. Knowledge of the most profound and intimate emotions is literally conveyed by touch" (p. 96).

Although Perricone's research deals with the tactile qualities of artwork and not with the ways that humans come to know animals through touch or photographs of touch, his article pulls together important ideas for understanding the ways that physical contact conveys meaning to viewers of contemporary photography. For social mammals like humans and domesticated animals, social bonding, facilitated by mutual grooming (petting your cat, who then licks you), plays a crucial role in the formation of relationships (see Olmert 2009). The photographs discussed in this chapter are in a unique position to activate the tactile imagination given their use of touch as a bridge from animal bodies to human bodies. According to Perricone (2007), "Chances are the tactile imagination is inspired most, given the allocation of brain space and the density of receptors, by what the hand touches" (p. 92). And, for the artists in this chapter, what the hand touches is the animal next to it.

Martha Casanave: *Beware of Dog*

In the book *Beware of Dog* California-based photographer Martha Casanave (2002) recorded her interactions with her pet whippet in her home. The images are black and white and are beautifully shot with lush natural lighting. In the introduction to the book, Casanave's written description of the dog and their life together is heavily grounded in perceptual descriptions of the dog and the dog's behavior, e.g. "My dog always smells what's different about me" (p. 9). At the same time, she also speaks about the limitations of the sensory information that a black and white photograph can convey to a viewer. She lists nuanced perceptual information about the dog she feels *cannot* be conveyed by the sight of a photograph, including smells, sounds and the feeling of touch, e.g., "the velvety feel of a whippet's belly skin" (p. 7). However, in the next paragraph she describes how the activity of photographing the dog —"the process of paying close attention to every detail of my dog and watching her behavior closely"—became (almost) more important than the photographs produced by this activity (p. 8). I argue that for Casanave the process of paying such close attention to the whippet allowed her to develop photographic strategies or tropes that, in contrast with her stated repudiation, *do* allow the viewer to experience an embodied tactility or felt-sense of the dog.

Instead of conforming to established conventions for photographing pets, the artist produces images that reflect intrinsic characteristics of the animal. Casanave notes that the nature of the whippet is to alternate between periods of quietude and then intense activity: "indolence and exuberance" (p. 8), and the two discrete photographic strategies that she develops to photograph the dog reflect these modes. Only one employs touch between herself and the animal, but both utilize a visceral, embodied connection that transfers information about the animal to the viewer of the photograph. Cleverly exploiting the inherent materiality of black and white film, Casanave uses motion blur to illustrate one aspect of the animal's persona: grainy swirls of silver trace the whippet's rapid bursts of frantic motion, conveying to the viewer a visceral sense of the dog's action. The dog is shown running or chasing a ball, full body, oscillating against the wide expanse of a backyard lawn and fence in streaky motion (figure 2.1). Then, in contrast with the first technique, but still communicating with the viewer through the felt-senses, the artist uses elegant, tightly composed, focused and lit images to convey the dog's complementary moments of stasis. Photographed inside Casanave's house, the home being the frequent site of human-companion animal encounters, the shots of the dog at rest often include the artist in contact with the animal. The artist captures minor, mundane, almost trivial moments of intimacy between dog and person, asleep, napping during the day on a sofa or bed. The tight spaces, sharp focus, accentuating the details of the dog's coat, bony spine, nails, and whiskers, and shallow depth of field combine with the minimal background detail of the images to invite the viewer to engage with the small moments of intimacy that exist between the two.

FIGURE 2.1. Martha Casanave, "Untitled," from *Beware of Dog*. Carmel, CA: Center for Photographic Art, 2002. Photo courtesy of the artist.

Although in the preface to her book Casanave writes in almost painfully tender detail about aspects of their life together, she does not refer to the dog by name. In somewhat similar fashion, she excludes her own face from the photographs, crops her body close enough that it remains relatively gender-neutral, and often fragments the dog's face and body as well. These tropes provide a visual context for the viewer to encounter the dog with a sense of parity to the human, an unusual situation for many pet photographs. By including herself in the photographs along with the dog, Casanave is similarly unable to maintain the typical unequal power relationship formed between artist and model, in which the artist is in complete control of her subject's representation.

In addition to creating a sense of visual equality or correspondence in the frame of the photograph, the implied motion, based on touch, generated by the pair's actions toward each other allow the viewer to respond in an embodied fashion. In one photograph Casanave appears to measure the dog's paw, comparing it with her own hand (figure 2.2). In other photographs she gently wiggles the bones in the dog's leg, strokes her smooth face, and cups her muzzle between her forefinger and thumb, all gestures of intimate and equalizing familiarity. The viewer can imagine Casanave

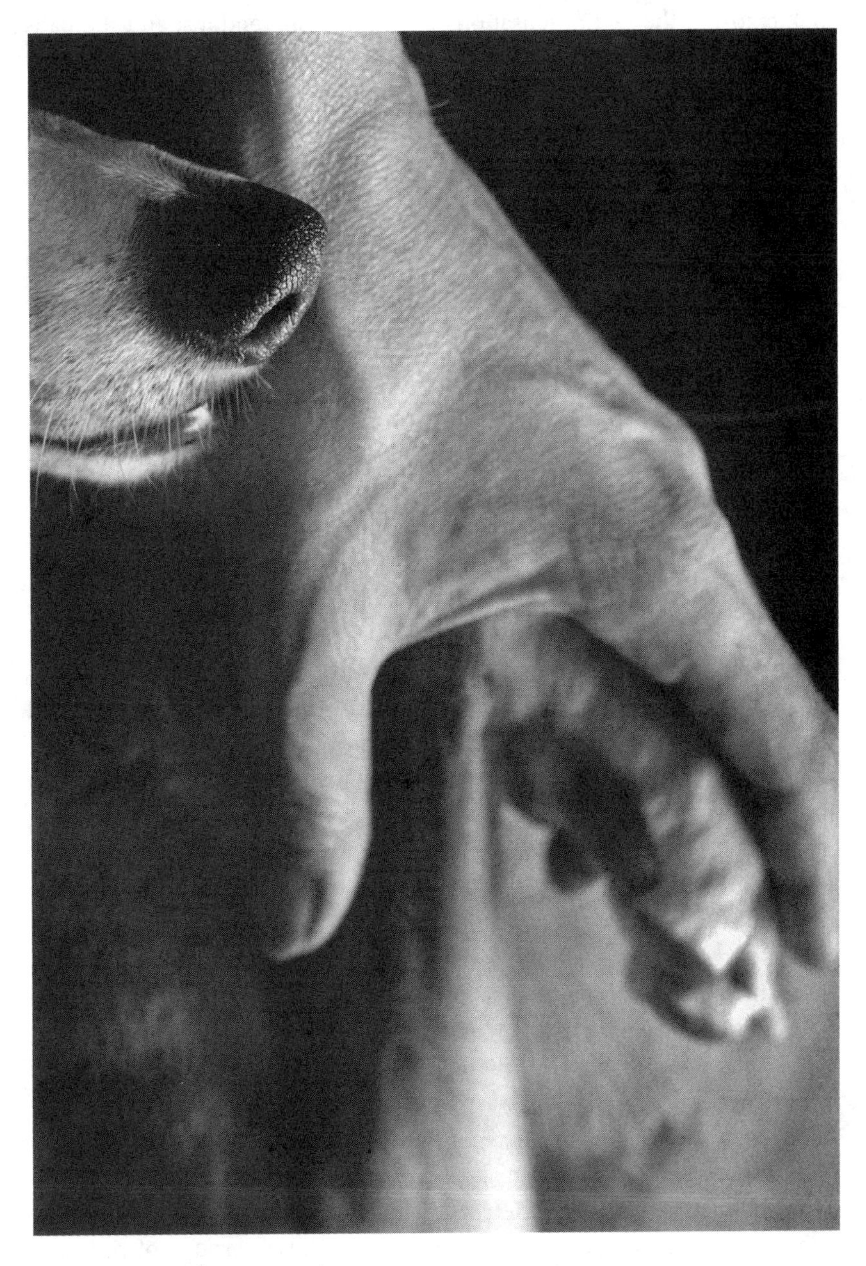

FIGURE 2.2. Martha Casanave, "Untitled," from *Beware of Dog*. Carmel, CA: Center for Photographic Art, 2002. Photo courtesy of the artist.

exploring the dog's body, absorbing embodied knowledge about the animal, enjoying the tactile sensations of her fur, tail, tendons, and the pads of her feet, and can experience the haptic sensations Casanave feels and transmits through her human presence in the photographs.

Casanave's trope of fragmenting her human and animal subjects creates meaning in the images in a number of provocative ways. The close-up, gender-neutral limbs of the dog and human subject provide a certain anonymity and thus allow the photographs to function as images about the relationship of human and companion animals, exploring ideas that are limited not just to this particular pair of subjects. In addition, the fragmentation of the bodies allows the artist to circumvent certain neotenous or nostalgic associations with photographs of pet animals in a vernacular context. In the images in which Casanave moves in close to her subjects and crops their bodies, the resulting abstraction of the forms creates a particular synchronicity for the viewer. In one image from the book, the close-cropped round of the whippet's back so closely mimics the curve of a human shoulder as to provide a visceral jolt of surprise when the viewer realizes the form is canine and not human. This technique creates a way for the viewer of the images to enter the space of the photograph by encouraging a kinesthetic merging with the fragmented bodies in the photographs. The fragmented limbs in the image can, in effect, fuse with the viewers' own and allow viewers to submerge themselves in the pair's insular world without unnecessary distraction. Casanave's images invite the viewer to focus on intimate details of the dog and human bodies and participate in the feeling of the larger wholeness of a composite human-dog being. Although perhaps counterintuitive given the fracturing of the bodies, this fragmentation and subsequent remerging ultimately give visual form to Weber's field model of touch and its underlying assumptions about the interconnected, inseparable nature of the physical universe and its inhabitants (1990:14).

In an untitled image from the book, the sleeping whippet enfolds Casanave's crossed leg with her own body, upturned head resting gently against her knee (figure 2.3). Often the assumption about this type of casual physical intimacy is that it is reserved exclusively for select groups of human counterparts. For example, in the book *The Caring Touch*, authors Pratt and Mason note: "In our society, actual physical contact as a full and free expression of intimacy is seen to occur only in two groups of people—between mothers and children and between lovers" (1981:4). Pratt and Mason point out that touch is perhaps the only form of sensory exchange that "requires some kind of contract [agreement] between the parties concerned" (p. 3). They also suggest that intimate body contact is the "starting

FIGURE 2.3. Martha Casanave, "Untitled," from *Beware of Dog*. Carmel, CA: Center for Photographic Art, 2002. Photo courtesy of the artist.

point of all loving relationships" (p. 4). Although the contact is between human and nonhuman animal, and not human and human as Pratt and Mason presume, Casanave and her dog have established this type of contract, and, for them, touch manifests as an expression of love and care. The images create an extraordinary sense of human and nonhuman animal together in embodied physicality, beyond the place where language is necessary. Images such as Casanave's, which implicitly value human and pet equally, help provide a framework for the articulation of ideas about our mutual importance and interrelatedness or, in the words of Alice Kuzniar, "reciprocity and a kind of intercorporeality" (Kuzniar 2006:126), which is so vital to many pet owners today.[2]

Barbara Dover: *Face to Face*

In *Face to Face,* a 2007 mixed-media installation and photographic series, Australia-based artist Barbara Dover creates photographic renderings of dyads of human and nonhuman animals. Dover (2007b) writes about human relationships with pets in a way that is reminiscent of Renée

Weber's ideas of relatedness, emphasizing "interdependent and interconnected" natures (1990:40). However, where Weber underscores touch rather than vision as the sense that activates this mutuality, Dover prioritizes the gaze—not our gaze at the animal, but a mutual gaze returned and completed by the eyes of the pet. In her photographs of people and their pets, Dover attempts to demonstrate reciprocal eye contact between the participants. Human and animals lay, sit, or stand comfortably together, situated in close proximity, one casually touching the other. However, it is touch rather than a gaze that provides a more or less constant point of contact; more often than not the gaze between human and animal is interrupted in some way, as when one of the participants turns his head or alters position. Certainly for the viewer, who may not even be able to see the eye contact between the two because of the camera angle, the perceptual link is maintained not by sight but by touch.

Arranged in a grid system, her large-scale black and white images are accompanied by objects and drawings on transparent tracing paper overlaying various sections of the photographs. Conceptually layered throughout Dover's images are ideas about the emotion, self-awareness, and sentience embodied by the animals and transmitted to the humans through the locus of their contact. The objects she includes provide a physical manifestation of her conceptual constructs by reinforcing ideas about relatedness and further enhancing the haptic quality of the images. The tracing paper drapes against the photographs, and the other objects, often extremely tactile in nature, are piled and stacked together and then further joined with lengths of twine or ribbon. For many of the *Face to Face* images, Dover presents the human-animal pairs in series of sequential photographs. The settings for the photographs are domestic: bedrooms, suburban backyards or decks, or, in the case of the horses, exterior landscapes with trees. The moments the artist captures are quiet, transitory, and private. The photographs demonstrate the sense of camaraderie and ease that exists between the two figures, and, in many ways, they overtly refuse the conventions that have been established in the history of pet photography. None of the figures is highly posed. Unlike conventional pet photography, which requires the animal to mug for the camera, neither human nor pet appears overly conscious of the camera at all. Filmic in nature, the serial imagery enables viewers to see the pairs interacting over time. Their responses—humorous, playful, whimsical, or warm—are conveyed to the viewer both by facial expressions and by the gentle, mutual, consensual nature of the touch, which activates the viewer's tactile imagination and allows her to share in the palpable, tangible connections between the two participants.

Other photographs in the series are large-scale, full-body representations of human and companion animals looking each other in the eyes. In these cases the artist floats the figures on a white background, removing any surrounding distractions, and allows the viewer to feel as if he is witnessing the private interaction between the figures (figure 2.4). Unlike Casanave, Dover includes the faces of both her subjects; and in the large-scale single images she titles the photographs with the subjects' names. Thus, although we have less information about each pair than Casanave provides to us about herself and her whippet, we are given images that are much more specific in nature: specific to a particular human-animal dyad and to a particular incident of contact. The photographs are approximately life size, and Dover provides access to a felt-sense or visceral quality through the tactility of the objects she includes with the photographs. In *(face to face) entwined* the girl and the horse appear on eye level, taking up equal shares of the frame, and look as if they are playfully nuzzling each other back and forth (figure 2.5). Nestled together on the floor in front of the photographs are long bundles of human and horse hair. The objects extend the installation into the physical space of the gallery and link the viewer to the tactility of the action taking place in the photographs above, just as the tracing paper loops off the wall

FIGURE 2.4. Barbara Dover, *(face to face) Taku and Victoria*, 2007. Giclee print, pencil, 1120 x 1800 mm. www.barbaradover.com.

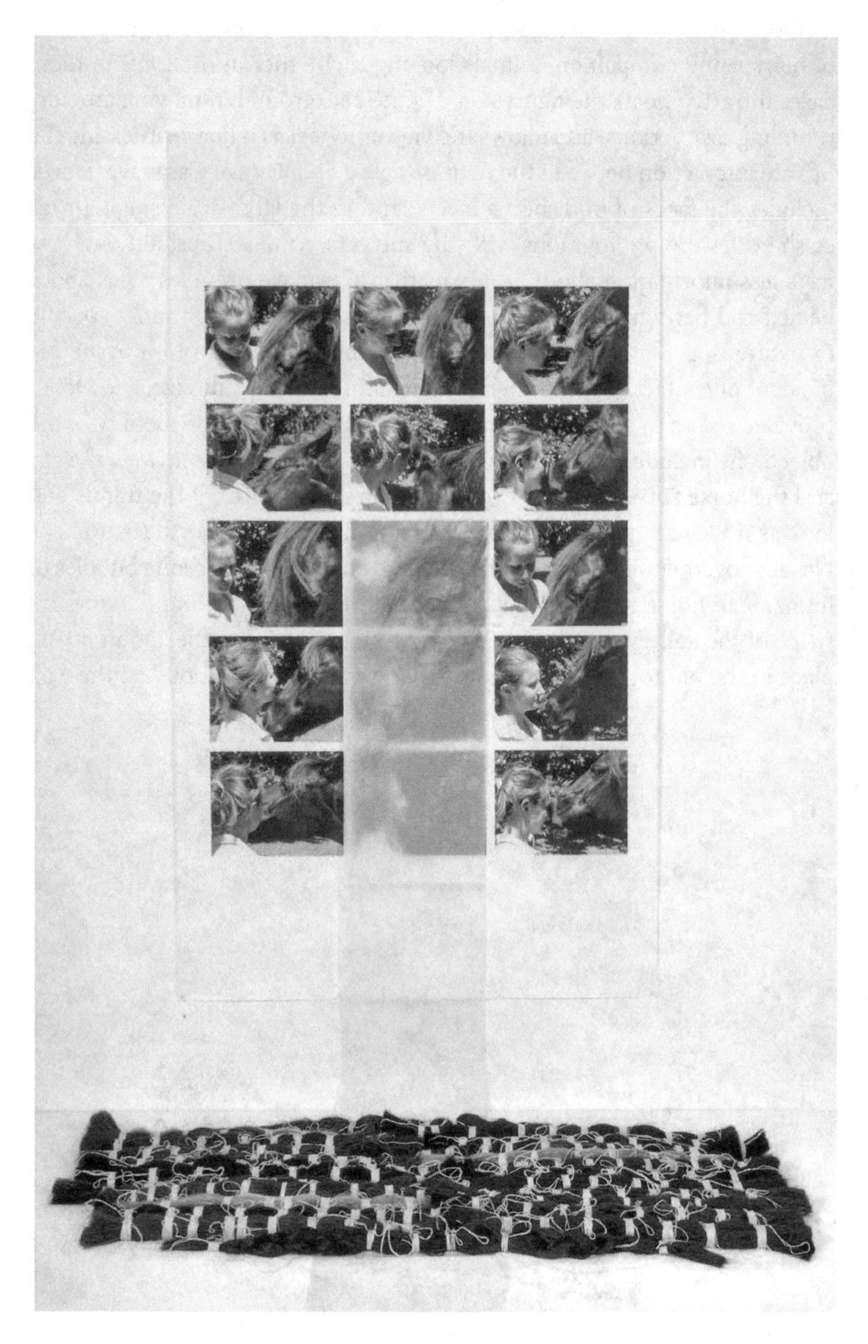

FIGURE 2.5. Barbara Dover, *(face to face) entwined*, 2007. Giclee print, tracing paper, pencil, human hair, synthetic hair, horse hair, 1800 x 1120 x 700 mm. www.barbaradover.com.

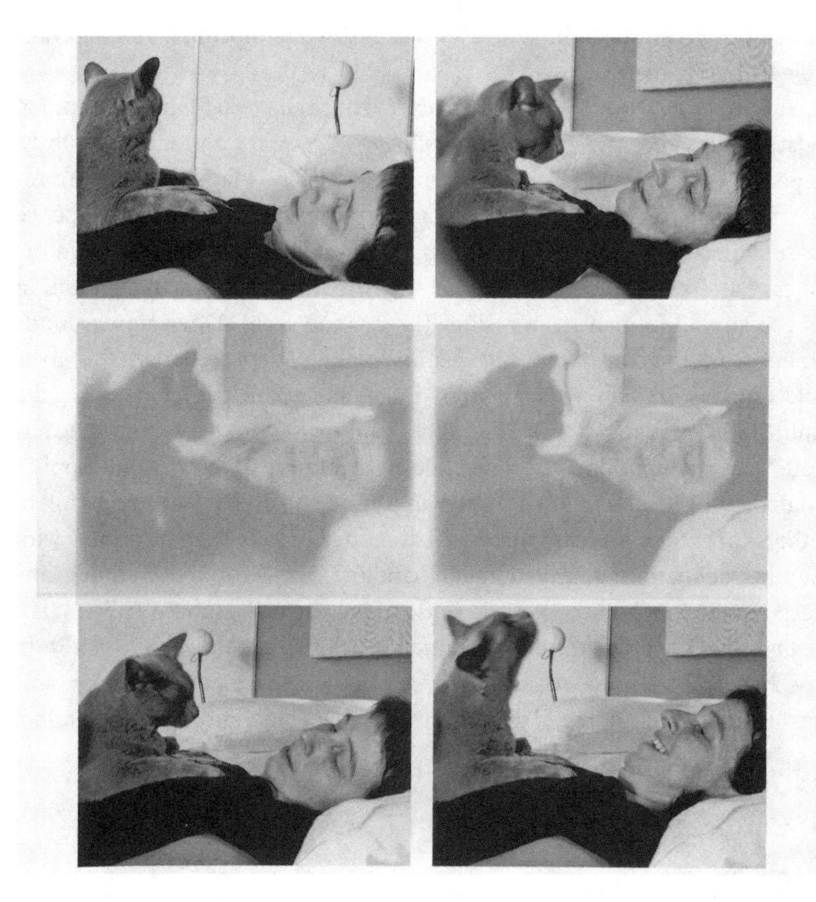

FIGURE 2.6. Barbara Dover, *(face to face) Betty and Julie—heart to heart*, 2007. Giclee print, tracing paper, pencil, 1800 x 1120 mm. www.barbaradover.com.

linking the photographs to the bundles of hair. The bundles extend Weber's "reach" from a simple touch to an entwined relatedness, binding horse and girl together metaphorically as well as physically.

In *(face to face) Betty and Julie—heart to heart* a cat lies across a woman's chest (figure 2.6). The viewer encounters the visceral sense of the weight of the cat on the woman's chest. The proximity between cat and woman, the fact that the cat is clearly comfortable enough to stay on the woman's chest for a protracted period of time with the photographer present, the smile on the face of the woman, and the body language of the cat all indicate a sense of ease, comfort, and reciprocal trust between human and companion animal. Thus the artist shares with the viewer intimate, private moments of contact between the animal and human that would normally not be available to an outsider.

In this series Dover locates in her art making a concern for human-animal relationships and "the moral responsibility we have for animals" (2007b:2), especially for companion animals, and is insistent that the visual strategies she employs be reflective of that sense of responsibility. "The essence of such mutuality, sharing, participation and connection creates a context of care and responsibility, a context appropriate to the visual exploration of the animal-human relationship" (Dover 2007a:22). Dover foregrounds in her work the conviction that relationships she depicts between animals and humans are "unmistakably reciprocal and participatory between the animal and human" (p. 22). We might again turn to Renée Weber to explicate Dover's intentions. Weber's philosophy of touch borrows philosopher Martin Buber's distinction between an I-Thou and an I-It way of relating to others (pp. 26–27). Weber links the I-It mode of interacting physically with another being to "manipulation and handling," while the I-Thou mode corresponds to touch in the holistic sense of "acceptance and mutuality. It is a living relationship, with its own dimensions of space and time" (p. 26). In the *Face to Face* series, Barbara Dover creates a space for the human-animal pairs she photographs to demonstrate their mutuality and connection.

Carolee Schneemann: *Infinity Kisses*

While pioneering feminist performance artist Carolee Schneemann has incorporated cats into her artwork since the 1960s, two bodies of work, *Infinity Kisses I* and *Infinity Kisses II*, in which she photographically documented her interactions with two cats, have gained her particular notoriety. During the 1980s and '90s, both her cats, Cluny and Vesper, woke Schneemann daily with extended periods of licking her lips and mouth. While Schneemann shocks viewers with the transgressive content of the images, she nonetheless opens up a new avenue for artists and viewers alike who are investigating animal psychology. Circumventing language altogether, Schneemann's photographs privilege the exploration of actions of touch initiated by the animals. These actions, recorded photographically, are experienced by the viewer through powerful sensory experiences of their own. Encountering the images, viewers involuntarily feel the cat licking the inside of their own mouths.

Schneemann kept a 35mm film camera next to her bed and photographed the cats and herself. She held the camera out without looking through the lens finder to include the two of them, and so the framing of the shots is variable and somewhat arbitrary (Schneemann 2002:264). Schneemann displays the photographs in the form of a grid of images, and in 2008 produced a short movie that juxtaposes the 35mm images with enlarged details of the photographs (figure 2.7). Each cat appears on top of the artist, straddling her chest with his paws; both usually have their eyes closed, and saliva runs down Schneemann's chin at times. Schneemann's nightclothes and the pattern of her bedsheets vary from image to image, and for the most part Cluny and then Vesper appear to take the place of a human lover. Schneemann was accused of bestiality and obscenity with this work, and certainly she allows her cats to act in a manner that is outside of generally accepted social bounds (Weintraub 1998:133). However, Schneemann accepts the actions of the cats and, in fact, welcomes them as indicative of the close interspecies bond she felt with both Cluny and Vesper (Weintraub 1998:130, 133). By doing so, Schneeman refuses to negate the part her cat is playing in directing certain aspects of their relationship, as she would have had she disallowed the cat's unconventional behavior (unconventional from a human standpoint). Schneemann says, "The cat is an invocation, a sacred being, profoundly devoted to communicating love and physical devotion, and the cat is self-directed" (2002:215).

For Schneemann's work, touch must be contextualized within her forty-year career as a feminist performance artist, a career in which the

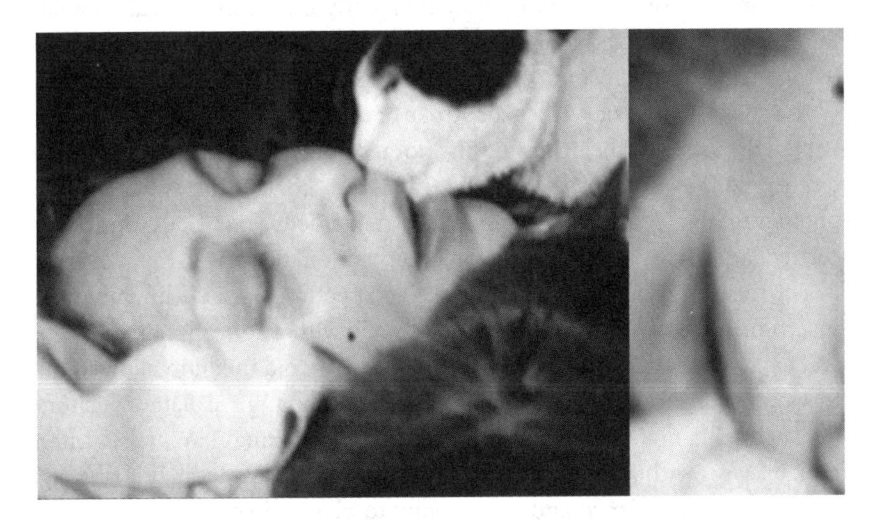

FIGURE 2.7. Carolee Schneemann, "Infinity Kisses—The Movie," 2008. Video still from self-shot photographic series. Copyright © Carolee Schneemann.

corporeal and the tactile have played highly significant roles and within which Schneemann has situated her relationships with her cat companions (Schneemann 2002:10). She credits the inspiration for her first performance piece to her cat Kitch (Schneemann 1997:7), and has incorporated cats as part of her performance pieces and films since that time. Schneemann uses the tactile, the corporeal to tell her stories, to convey her ideas. Because language separates human and animals, she has always been open to modes of communication that replace language: psychic connections, messages in dreams, corporeal and sensual transfers of mental or emotional content (Schneemann 2002:273, 279). The taboo has been part of the ground Schneemann has worked since her career began, and from the first her cats have been part of her creative milieu, as has her sexuality.

Julia Schlosser: *Contact*

I end this chapter with a discussion of a body of my own photographic work, which, like that of the artists previously discussed, explores the complex relationships between pets and humans in the space of the home, both my home and those of others. In the series *Contact* I investigate the nature of touch as it occurs between human and companion animals on a daily basis (figure 2.8). This kind of touch often takes place in the course of daily tending activities. As people feed, groom, play with, or move their animals from one location to the next, they frequently exchange mutual contact, developing daily patterns of habitual interaction. Because I was interested in the transient touch connections that happen many times daily, such as a pat on the head, rub on the leg, or stroke of the tail, I wanted to be as unobtrusive as possible when I visited the homes of my subjects. It was important to me that I did not knowingly frighten or control the animals I was photographing. I neither scripted their movements nor body positions, carried large lights nor intrusive equipment, nor manipulated them overtly.

For this series of photographs, I used a Spectra Polaroid camera with a macro or close-up lens. The camera is small, with a soft on-camera flash, and can be operated with one hand, and the Polaroid medium allowed me to view in the moment what I was photographing. To isolate transitory instances of contact between people and their companion animals, I often took pictures without looking through the lens of the camera. This strategy allowed for coincidence and spontaneity to play a large role in the images. Rarely does the face of the human or animal show in the picture, and,

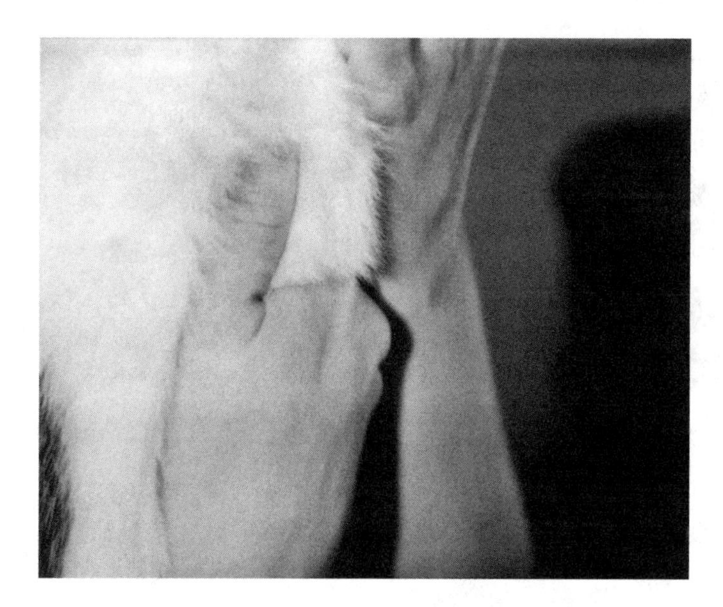

FIGURE 2.8. Julia Schlosser, "Untitled (Alice with Sam)," from the series *Contact*, 2005. Archival pigment print scanned from Polaroid, 6¾" x 5½". Photo courtesy of the artist.

because of the nature of the close-up lens, body parts are fragmented and mixed together. The space within the photographs is often ambiguous, as is the particular action taking place in the image (figure 2.9). These momentary glimpses of contact between pet and human allow the viewer to imagine the entirety of the scene and form her own open-ended sense of meaning about the animal and human subjects in the images. The small, grainy photographs, muted color palette, fragmentary nature of the figures, and spatial ambiguity all develop a unique visual vocabulary for the depictions of animals. This way of seeing animals offers an alternative to the overly processed images of commercial photographers, which often reflect less of the nature of the animals being photographed and more of the desires of the humans for conventional representations.

Pet owners come to know the minds of the animals they live with, and the animals come to know their owners' minds, in myriad ways: by paying attention to the everyday behavior patterns of each other, by watching and listening to each other's reactions when confronted with new situations (for example, those created by other people and animals), and by engaging in

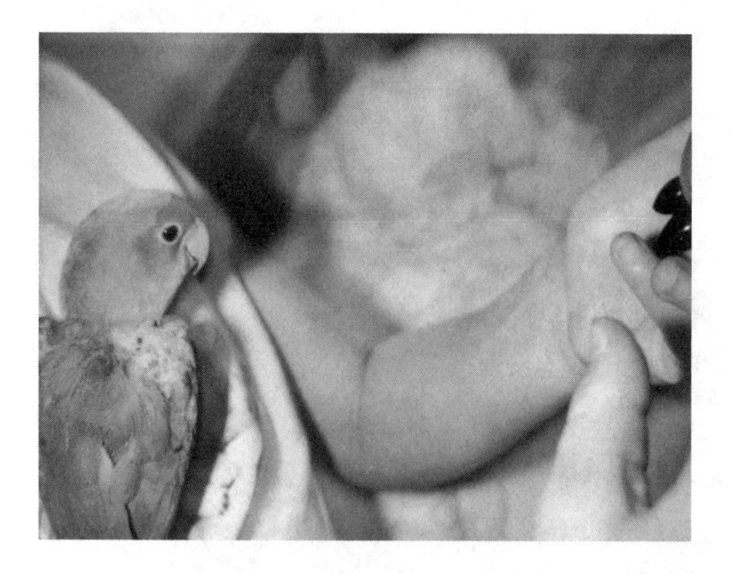

FIGURE 2.9. Julia Schlosser, "Untitled (Pepy with Hazel)," from the series *Contact*, 2005. Archival pigment print scanned from Polaroid, 6 ¾" x 5 ½". Photo courtesy of the artist.

touch. My emphasis has been on knowledge gained through touch, touch through play, mutual social grooming, tending, and care (figure 2.10). The sensory functions of touch and sight are profoundly interconnected, facilitating the formation of social connections or bonds between humans and domesticated animals. I believe that the intertwined nature of touch and sight, and the strong social connection that humans share with companion animals, allow photographs of humans and animals touching to create a special nexus of embodied perception for the viewer.

Using photography, these four artists have explored the possibilities of understanding animals through the vehicle of the body. Each captures a point of contact between animal and human subjects, an intersection between animal and human bodies. The act of contact, this moment of embodied knowledge, is then transferred to the viewer through the viewer's own felt-senses. Absent communication with language, touch is one avenue by which humans and animals learn about each other (and their worlds) and form relationships. Aware that domesticated pet animals often welcome human contact, these artists are free to explore the rich

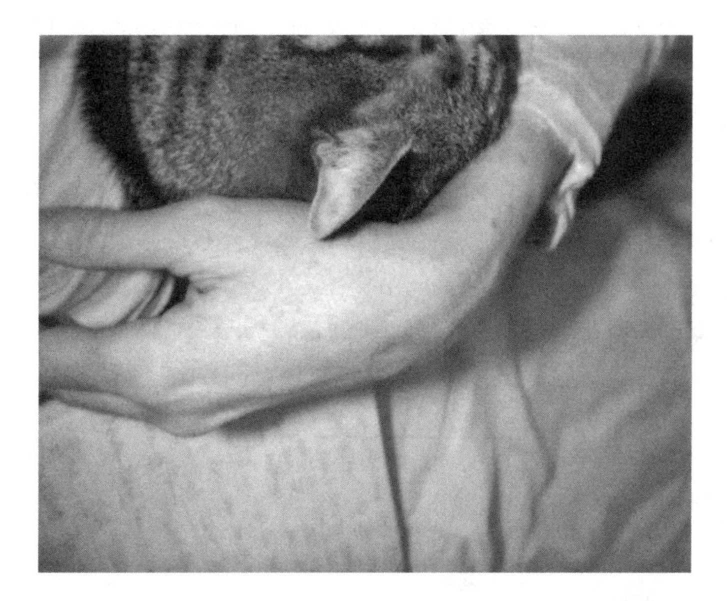

FIGURE 2.10. Julia Schlosser, "Untitled (Sebastian with Julia)," from the series *Contact*, 2005. Archival pigment print scanned from Polaroid, 6 ¾" x 5 ½". Photo courtesy of the artist.

and complex data communicated from established and incidental patterns of contact with animals. This knowledge, intuitive and subtly nuanced, is transmitted directly to viewers through their felt-senses. For companion animals, touch is one of the primary methods with which they can communicate with humans, and more focused observation of the ways they express themselves through touch provides for wide-ranging and perhaps validating evidence of their minds.

Notes

This chapter is dedicated to Claire, who touched my life. I am very grateful to Bob and Julie for their discerning editorial skills and gently relentless encouragement. Thanks also to the artists for their generosity in allowing reproduction of their artwork in this publication.

1. Since the 1970s the feminist movement has made great strides in reclaiming these once maligned areas for artistic production. However, especially in the photo-

graphic community, where modernist conventions have been slow to give way to more inclusive practices, ingrained prejudices about the acceptability of certain subjects still remain in force.

2. Kuzniar uses these words to describe Rhoda Lerman's relationship with her Newfoundland dogs.

References

Baker, S. 2000. *The Postmodern Animal*. London: Reaktion.

——. 2001. *Picturing the Beast: Animals, Identity, and Representation*. Champaign: University of Illinois Press.

Candlin, F. 2006. "The Dubious Inheritance of Touch: Art History and Museum Access." *Journal of Visual Culture* 5, no. 2: 137–154.

Casanave, M. 2002. *Beware of Dog*. Carmel, CA: Center for Photographic Art.

Dover, B. 2007a. *Eye to Eye*. Dubbo, AUS: Western Plains Cultural Centre.

——. 2007b. *Interrogating Gaze*. Townsville, AUS: Pinnacles Gallery.

Kuzniar, A. A. 2006. *Melancholia's Dog: Reflections on Our Animal Kinship*. Chicago: University of Chicago Press.

Levinson, B. M. 1972. *Pets and Human Development*. Springfield, IL: Thomas.

Merleau-Ponty, M. 1962 [1945]. *The Phenomenology of Perception*. Trans. C. Smith. New York: Humanities.

Olmert, M. D. 2009. *Made for Each Other: The Biology of the Human-Animal Bond*. Cambridge: Da Capo.

Perricone, C. 2007. "The Place of Touch in the Arts." *Journal of Aesthetic Education* 41, no. 1: 90–104.

Pratt, J. W., and A. Mason. 1981. *The Caring Touch*. London: HM+M.

Schlosser, J. 2007. "The Postmodern Pet: Images of Companion Animals in Contemporary Photography and Video." Masters thesis. California State University, Northridge.

Schneemann, C. 1997. *More Than Meat Joy: Performance Works and Selected Writings*. Ed. B. R. McPherson. Kingston, NY: McPherson.

——. 2002. *Imaging Her Erotics: Essays, Interviews, Projects*. Cambridge: MIT Press.

Serpell, J. 1986. *In the Company of Animals*. Cambridge: Cambridge University Press.

Ullrich, P. 1996–1997. "Touch This: Tactility in Recent Contemporary Art." *New Art Examiner* 24 (December/January): 10–15.

Weber, R. 1990. "A Philosophical Perspective on Touch." In K. E. Barnard and T. B. Brazelton, eds., *Touch: The Foundation of Experience*, pp. 11–43. Madison, CT: International Universities Press.

Weintraub, L. 1998. "Interspecies Eros." In M. Seppälä, J. P. Vanhala, and L. Weintraub, eds., *Animal. Anima. Animus*, pp. 128–137. Pori: Pori Art Museum.

Wyschograd, E. 1981. "Empathy and Sympathy as Tactile Encounter." *Journal of Philosophy* 6:25–42.

3

Beaver Voices

Grey Owl and Interspecies Communication

ALBERT BRAZ

> Animals have only their silence left with which to confront us. Generation after generation, heroically, our captives refuse to speak to us.
>
> (COETZEE 2003:70)

Excluding perhaps his still controversial attempt to pass as a Scottish-Apache halfbreed, Grey Owl today is best known as the trapper who experienced a life-changing conversion in the 1930s and became an internationally famous conservationist and nature writer. More specifically, he emerged as the great champion of the animal on which he had relied the most to make a living, Canada's own national emblem, the beaver. However, what is not generally known about Grey Owl is that he often attributes his identification with the beaver to their language, which he describes as humanlike. Grey Owl's attitude toward the beaver, as that toward other wild animals, is paradoxical. On the one hand, he asserts that he does not anthropomorphize them. On the other hand, he maintains that the "kinship between the human race and the rest of our natural fauna" is real and "very apparent to those of us who sojourn among the latter for any length of time" (Grey Owl 1990 [1935]:preface, para. 9; see also Dawson 2007:119–121). These interspecies affinities, argues Grey Owl, become most evident in the similarities between human speech and the languages of wild animals. As he says of what he terms the "queer language" of the beaver, it is so akin to ours that "it seemed we could almost understand [it], so human did it sound at times" (p. 93). Indeed, as I will attempt to show in this chapter, Grey Owl comes to believe that the beaver are "small ambassadors from a hitherto unexplored realm" (p. 129) and that it is the responsibility of humans to strive to decipher their speech so that we can communicate with them.

The question of what differentiates human from nonhuman animals has long captivated writers and other thinkers. For someone like Margaret Atwood (2008), the main difference between the two is technological. As she states, somewhat humorously, "All animals eat, but only human beings cook" (p. 9). However, the critical distinction between human and nonhuman animals is much more frequently ascribed to our possessing some kind of consciousness or reasoning, notably the mechanism that is believed to make thinking possible, language. According to Oscar Wilde (1972), for one, there is "no mode of action, no form of emotion, that we do not share with lower animals" (p. 20). The sole exception is speech. "It is only by language," he asserts, "that we rise above [other animals], or above each other—by language, which is the parent, and not the child, of thought" (p. 20). In short, to be human is to speak. Or, to phrase it differently, language is us.

Although broadly embraced, the thesis that it is speech that distinguishes human from nonhuman animals is highly problematic. To begin with, it is hard not to notice that it is humans, notably writers and philosophers, who appear most determined to prove that "the [nonhuman] animal is without language" (Derrida 2002:400). Moreover, even among humans, the idea that the possession of speech defines the human is suspect. As Erica Fudge (2008) has cogently argued, if that were true, then the deaf would not be considered human. Likewise, when one suffers a stroke and loses the ability to speak, one would cease to be part of the human family. The divide between human and nonhuman animals is a fluid one, and, historically, "many members of the human species have been relegated to the 'animal' side of the line," usually because of gender, national, or racial differences (Donaldson and Kymlicka 2007:192–193). Still, the equation of speech with the human is so transparently arbitrary that it begs to be challenged, something that Grey Owl does throughout his writings.

Grey Owl, who lived between 1888 and 1938, was born Archibald Stansfeld Belaney to an upper-middle-class Scottish-English family in Hastings, England. He migrated to Canada in his late teens and moved to northern Ontario, where he eventually assumed an Indigenous identity, adopted the name Grey Owl (Wa-Sha-Quon-Asin, or He Who Walks By Night), and became a trapper, forest ranger, and river guide. However, in the early 1930s he underwent a major intellectual crisis, which led him to abandon trapping and become a conservationist. Soon after, while claiming to be partly Indigenous, he began to metamorphose himself into one of the foremost nature writers and conservationists in the world. Grey Owl owed his renown to four books—two collections of essays about life in the wilderness, a much fictionalized memoir, and a children's novel—as well as to his extensive lecture tours of both North America and Great Britain. The impact of his lectures, which were often accompanied by short documentary films about the beaver he was raising, cannot be overestimated.[1] Grey Owl was not only a magnetic speaker but, as befits someone who wished to be perceived as Indigenous, also tended to address his audiences clad in buckskins and sporting an eagle feather in his long, braided hair. This may explain why people could not get enough of him and his message that the main "difference between civilised man and the savage" is that "civilised people try to impose themselves on their surroundings, to dominate everything" (as cited in Smith 1991:120), and that the only way the wild would be preserved would be if "civilised" people embraced the Indigenous ethos. For example, during his second British tour, in 1937, in just under three months, he gave "138 lectures" (Dickson 1938:16), speaking to "close to half a million people in audiences that sometimes exceeded three thousand at a time" (Grey Owl 1938a:38; Dickson 1938:20). Partly as a result of his talks and interviews, his books sold extremely well, with translations appearing in "all the main European languages" (Dickson 1973:242) and in several Asian ones (Smith 1991:266–267). Most importantly, his celebrity enabled him to undertake a crusade to save the beaver from imminent extinction.

Grey Owl's interest in the beaver is evident in all his works, beginning with the first one, *The Men of the Last Frontier*. Originally entitled *The Vanishing Frontier*, and renamed by the publisher without the author's permission (Grey Owl 1990 [1935]: 209, 253), the collection of essays is fascinating, not the least because, in the course of the book, the central subject appears to go from being a trapper of European descent to one of Indigenous ancestry, without any explanation for the change (Grey Owl 1973 [1931]:1, 252; see also Braz 2005:55–59). However, the text is also significant because of the way it reveals Grey Owl's growing identification with the beaver, an animal he eventually comes to see as his "patron beast" (Grey Owl 1990 [1935]:47). From the outset, Grey Owl (1973 [1931]) is struck by the beaver's humanlike "sagacity" (p. 23), which he states has led "those who know most about them" to believe that "they are endowed to a certain extent with reasoning powers" (p. 155). He acknowledges that building dams and houses and collecting feed does not make beavers unique among wild animals, since "muskrats also erect cabins and store food in much the same manner"

(p. 155). Yet, asks Grey Owl, "where do you find any other creature but man who can fall [*sic*] a tree in a desired direction, selecting only those which can conveniently be brought to the ground?" While he accepts that instinct may explain why the beaver "build their dams in the form of an arc," he fails to see how, other than through their native intelligence, they can "gain the knowledge that causes them to arrange that curve in a concave or a convex formation, according to the water-pressure" (p. 155). Interestingly, Grey Owl underscores that the beaver's acumen does not always have positive consequences. On the contrary, "by the very nature of his work," the beaver "signs his own death warrant. The evidences of his wisdom and industry, for which he is so lauded, have been after all, only sign-posts on the road to extinction" (p. 160), since they make it extremely easy for predators, particularly trappers, to locate the animal.

In any case, for Grey Owl, the most concrete demonstration that the beaver are intelligent is that they have speech. The United States conservationist Enos Mills (1913), whose ideas on the beaver are remarkably similar to Grey Owl's, once noted that he often wished "an old beaver neighbor of mine would write the story of his life" (p. 175). He also contended that when beaver become "separated from one another" when traveling, "they give a strange shrill whistle or call," which "appears to be a call of alarm, suspicion, or warning" (pp. 26, 27). Grey Owl goes even further than Mills. He claims that the beaver not only make sounds like "the moaning of a child" (1973 [1931]:148), but also that they have the ability to recognize human voices, such as his and that of his partner. "Their voices," explains Grey Owl, are "really the most remarkable thing about them, much resembling the cries of a human infant, without the volume but with a greater variety of expression, and at all hours of the day and night there was liable to be some kind of new sound issuing from the interior of the box" where they were being kept (Grey Owl 1990 [1935]:33). According to Grey Owl, "A human voice easily takes on a beaver call because the inflections are very like our own. It's wonderful, how close you can get to the animal kingdom, if you fetch a sympathetic heart with you" (1936:282). As he describes the beaver: "They have a large range of distinctly different sounds. The emotions of rage, sorrow, fear, joy, and contentment are expressed quite differently, and are easily recognized after a short period of observation. Often when a conversation is being carried on they will join in with their vocal gymnastics, and the resemblance to the human voice is almost uncanny to those not accustomed to hearing it, and has been partly the cause of their undoing, as they are a very easy animal to imitate. When in trouble they whimper in the most dolorous fashion, and become altogether disconsolate" (1973 [1931]:197). In fact, Grey Owl maintains that one of the reasons people should ensure the beaver do not become extinct is that "this little beast who seems almost able to think, possesses a power of speech in which little but the articulation of words is lacking" (p. 153). Again, the beaver's language makes them distinct.

Actually, it is his gradual awareness of "the almost human mentality" of the beaver that leads Grey Owl "to quit the beaver hunt altogether" (p. 191), a fateful decision he relates in his most influential book, *Pilgrims of the Wild*. In his memoir Grey Owl credits his transformation from a trapper to a conservationist largely to the influence of his Mohawk wife Anahareo. The fourth of his five wives or common-law partners, and eighteen years his junior, Anahareo was a compelling figure on her own. An urban Mohawk, who ironically tried to rediscover her ancestral roots through the man she called her "Jesse James, that mad, dashing, and romantic Robin Hood of America" (Anahareo 1972:2), she joined Grey Owl on the trapline soon after they married. Her fascination with life in the wild was short-lived, though (Braz 2007:210–213). Anahareo became disturbed both by the wholesale slaughter of fur-bearing animals and by "the great numbers of harmless birds and squirrels caught accidentally" in the traps, "often still alive, some screaming, others wailing feebly in their torment" (Grey Owl 1990 [1935]:23; Anahareo 1972:83). Because of her reaction, Grey Owl began to become conscious of "the cruelties" of his "bloody occupation" (1990 [1935]:23, 24). That is, by her presence on the trapline, Anahareo led Grey Owl to see himself as he was—or at least as he imagined she saw him—prompting him to reconsider his right to kill other animals in general and the beaver in particular.

The pivotal incident in the alteration of Grey Owl's attitude toward wild animals involves a family of beaver. It is late in the season, and the only reason he is still trapping is that, because of a combination of missing game and low fur prices, his financial circumstances are so dismal that he has no other option. Grey Owl (1990 [1935]) relates that, while setting a trap at an old beaver house, he hears "faintly the thin piping voices of kitten beavers." He tries to camouflage the sound from Anahareo, but she has also heard it and implores him "to lift the trap, and allow the baby beaver to have their mother and live" (p. 27). As he describes his response, "I felt a momentary pang myself, as I had never before killed a beaver at this time on that account, but continued with my work. We needed the money" (pp. 27–28). However, the next morning, Grey Owl changes his mind. After he realizes that the mother is missing, he decides to look for the young beaver and, when he finds the kittens, he agrees with Anahareo's suggestion to "save them" as some kind of "atonement" (p. 29). Indeed, it is primarily because of this episode that Grey Owl comes to believe that it is "monstrous" to hunt such creatures and determines to "study them," instead of killing them or "persecuting them further" (p. 53). Consequently, he and Anahareo start a beaver colony, a development that will lead him to write books, lecture, and even join Canada's National Parks Service as a naturalist (Smith 1991:89–92).

In addition to Anahareo, Grey Owl (1990 [1935]) usually credits his "newly awakened consideration" of wild animals and their habitat to what he terms his "imaginative ancestry" (pp. 24, 25). He claims that Indigenous people have a special affinity with wild animals, notably the beaver. Although Indigenous people "must kill" some beaver in order to support their families, Grey Owl writes, they have much respect for the rodents, and perceive them "almost as separate tribes of people, of a kind little different from themselves" (1977 [1935]:5). "Beavers are especially respected," he elaborates, "and some Indians can understand to a certain extent what they are saying to one another, as their voices are not unlike those of human beings." Such is the kinship between Indigenous people and the beaver that the former call the latter "Little Brothers" (pp. 5–6). In fact, it is because of his ostensible Indigenous heritage that Grey Owl develops "a guilty feeling" about his role in the near annihilation of so intelligent a creature (1973 [1931]:142). It is also purportedly because of his broad knowledge of Indigenous history that he resolves to ensure the beaver will not suffer the same fate as the bison, drawing a parallel between himself and the bison hunters of the second half of the nineteenth century. As he confides, while he has "ably assisted in the destruction" of the beaver, now that the animal is threatened with extinction he has "a sudden feeling of regret, something of that vacant feeling of bereavement that comes upon us on the disappearance of a familiar land mark, or on the decease of some spirited, well-respected enemy. Thus the hide hunters must have felt as the last buffalo dropped, so that some of them abjured forever the rifle and the knife, and strained every nerve to bring them back again" (1990 [1935]: 47). He simply does not want what happened to the bison to befall the beaver.

Grey Owl's transition from a trapper to a conservationist was considerably facilitated by the fact he became a writer and, subsequently, a lecturer. Like other "rugged" outdoorsmen-turned-authors (Glotfelty 2004:128),

Grey Owl experienced serious anxieties about his new profession, wondering "at times if it was quite manly to feel as I did toward these small beasts" (1990 [1935]:36). Yet he could not help but discern that there might be some tangible benefits to his being a conservationist and nature writer, even from an economic perspective. As he and his fellow trappers were only too well aware, their industry was in a deep crisis because of overhunting. In his apocalyptic description of the situation, when he meets veteran trappers on the trail, there is "a vague foreboding in their speech," for everyone realizes that the "fur [is] gone" (pp. 12, 13; see also Braz 2007:217–220). So he starts looking for another way to support himself and Anahareo, a process that leads him to surmise that he may be able to do it through writing. Moreover, in an epiphanic moment, it dawns on him that wild animals are "more fun alive than dead, and perhaps if I could write about them they would provide many times over the value of their miserable little hides, beaver included, and still be there as good friends as ever" (p. 141). As he adds, "I inwardly rejoiced that the bloodless happy hunting ground of my imagination was now within the bounds of possibility" (p. 150). That is, in a way, Grey Owl continues to make a living from the beaver. The crucial difference is that, like other naturalists in the early decades of the twentieth century who staged a vigorous campaign "to substitute animal photography for hunting" (Altmeyer 1995:108), he now can do so without having to kill the animals.

That said, the main reason Grey Owl becomes such a passionate defender of the beaver has to do with his evolving attitude toward the beaver themselves, particularly their language. Anahareo asserts that it is the two orphaned kittens they rescue, which they later name McGinnis and McGinty, that are "responsible for the change" in her partner (Grey Owl 1990 [1935]:90). This is a view that is supported by Grey Owl in *Pilgrims of the Wild* and other writings. As he interacts with the kittens, Grey Owl comes to see the beaver not as antagonists but as his "co-dwellers in this wilderness" (p. 49). They are his "comrades-in-arms" as well as his "unarmed fellow-country men," whom he must protect from the unscrupulous part-time trappers who have recently flocked to the North, those "alien interlopers who [have] nothing in common with any of us" (p. 50). For Grey Owl, the beaver are "the Wilderness personified, the Wild articulate, the Wild that was our home" (p. 203). Consequently, he develops "a transhuman ethic" (Braz 2007:207, 222), an idea of citizenship that includes all the legitimate denizens of the wild, both human and nonhuman, not least the beaver.

Grey Owl's identification with the beaver, as noted, reflects the fact that he is profoundly affected by their voices, which both he and Anahareo perceive as "almost" human, "Just like ours" (1990 [1935]:111). Yet ultimately the voices of the beaver create a major dilemma for Grey Owl, since he does not know exactly how to classify the rodents. Sometimes he states that the beaver seem "almost like little folk from some other planet, whose language we could not yet quite understand" (p. 53). They are strangers with a peculiar accent, which is the reason he refers to McGinnis and McGinty as "Immigrants" (pp. 93, 119)—as well as "Little Indians" (p. 42; see also Dawson 2007:121). Other times, though, he contends that the beaver are congenitally "different from other animals; they are very like persons" (1977 [1935]:176). As one of his characters says in the children's novel *Sajo and the Beaver People*, the beaver often appear "to be, not wild things at all, but hopeless, unfortunate little people who could not speak" (p. 125). In other words, it is not always clear whether humans do not understand the beaver because they come from some other celestial body, and we do not yet grasp their language, or because we deem them too insignificant to warrant listening to their voices.

What Grey Owl's fascination with the language of the beaver also illustrates, of course, is that he tends to favor them over all other wild animals. As other scholars have pointed out, there is an inherent "slipperiness" in the line differentiating humanity from animality. This instability has allowed individuals and groups "to suggest, in effect, that some animals have been mistakenly relegated to the animal category, when in fact they are sufficiently human-like to be placed in the human category" (Donaldson and Kymlicka 2007:194). To a certain degree, this is the strategy employed by Grey Owl. It is true that he advocates for the protection of other animals, even stating that animals other than the beaver "too might have qualities, which whilst not so spectacular perhaps, might be worth investigating" (1990 [1935]:129). However, there is no avoiding the fact he places the beaver in a special category of their own, as befits a creature that he says has "developed a degree of mental ability superior to that of any other living animal" (1973 [1931]:155). For instance, when describing the porcupine, Grey Owl writes that it is the "dumb cousin to the beaver, whom he resembles very closely except for the tail, the webbed hind feet, and his bristles. But it seems that when the brains were handed out between the two of them, the porcupine was absent and the beaver got them all" (p. 39). Similarly, he calls the fox "the evil genius of the north country" (1930:574) and the weasel "the gangster of the beaver country, a murderous, slinky no-account" (1936:270, 282). For Grey Owl, the porcupine, the fox, and the weasel clearly should not be equated with the beaver and one of the reasons for this is the beaver's purported intelligence, which is manifest in their ability to build beaver damns and, above all, in their speech. As he responds to a correspondent who inquires if the beaver can reason like humans, "Beaver are able to experience a great variety of emotions and can express them very well both by voice and action" (Grey Owl 1935:3). Once more, one of the pivotal elements that make the beaver special is that, like humans, they possess speech.

Needless to say, the human desire to communicate with nonhuman animals is not devoid of either political or psychological complexities, if not outright contradictions. Many writers and philosophers actually believe that it reflects less our empathy toward other species than our hubris. For someone like James Fenimore Cooper (2004 [1841]:568), "It would exceed all the means of human knowledge to pretend to analyze the influences that govern the acts of the lower animals," such as brown bears. Atwood (2008), in contrast, perceives it as one of those profoundly paradoxical human longings not only to have it all, but to have it both ways. In her words, "We want to drink a lot without having a hangover. We want to speak with the animals. We want to be envied. We want to be as gods" (p. 11). Atwood, in fact, seems to agree with Wilde when it comes to the critical differences between human and nonhuman animals. She contends that, besides fire, which made cooking possible, the only other technology that was "developed by human beings, and by humans alone," is "grammar." While most "animals have methods of communication," and some even have "systems that can be called languages," she writes, "only human languages have complex systems of grammar—systems that allow us to formulate thoughts that dogs, for instance, most likely don't bother with" (p. 9). Yet Atwood illustrates why humans find it difficult to dissociate ourselves from other animals. As she says of trappers:

> I can understand
>
> the guilt they feel because
> they are not animals
>
> the guilt they feel
> because they are
>
> (Atwood 1968:35)

The predicament faced by trappers, which easily might also have been that of the trapper-turned-conservationist Grey Owl, is the human quandary. The reality is that we cannot help but sense that our responses to nonhuman animals are always conditioned by the fact we know that we are not like them—and yet we are.

In the end, the one solid conclusion one can reach about Grey Owl and the beaver is that he is deeply affected by their utterances. Whether those utterances constitute speech or are merely guttural noises remains open to debate. For instance, no less a figure than the animal studies scholar Temple Grandin (2008) maintains that "animals do not have verbal language" (p. 0212). Considering that Grandin is autistic and is fully aware that a significant number of "philosophers think that if you don't have language, you don't have true thought" (as cited in Allemang 2009:F4), Grey Owl's view on animal language is obviously not the dominant one. Yet given the way the beaver were able, through their utterances, to influence the mind of a professional trapper, it does not seem wise to dismiss those utterances as nothing more than purposeless cacophony.

Furthermore, whatever one may think of his views on the beaver and their language, it is evident that Grey Owl does not subscribe to the idea that nonhuman animals are responsible for the lack of communication between themselves and humans. He certainly would not share the sentiments expressed by J. M. Coetzee's character Elizabeth Costello, which serve as the epigraph to this essay, that the only weapon nonhuman animals have "left with which to confront" humans is "their silence" and that, for generations, they have employed it by "refus[ing] to speak" to their captors (2003:70). For Grey Owl, if there has been no interchange between humans and other animals, it is because humans have not made a sustained effort to study the latter's speech. Another significant aspect of Grey Owl's writings on the language of the beaver is that he tends not to reach categorical conclusions. Unlike someone like Grandin, who claims that her autism gives her special insights into the minds of domesticated animals and enables her to determine why they do "the things they do" (Grandin and Johnson 2005:7), Grey Owl is much more tentative about the meaning of what he observes. He has no doubt that humans have numerous affinities with nonhuman animals. Thus he argues that it is "utterly unsportsmanlike" to hunt "other living creatures by means of a horde of dogs, causing pain and terror for amusement," because we are all related. As he reminds his listeners and readers, nonhuman animals "are your fellow-dwellers on this earth; and if you met on Mars you would be inevitably drawn to one another like fellow-townsmen meeting in a foreign country" (Grey Owl 1938b:84). Still, even though Grey Owl is positive that the beaver speak, he is never quite certain what their language means. This is the reason that he exhorts us to study the beaver, and presumably other wild animals, the way we study human groups.

Note

1. Short segments from these films can be watched at http://www.nfb.ca/film/Beaver_People/ and http://www.nfb.ca/film/Beaver_Family/.

References

Allemang, J. 2009. "Autistic Author Beats Dr. Doolittle—by Not Talking to the Animals." *Globe and Mail,* February 7, F1, F4.

Altmeyer, G. 1995. "Three Ideas of Nature in Canada, 1893–1914." In C. Gaffield and P. Gaffield, eds., *Consuming Canada: Readings in Environmental History,* pp. 96–118. Toronto: Copp Clark.

Anahareo. 1972. *Devil in Deerskins: My Life with Grey Owl.* Toronto: New Press.

Atwood, M. 1968. "The Trappers." *The Animals in That Country,* pp. 34–35. Toronto: Oxford University Press.

——. 2008. "Why Poetry?" *Prairie Fire* 29, no. 2: 6–11.

Braz, A. 2005. "The Modern Hiawatha: Grey Owl's Construction of His Aboriginal Self." In J. Rak, ed., *Auto/biography in Canada: Critical Directions,* pp. 53–68. Waterloo, ON: Wilfrid Laurier University Press.

——. 2007. "St. Archie of the Wild: Grey Owl's Account of His 'Natural' Conversion." In J. Fiamengo, ed., *Other Selves: Animals in the Canadian Literary Imagination,* pp. 206–226. Ottawa: University of Ottawa Press.

Coetzee, J. M. 2003. *Elizabeth Costello.* New York: Viking.

Cooper, J. F. 2004 [1841]. *The Deerslayer or the First Warpath.* New York: Signet.

Dawson, C. 2007. "The 'I' in Beaver: Sympathetic Identification and Self-representation in Grey Owl's *Pilgrims of the Wild.*" In H. Tiffin, ed., *Five Emus to the King of Siam: Environment and Empire,* pp. 113–129. Amsterdam: Rodopi.

Derrida, J. 2002. "The Animal That Therefore I Am (More to Follow)." Trans. David Wills. *Critical Inquiry* 28, no. 2: 369–418.

Dickson, L. 1938. "The Passing of Grey Owl." In L. Dickson, ed., *The Green Leaf: A Tribute to Grey Owl,* pp. 9–33. London: Lovat Dickson.

——. 1973. *Wilderness Man: The Strange Story of Grey Owl.* Toronto: Macmillan.

Donaldson, S., and W. Kymlicka. 2007. "The Moral Ark." *Queen's Quarterly* 114, no. 2: 187–205.

Fudge, E. 2008. "Gestural Language and Ideas of Animal Minds in Early Modern Thought." Paper presented at The Minds of Animals: Conceptions from the Humanities, Sciences, and Popular Culture, August 12. University of Toronto, Canada.

Glotfelty, C. 2004. "'Once a Cowboy': Will James, Waddie Mitchell, and the Predicament of Riders Who Turn Writers." In M. Allister, ed., *Eco-man: New Perspectives on Masculinity and Nature,* pp. 127–140. Charlotte: University of Virginia Press.

Grandin, T. 2008. "Response by Temple Grandin to the Essay 'Are Animals Autistic Savants?'" *PLoS Biology* 6, no. 2: 0212.

Grandin, T., and C. Johnson. 2005. *Animals in Translation: Using the Mysteries of Autism to Decode Animal Behavior.* Orlando: Harvest.

Grey Owl. 1930. "Little Brethren of the Wilderness, part 2." *Forest and Outdoors* 26, no. 10 (October): 573–574.

——. 1935. Letter to John H. McCallum, July 14. Donald B. Smith Papers, M 8102, box 4, file 48. Glenbow Museum Archives, Calgary, Alberta.

——. 1936. "Grey Owl Speaks His Mind." *Illustrated Canadian Forest and Outdoors* 32, no. 9 (September): 269–270, 282.

——. 1938a. "My Mission to My Country." As told to Robson Black. *Forest and Outdoors* 34, no. 2 (February): 37–38, 52.

——. 1938b. "Grey Owl's Precepts." In L. Dickson, ed., *The Green Leaf: A Tribute to Grey Owl,* pp. 83–85. London: Lovat Dickson.

——. 1973 [1931]. *The Men of the Last Frontier.* Toronto: Macmillan.

——. 1977 [1935]. *Sajo and the Beaver People.* Toronto: Macmillan.

——. 1990 [1935]. *Pilgrims of the Wild.* Toronto: Macmillan.

Mills, E. A. 1913. *In Beaver World.* Boston: Houghton Mifflin.

Smith, D. B. 1991. *From the Land of the Shadows: The Making of Grey Owl.* Saskatoon: Western Producer Prairie.

Wilde, O. 1972. "The Critic as Artist." In Isobel Murray, ed., *Plays, Prose, Writings, and Poems,* pp. 3–65. New York: Dutton.

PART II

Anthropomorphisms

4

The Historical Animal Mind

"Sagacity" in Nineteenth-Century Britain

ROB BODDICE

I

Sagacity was the prevalent term in nineteenth-century Britain for the intelligence of animals. On this subject there is an orthodoxy of opinion, which this chapter sets out to nuance. Briefly stated, the orthodox opinion is that sagacity helped to order creation hierarchically, with human beings on top. Harriet Ritvo (1987) suggests that sagacity distinguished animal intelligence from human intelligence, noting that should it be "attributed to human beings it often had an ironic or less than flattering connotation" (pp. 37–38). For Ritvo, the "concept of sagacity actually reinforced human dominion. It could be defined so that the animals that exemplified obedient subordination had the largest measure." Under this maxim, domesticated animals, especially dogs, were the most *sagacious*. Rod Preece (2005) asserts that *sagacity* has been used for animals where *reason* would be employed "if the discourse were about humans," denying animals access to "the favoured category of reason" (p. 56). Diana Donald (2007) notes that animal sagacity "was measured by the docility of tame or tameable species; or, failing that, by the impulses of familial tenderness that were thought to offer a moral example to humanity" (p. 106). Animal intelligence therefore served to reflect an anthropocentric view. This intelligence, however, differed in *kind*, rather than by *degree* from human intelligence, a fact ascribed to the era's distinction between instinct and reason (p. 107). Though Donald notes the increasing fuzziness of this distinction, the superiority of human intellect was still retained (pp. 109–110).

II

Sagacity has etymological roots in the Proto-Indo-European base **sag-*, meaning "to track down, trace, seek," and, indeed, in seventeenth- and eighteenth-century England the term generally referred to a keen sense of smell (see Sagacity n.d.). Its Latin equivalent, *sagacitas*, stands for "quickness of perception," and the English usage clearly has ties to the French *sagacité*, which resembles "shrewdness." *Sagacity* is therefore ambiguously defined. Samuel Johnson (1755–1756) represented this confusion in his *Dictionary*, defining the term as relating to "quickness of scent," to "acuteness of discovery" and to being "quick of thought"; it was synonymous with "penetration." Johnson took special care to show where *sagacity* referred to thought (in humans) and where to scent (in animals).[1] Yet there was a blurring of the lines. The "sagacious" actions of scenting animals were colloquially celebrated as much for intellectual display as for olfactory prowess.

To fully grasp what was meant by sagacity in animals, we must first understand its bearing on human intellect. Sagacity was political more often than zoological. "Political sagacity" was commonly ascribed to those public figures who were bestowed with a natural gift in debate or in diplomacy. It served sometimes to satirize but more frequently to honor the intellectual cunning of "great men."[2] In one notable and useful passage, William Minto (1881) tried to weigh the *nous* of Edmund Burke, whose political sagacity he deemed not "of the first rank" (p. 440). Yet he noted that it is "of course, unprofitable to argue regarding a term so vague" (p. 440). Even among those dealing in human intelligence, *sagacity* was a slippery term. But it was without doubt a natural intellectual quality to be valued.

With this usage in the cultural background, observers and gatherers of anecdotal evidence on animal sagacity saw no clear distinction between scenting and thinking, since finding a scent and following it betrayed discriminatory powers of mind. An end was apprehended through an apparent process of deduction: Scent equals the previous presence of another animal; it went *that* way; if I want (to eat, kill, or mate with) it I must follow the scent. Sagacity therefore referred to processes of deduction, regardless of the capacity to follow scent. The elimination of the distinction between scenting and thinking made ascriptions of human sagacity equivalent to ascriptions of animal sagacity. As one purveyor of anecdotes remarked in 1824, animals could "outstep that faculty which forms part of their nature" and "display an extraordinary sagacity which . . . seems to approach reason" (Anonymous 1824:5–10). This elision, from scenting to thinking, broadened the category of sagacious animals from its most obvious association with dogs to include potentially any "intelligent" animal.

Anecdotal stories for children and adults alike made the most of sagacity's greater import. In *Stories of Animal Sagacity*, the author undertook to prove to his young readers that animals were capable "of exercising a kind of reason, which comes into play under circumstances to which they are not naturally exposed" (Kingston 1874:13). This was sagacity, which was largely reserved for "those animals more peculiarly fitted to be the companions of man," but probably existed "in a certain degree among wild animals" (p. 13).For many writers, the concept of sagacity allowed them to endow animals with human-like reason without going so far as saying so explicitly. As early as 1742, Dennis de Coetlogon (pp. 1–2) proclaimed that "all learning," without "natural sagacity" was "in *effect* nothing." In keeping with orthodox ideas, the sagacity of animals provided examples from which men might learn: "It is this great Sagacity which teaches Men and Beasts, to take care of themselves; it is thus that Men were instructed and learn'd from the *Sagacity of Animals*, many useful particulars for the cure of *human Maladies*."

Even among those more sceptical that sagacity could be defined as reason, this capacity was nevertheless "superior to mere instinct as ordinarily displayed" (Pardon 1857:180). George Frederick Pardon took great lengths to demonstrate how apparent animal cognition was not the same as human reason and how in most cases animal intelligence could be explained by instinct. Man, after all, was, "from his superior intellectual organization," "the natural protector and ruler over brutes" (p. 207). Nevertheless, Pardon had to err on the side of caution, since the "intelligence of animals seems sometimes to lead to very remarkable results, and occasionally to the formation of plans that the reason of man would probably have failed to adopt under circumstances of a similar nature" (p. 179). Likewise, Priscilla Wakefield (1811), in a book whose title seemed to conflate sagacity with instinct, sounded a note to the contrary with regard to the most intelligent of animals. "Quadrupeds, after man, are the most intelligent of the lower world," she said, "and the most capable of deviating from the instinctive impulse." This was "evinced by innumerable well-attested instances of sagacity, that seem to be the result of reflection and experience, in the horse, the dog, and the half-reasoning elephant" (pp. 277–278).

The notion of contiguity in nature was taken to extreme lengths by J. E. Taylor (1884) who wrote a treatise on the *Sagacity & Morality of Plants*. Taylor (pp. 4–5) thought that the "intelligent acts" of animals proceeded from "cerebration, exactly in the same way as the intelligent and rational actions of men." If *sagacity* had previously been thought to denote in animals a "different kind of mentality" to that in humans, then this usage was "old-fashioned" and would "soon become extinct." In a similar vein, John Selby Watson (1867) wrote a pseudo-scientific treatise with the singular object of showing that "inferior animals . . . have a portion of that reason which is possessed by man" (p. 1). The refusal to admit this was borne "from fear of admitting them to be on a level with ourselves" (p. 2). For this classically educated clergyman such fears were unwarranted. He made this clear in his precise definition of reason, which included not the "higher" qualities of "abstraction and generalization," but rather the "power of understanding . . . what is presented to the observation, and of forming conclusions from experience, so as to conceive of consequences, and to expect that what has happened under certain circumstances at one time will happen under like circumstances at another" (pp. 1–2). Watson clearly thought sagacity to be synonymous with reason, so defined. Most explicitly, anecdotes of the sagacity of dogs proved them "to have some share of the reason which pre-eminently distinguishes man" (p. 60, see also pp. 459–460).

Charles Darwin did not radically alter the debates about the status of animals (Boddice 2009:279–284, 317–324; Preece 2003). Age-old disagreements about their rationality continued as before, but with an altered vocabulary and, admittedly, in much greater volume. For all those, like Watson, and later George Romanes, who argued for the logical connectedness of human and animal minds, there were other prominent voices who dissented. William James (1878) argued that animal sagacity involved mental processes that "may as a rule be perfectly accounted for by mere contiguous association, based on experience," unlike higher human reasoning, which was unempirical and based on applied theories and prior knowledge in connection (pp. 237–249, 260). Similarly, C. Lloyd Morgan (1890–1891), lamenting tendencies to anthropomorphize, plainly stated that "in man alone, and in no dumb animal, is the rational faculty": "the pity of it is that we cannot think of [animals] in any other terms than those of human consciousness. The only world of constructs that we know is the world constructed by man" (pp. vi, 335).

Darwin himself struggled to map the theory of evolution onto the mind and encouraged his young disciple Romanes to push the field. Prominent scientists grappled unsatisfactorily with the concept of intelligence—not to mention sagacity—in man as well as in animals. T. H. Huxley, for example, was not above name calling in terms running contrary to the principles of Darwinian contiguity in human and animal intelligence. He even wondered about the mental capacities of Darwin himself: "Exposition was not Darwin's *forte*—and his English is sometimes wonderful. But there is a marvellous dumb sagacity about him—like that of a sort of miraculous dog—and he gets to the truth by ways as dark as those of the Heathen Chinee" (quoted in Cock and Forsdyke 2008:129). Sagacity was clearly not quite human reason (or at least not quite civilized enough) for Huxley, but it did nevertheless facilitate arriving at the right answers to rational questions! Darwin himself wrote Romanes in 1881, with regard to the manuscript of Darwin's work on worms (see Darwin and Seward 1903). "In the middle," he said, "you will find a few sentences with a sort of definition of, or rather discussion on, intelligence. I am altogether dissatisfied with it." He could not pin down what he meant by "intelligent," but knew that "it will hardly do to assume that every fool knows what 'intelligent' means." He expressed pity for Romanes, whose work dealt precisely with this issue (pp. 213–214).

Romanes (1882:2–17) himself, in his most famous work, *Animal Intelligence*, used the term *sagacity*, but left its ambiguity intact. Although he went to some lengths to establish the distinctions between reflex action, instinct, and reason, he did not mention *sagacity* in his introduction. This makes his subsequent use of the term irritatingly unclear. Thus we are left to wonder at the meaning of "an instance of sagacity—indeed, amounting to reason—in a dog" (p. 466), as well as the displays of "remarkable sagacity" in elephants, which were "probably fabulous" (p. 386). He detailed "unusually high" displays "of sagacity" in elephants, juxtaposing the term with references to "higher mental faculties" and "docile intelligence" (pp. 396–397). Thus he elevated the concept of sagacity, associating it directly with reason, as opposed to some form of naturally occurring intelligence between instinct and reason. Canine intelligence, in particular those manifestations of it in a dog's powers of communication, is given by Romanes as "sagacity." Indeed, Romanes's usage would make canine sagacity akin to "a high degree of intelligence" (p. 447). After dealing with dog emotions and capacities to communicate, Romanes pursued "cases showing the higher and more exceptional developments of canine sagacity," deploying anecdotes to make his case (pp. 447–470).

As Romanes's work on animal and human psychology developed, his usage of the word *sagacity* and his dependence on anecdotal evidence declined. The word does not appear in the more complex if less widely read sequels to *Animal Intelligence*, entitled respectively *Mental Evolution in Animals* (1883) and *Mental Evolution in Man* (1888). Romanes worked tirelessly to establish that animal and human psychology differed only by degree and not in kind. As he strove for analytical precision, sagacity fell away as a useful concept. Nevertheless, Romanes never ceased to admit that his science was necessarily anthropomorphic: "We must always remember that we can never know the mental states of any mental beings other than ourselves as *objects*; we can only know them as *ejects*, or as ideal projections of our own mental states," he stated, continuing that "it is from this broad fact of psychology that the difficulty arises in applying our criterion of mind to particular cases—especially among the lower animals" (1883:22). Note that "higher" and "lower" are established a priori, on the basis of similarity to human physiology.

Analogies of mind depended on inference rather than on fact. Judging animal minds was by Romanes's own definition a reflexive activity without any objective basis. This inherent anthropomorphism in studies of animal minds was ultimately rejected by Lloyd Morgan, who ushered psychology toward behaviorism. Yet anthropomorphism essentially defined sagacity's meaning, for it placed animal cognition into human categories of understanding. Discussions of the intelligence of animals were rooted in a desire to explain, justify, or apologize for the intelligence of humans, whether distinct or not.

The fine distinctions employed to explain the origins or ends of thinking in animals are crucially important to the understanding of sagacity and also for grasping the richness of anthropomorphism in nineteenth-century discourse. The aforementioned orthodoxy can thus be modified: Sagacity could earmark an intelligence as lower than, and different in kind to, that of humans, or it could distinguish the human and the animal mind by degree only, or it could establish human-animal contiguity through a form of "natural" ingenuity (distinct from instinct), or it could even mean "reason." Applied to humans it was more likely to be honorific than derogatory and referred in both humans and animals to a shrewdness in problem solving. Sagacity's meaning, ultimately, was dependent on context. Turning now to the rat, fox, and dog, it will emerge how the concept of sagacity could be played out in varying contexts.

III

In accordance with Jonathan Burt's (2006) claim that rats hold a "central, sometimes disturbing, role in human culture," serving as dark "mirrors" of human beings (pp. 7, 13), nineteenth-century stories of rat sagacity typically style the rat as either thief or villain. The *Quarterly Review* ("Rats" 1857) pronounced that "the sagacity of the rat in the pursuit of food is so great, that we almost wonder at the small amount of its cerebral development. Indeed he is so cunning, and works occasionally with such human ingenuity, that accounts which are perfectly correct are sometimes received as mere fables" (p. 130). If rats stole with human ingenuity then they were guilty of the crime of theft.

Two story forms are oft repeated by different writers, involving the stealing of eggs and the consumption of oil or other liquids in bottles. The method of stealing eggs involved the placing of an egg on the belly of one rat lying on its back, which then allowed other rats to pull it along by its tail. The safety of the egg was thereby ensured, and the tellers of anecdotes wondered with some awe at the collaboration, communication, and intelligent problem solving of the animals involved. Similar accounts note the effective teamwork of rats moving an egg up or down stairs, handing the precious food to each other to assure its safe passage. Rats consumed cooking oil (or sometimes treacle or jam) by removing corks from bottles and dipping in their tails. They could then easily lick their tails clean and repeat until the oil was consumed ("Sagacity of Rats" 1825; "Rats" 1857:130; Anonymous 1862:262–263; Jennings 1885; Morris 1885; D. 1896:120–124; Lee 1852:264; Rodwell 1858:93–95; Romanes 1875:515, 1882:361–363; Pardon 1857:180; Kingston 1874:224; Sax 1992:101–109; Watson 1867:289–297).

The reaction to these acts of sagacity was usually to the detriment of the rat. James Rodwell (1858), perhaps the chief expert on (destroying) rats in the nineteenth century, thought that "if rats could by any means be made to live on the surface of the earth instead of in holes and corners... there would not be a man, woman, or child but would have a dog, stick, or gun to effect their destruction, wherever they met with them" (p. 2). The rat's ability to evade human traps and death at human hands made it a threat. Its intelligence added to its reputation of malignancy. The *Quarterly Review* ("Rats" 1857) referenced its "wonderful presence of mind" when in danger, remarking on its "reasoning power": "The sagacity of the rat in eluding danger is not less than his craftiness in dealing with it when it comes" (p. 134). "Man" was the rat's "most relentless and destructive enemy," due to "the repulsive idea which attaches to this animal under every form" (p. 137). The reputed anecdotist Mrs. Lee (1852) filled her book with stories of killing rats. Rat sagacity lacked morality, leading to "mischief": the eating of human food, the undermining of houses, the burrowing through dams, the destruction of drains and the committal of "incalculable havoc, in every place and every thing" (p. 262). A ferocious, immoral rapacity defined their inferiority to humans. They partook "of the character of the wolf, and in their cunning, of that of the fox" (Kingston 1874:223; see also Watson 1867:289, 299; Romanes 1882:361). The rat combined the qualities of what were colloquially considered to be the most vicious, noxious, and intelligent of animals. Its emotions did not appear "to be of an entirely selfish character," rats having been "frequently known to assist one another in defending themselves from dangerous enemies" (Romanes 1882:360). This combination of qualities, together with the absence of conscience, was a nightmarish proposition to be extinguished at all costs. Rat sagacity threatened human dominion.

Foxes are proverbially clever, cunning, and sly: qualities synonymous with sagacity. The fox's proverbial cleverness has been endlessly documented (e.g., see chapter 3, Wallen 2006). The nineteenth century's unanimous acclaim for the superior intelligence of the fox was coupled with an overwhelming desire to kill it. Stories of the sagacity of foxes generally took two forms. Either the fox demonstrated its cunning as a rogue, through the stealing of food from humans (this is generally given as a reason for killing it), or the fox showed its sagacity in escaping from those trying to kill it. A significant ritual element of foxhunting was the concept of fair play, and nothing appealed more to the minds of nineteenth-century sportsmen than the prospect of a worthy opponent (Boddice 2009:290–291, 298–299). Unlike thoroughbred hounds, designed by man to pursue, the fox's mind was a product of nature (Marvin 2001). Its cunning made it fair game, but it also murkily reflected human reasoning. The depraved fox served as an analogy of humanity's inner beast.

W. H. G. Kingston (1874) demonstrated the lessons for humanity to be drawn from animal intelligence in his *Stories of Animal Sagacity*. Of the fox he said that "no other animals so carefully educate their young in the way they should go. . . . He is a good husband, an excellent father, capable of friendship, and a very intelligent member of society; but all the while . . . an incorrigible rogue and thief." The moral was immediately forthcoming: "Do not pride yourself on being perfect because you possess some good qualities. Consider the many bad ones which counteract them, and strive to overcome those" (pp. 194–207, especially 196–197). All very well for humans, but the fox's moral character—encapsulated in its sagacity—was beyond redemption. The example bolstered the fox's status as malignant, destined to be routed out and killed. Mrs. Lee (1852) illustrated this tendency well. "The fox is generally a suspicious animal," she said. Regardless of the degree to which the animal was habituated to the company of man, "he seems to think he is going to be deceived and ill-treated: perhaps he judges of others by himself." Accordingly, Mrs. Lee's anecdotes tend to end with the killing of foxes, despite the fact that they are "much too coy and clever to be easily entrapped" (pp. 174–176). Watson (1867) too noted the fox's "amiableness," despite "his craftiness," and he provided numerous examples of the fox's sagacity in escaping its fate at human hands (pp. 259–263).

Wonderment at fox intelligence—its stark refusal readily to submit to its "fate" without putting man "to all his shifts" ("Morality" 1870, quoting Earl Winchilsea)—was heightened by foxes' independence of human influence. Romanes, recognizing this, dedicated a separate chapter of *Animal Intelligence* to foxes, wolves, jackals, etc., on the basis that "from never having been submitted to the influences of domestication, their mental qualities present a sufficient number of differences from those of the dog" (1882:426). The qualities of these animals were determined precisely by the lack of mankind's civilizing influence. To arrive at an understanding of their character, one must "subtract from the domestic dog all the emotions arising from his prolonged companionship with man, and at the same time intensify the emotions of self-reliance, rapacity, &c." (p. 426). Relying on well-rehearsed anecdotes, Romanes concluded that the sagacity of the fox (comparably with the rat) was "of a very remarkable order," nothing short of *reason* (pp. 426–429). But the assumption of this sagacious character's "rapacity" provided justification for the human proclivity for killing foxes.

Anecdotal accounts of canine sagacity are the most difficult neatly to summarize because of their sheer volume. F. O. Morris's *Records of Animal Sagacity and Character* (1861), for example, devotes 109 pages to the first chapter, on the sagacity of dogs, out of a total of 302 pages including 27 additional chapters. The dog, clearly, was considered the foremost of sagacious animals. Watson (1867) set out to prove that dogs "have some share of the reason which pre-eminently distinguishes man" and used sagacity interchangeably with "reason" and "intelligence" (p. 60). Kingston (1874) noted the "staunch fidelity . . . courage . . . devotion and generosity" exhibited by the dog, marvelling at "his wonderful powers of mind," and ranking it "the most sagacious of all animals" (p. 52). The dog's closeness to its human "owner" defined its superior intelligence. More than any other animal, the dog demonstrated "delight in the companionship of man" (D. 1896:60). In a stronger sense, it was the human's intelligent crafting of the dog's character that defined dog intelligence as an analogue of that of humans.

Canine sagacity was widely, and anthropomorphically, understood in these terms. Priscilla Wakefield (1811), for example, noted that dogs "seem highly favoured" with regard to intelligence, possessing "more sense than most of their fellow-brutes" (p. 90). She reasoned that this was because they lived "on such a very familiar footing with man" and that they "owe part of their superiority to that circumstance." Charles Hamilton Smith (1839) thought that a dog's sagacity, among other qualities, made it impossible for humans to withhold their "admiration and affection" (p. 78). All the more so since dog intelligence directly aided the "higher purposes" for which man was "created." "Subordinate creatures," he thought, were "so constituted as to be important elements of co-operation," and "were called into existence to further [human] design, and to facilitate his intellectual development" (p. 103). In other words, if dogs were considered sagacious, then they were so *for* man; they were *designed* to be understood in human terms. If *Canis familiaris* could be traced to several distinct species of wild dogs, then variations were caused by man for his own ends.

This point was taken to its logical conclusion as evolutionary theory advanced. Romanes (1882) thought that dog intelligence was "special," having been "from time out of record . . . domesticated on account of the high level of its natural intelligence." What began with "natural" sagacity was "greatly changed" by "persistent contact with man, coupled with training and breeding." This contact caused "general modification in the way of dependent companionship and docility" (i.e., the capacity to learn or be educated) as well as "a number of special modifications, peculiar to certain breeds, which all have obvious reference to the requirements of man" (p. 437). The implications of human intervention in the selection and breeding of dogs were realized by Romanes with all the grand eloquence he could muster in support of Darwin's theory of natural selection: "The whole character of the dog may therefore be said to have been moulded by human agency with reference to human requirements." Elaborating, he observed that "for thousands of years man has here been virtually, though unconsciously, performing . . . a gigantic experiment upon the potency of individual experience accumulated by heredity; and now there stands before us this most wonderful monument of his labours—the culmination of his experiment in the transformed psychology of the dog" (pp. 437–438). If sagacity in the case of the dog can be demonstrated to mean exactly what Ritvo stated it meant—obedient subordination—then this in large part was because humans had themselves *created* it. Dog sagacity was not threatening, malignant, or immoral—as in the case of the rat and the fox—because it was understood to be a human form of intelligence, instilled and controlled by human hands.

IV

John Selby Watson (1867) summarized speculative thought on the reasoning capacity of animals, incorporating the concept of sagacity: "We have now sought for indications of reason and understanding through the inferior animal creation from the elephant to the ant and the beetle," he said, finding "signs of intelligence in various creatures . . . quite distinguishable from the mere promptings of instinct." Watson pointed out "how dogs distinguish themselves by their general sagacity and perception of things," and he paid attention "to the exhibitions of sagacity and artfulness in monkeys and rats, cats and foxes" (p. 459). "With all these particulars before us," Watson (p. 460) said, "may we not say of many of the animals of the present day, as Milton made the angel say of those in Paradise,

> They also know
> And reason not contemptibly?"

The quality of sagacity was widely distributed, not particularly favoring animals that associated closely with humans over those deemed wild and untameable. Yet the meaning of sagacity varied according to the context of human-animal relations. The mind of a dog represented the mind of man, since its intellect had been developed by and for man; the minds of rats and foxes could only be understood through human standards and categories of intelligence. Sagacity could be either allotted in greatest measure to those animals most obedient to man, since man had a great share in its being there *at all*, and therefore to animals that were analogously *most like man by contrivance*, or allotted in greatest measure to those animals most fiercely independent of man, and, in fact, in direct competition with man's means of survival, and therefore to animals that were analogously *most like man by nature*. In either case, sagacity was defined anthropomorphically. The character of an animal was determined by the way in which its sagacity was presumed to have been acquired. An animal whose intelligence was independent of man often commanded respect but nevertheless was subject to elimination. Conversely, those intelligent animals that assisted man and depended on him tended to be honored and protected. Leonard Larkin saw this as early as 1879. Responding to a lecture by Romanes given in 1879 (see Romanes 1877–1879), Larkin pointed out that animal intelligence was "only our own reflected by them. . . . They have nothing of it themselves and would never have it unless they were brought into connection with man" (p. 15).

Animal sagacity was significant principally in its human implications. In partial accord with the assertions of Ritvo and Preece, I have argued that sagacity in *some* animals denoted animal subordination, even faithful obedience. In dogs, especially, sagacity was taken to be a human accomplishment. In other cases, sagacity in animals could not be ascribed to human intervention, and, where animals competed directly with humans, this intelligence was maligned, however remarkable. Precisely, therefore, when these animals were most intelligent, they were the least subordinate and the least obedient. In an anthropocentric world such a sagacious animal sealed its own fate.

Notes

This chapter presents a development of the argument, using largely new material, of Boddice 2009:284–304.

1. See the entries for *sagacious, sagacity, penetration.* Literary references under *sagacious* were, in relation to animals and scent, by Milton and Dryden; references for the same, for human thought, were by Locke.
2. An impressionistic measure of the prevalence of this term can be apprehended through a search for the exact phrase *political sagacity* in all books published between 1800 and 1899 currently listed on books.google.com. The return of 1006 hits (as compared with 651 for *animal sagacity*) is telling by itself (https://books.google.com, accessed February 24, 2009; search terms: "political sagacity" date: 1800–1899; "animal sagacity" date: 1800–1899.

References

Anonymous. 1824. *Animal Sagacity, Exemplified by Facts: Shewing the Force of Instinct in Beasts, Birds, &c.* Dublin: W. Espy.

Anonymous. 1862. *The Children's Picture-Book of the Sagacity of Animals.* London: Sampson Low.

Boddice, R. 2009. *A History of Attitudes and Behaviours Toward Animals in Eighteenth- and Nineteenth-Century Britain: Anthropocentrism and the Emergence of Animals.* Lewiston, NY: Mellen.

Burt, J. 2006. *Rat.* London: Reaktion.

Cock, A. G., and D. R. Forsdyke. 2008. *Treasure Your Exceptions: The Science and Life of William Bateson.* New York: Springer.

D., D. J. 1896. *Stories of Animal Sagacity.* London: S. W. Partridge.

Darwin, F., and A. C. Seward, eds. 1903. *More Letters of Charles Darwin: A Record of His Work in a Series of Hitherto Unpublished Letters.* Vol. 1. London: John Murray.

de Coetlogon, D. 1742. *Natural Sagacity the Principal Secret, If Not the Whole in Physick; All Learning, Without This, Being in* Effect *Nothing. Which Is Contrary to the Assertion of a Pamphlet, Lately Publish'd, Call'd, "One Physician Is as Good as T'other, &c."* London: T. Cooper.

Donald, D. 2007. *Picturing Animals in Britain, 1750–1850.* New Haven: Yale University Press.

James, W. 1878. "Brute and Human Intellect." *Journal of Speculative Philosophy* 12:236–276.

Jennings, S. 1885. "The Rat." *Times,* July 9, p. 5B.

Kingston, W. H. G. 1874. *Stories of Animal Sagacity.* London: T. Nelson.

Larkin, L. 1879. *Animal Intelligence; Being a Reply to a Lecture Given by George J. Romanes.* Manchester: Chas. Sever.

Lee, Mrs. R. 1852. *Anecdotes of the Habits and Instincts of Animals.* London: Grant and Griffith.

Marvin, G. 2001. "Cultured Killers: Creating and Representing Foxhounds." *Society and Animals* 9:273–292.

Minto, W. 1881. *A Manual of English Prose Literature, Biographical and Critical: Designed Mainly to Show Characteristics of Style.* London: William Blackwood.

"Morality of Fox-Hunting." 1870. *Manchester Guardian,* January 5, p. 7F.

Morgan, C. L. 1890–1891. *Animal Life and Intelligence.* London: Edward Arnold.

Morris, F. O. 1861. *Records of Animal Sagacity and Character: With a Preface on the Future Existence of the Animal Creation.* London: Longman, Green, Longman, and Roberts.

——1885. The Rat. *Times,* July 29, p. 13C.

Pardon, G. F. 1857. *Dogs: Their Sagacity, Instinct, and Uses; with Descriptions of Their Several Varieties.* London: James Blackwood.

Preece, R. 2003. "Darwin, Christianity and the Great Vivisection Debate." *Journal of the History of Ideas* 64:399–419.

——. 2005. *Brute Souls, Happy Beasts, and Evolution: The Historical Status of Animals.* Vancouver: UBC Press.

"Rats." 1857. *Quarterly Review* 101:123–41.

Ritvo, H. 1987. *The Animal Estate: The English and Other Creatures in the Victorian Age*. Cambridge: Harvard University Press.

Rodwell, J. 1858. *The Rat: Its History and Destructive Character; with Numerous Anecdotes*. London: G. Routledge.

Romanes, G. J. 1875. "Tails of Rats and Mice." *Nature* 12 (October 14): 515.

——. 1877–1879. "Animal Intelligence: A Lecture Delivered in the Hulme Town Hall, Manchester, March 12, 1879." In *Science Lectures Delivered in Manchester*, pp. 152–170. 9th and 10th series. Manchester: John Heywood.

——. 1882. *Animal Intelligence.* 3d ed. London: Kegan Paul, Trench.

——. 1883. *Mental Evolution in Animals.* London: Kegan Paul, Trench.

——. 1888. *Mental Evolution in Man: Origin of Human Faculty.* London: Kegan Paul, Trench.

Sagacity. n.d. In Online Etymology Dictionary. Retrieved July 10, 2008 from http://www.etymonline.com/index.php?search=sagacity&searchmode=none.

"Sagacity of Rats." 1825. *Times*, December 27, p. 3E.

Sax, B. 1992. *The Parliament of Animals: Anecdotes and Legends from Books of Natural History, 1775–1900*. New York: Pace University Press.

Smith, C. H. 1839. *Dogs, Canidæ, or Genus Canis of Authors: Including also the Genera Hyæna and Proteles*. Edinburgh: W. H. Lizars.

Taylor, J. E. 1884. *The Sagacity & Morality of Plants: A Sketch of the Life & Conduct of the Vegetable Kingdom*. London: Chatto and Windus.

Wakefield, P. 1811. *Instinct Displayed, in a Collection of Well-authenticated Facts, Exemplifying the Extraordinary Sagacity of Various Species of the Animal Creation.* London: Darnton, Harvey, and Darnton.

Wallen, M. 2006. *Fox.* London: Reaktion.

Watson, J. S. 1867. *The Reasoning Power in Animals.* London: Reeve.

5

Science of the Monkey Mind

Primate Penchants and Human Pursuits

SARA WALLER

In this chapter I will give textual evidence for several claims. First, I will argue that there is a pop culture paradigm that demands *surprising similarity* between humans and nonhuman primates in terms of their mental lives, especially in terms of cognition and emotionality. As I discuss headlines from popular media,[1] I will review the originating scientific studies of nonhuman primate cognition and note that they too aim at similarity finding and are actually structured to reveal similarity rather than to uncover disparity or profound difference.[2] I will finally suggest that similarity seeking ultimately narrows the way science is being done and propose, with Uexküll (1985 [1909]), that a good cognitive ethology should also incorporate studies that concentrate on a broader, less anthropocentric range of cognition and emotional states and that such a science will prove to be more useful than we might immediately imagine.

Surprising Similarity in Scientific Studies and the Media

Which similarities between nonhuman primates and humans are surprising, and which are not? Evolutionarily speaking, all higher organisms must be able to navigate using the senses and avoid danger. So we would not be surprised if we were told that cotton-topped tamarins can see or feel or that orangutans can sense heat and move toward the shade on a sunny day. We are also not surprised if we hear that rhesus monkeys express fearful behaviors or that great apes can demonstrate aggression (Balcombe 2006:19).[3] Why not? First, we come from a long Western tradition that expects certain things of animals and not others. Living a life based on sensations and primal emotions such as fear and aggression is the paradigmatic case of living an animalistic life rather than a thoughtful, reflective human life. René Descartes (1999 [1637]) claimed, in the *Discourse on Method*, that animals do "not act on the basis of knowledge, but merely as a result of the disposition of their organs" (p. 40). To be governed by the physical drives and bodily passions is the Cartesian definition of what it is to be animal and to be unconscious. But Descartes did not found this dividing line. Aristotle (1941 [322 bc]) writes in *De Anima,* "In all animals other than man there is no thinking or calculation" (book 3, part 10, p. 597). Immanuel Kant (1981 [1785]), in the *Groundwork*, dismissed the moral status of animals: "If they are irrational beings, [they have] only a relative value as means, and are therefore called things; rational beings, on the contrary, are called persons, because their very nature points them out as ends in themselves, that is as something which must not be used merely as means" (pp. 35–36). The dividing line between humans and animals continues in the present day, where we find Daniel Dennett (1996) arguing that animals often have "Skinnerian" or "Popperian" but not "Gregorian" minds. He claims they can learn through stimulus and response (Skinnerian), as well as can model reality internally (Popperian—so their hypotheses can die in their stead), but they are fundamentally unable to incorporate their tools into their conceptual structure and thus modify their mental landscapes through the use of technology. He claims that "we, and only we, have developed [a technique of] deliberately mapping our new problems onto our old problem-solving machinery" (p. 145). For Dennett, animals simply do not have the cognitive equipment that humans do. From this brief survey, we can see that major philosophers across several time periods have argued that there is a clear dividing line between human and beast, and that dividing line has something to do with rationality, cognition, thoughtfulness, self-reflection, and the higher, more esoteric forms of mentality. As a result of participating in this culture of teachings, we tend to expect basic similarity between human and animal in the ability to perceive the world and feel primitive emotions and drastic difference in cognition. That which violates these expectations, then, is what I will take to be surprising.

Headlines are designed to grab our attention, and thus the familiar line mocking the news media "if it bleeds, it leads" is well-known to all. Headline makers are alert for discoveries that somehow bump up against our traditional beliefs and expectations. This is as true for news in the science of animal minds as it is for news in the human world, and headlines often reveal the traditional presuppositions of human-animal division: rationality, cognition, self-reflection, etc. For example, note these headlines about parrots who can do mathematics, "Parrot Prodigy May Grasp the Concept of Zero" (Pickrell 2005), or speak as humans do, "Counting, Speaking Parrot Alex Dies" (2007), or apparently join the ranks of MENSA, "Alex, Genius Parrot Who Dazzled Scientists Dead at Thirty-one" ("That Damn Bird" 2007). These headlines show we have not changed much in what seduces our attentions since the social sensation of the horse, Clever Hans, who was purported to count, add, and subtract. Likewise, reports on New Caledonian crows who make their own tools lead with headlines such as "Crow Reveals Talent for Technology" (Pain 2002) and "Crows Match Great Apes in Skilful Tool Use" (Randerson 2007).[4] But birds are not the only ones that can surprise and excite us, and the abilities of mammals everywhere have a full share of the media. Headlines spotlight the symbolic skills of vervet monkeys who practice gestural communication in "Monkeys Terrorize Village in Kenya" (Njeri 2007), "African Monkeys Sexually Harass Women" (Kalla 2007), "Sexist Monkeys Terrorize Village" (Rolen 2007), and the very clever "Monkeys Ape Sex Harassment" (Neighbour 2007). The media is likewise alert to dolphins, those cooperative social predators of the sea who understand syntax: "Dolphin Whistles Offer Signs of Language Ability" (2000). And the spotlight falls on creatures from chimpanzees to elephants who gaze at their own mirror reflections: "Mirror Test Reveals Elephants Are No Dumbos" (2006) and "Dolphin Self-Recognition Mirrors Our Own" (Wong 2001). We can see that surprising similarities across the animal kingdom have caught the attention of reporters in a variety of popular venues.

Of course, there are headlines based on differences, specifically, on astounding abilities some animals have that far outstrip those of humans.[5] After all, if headlines sell media subscriptions based on generating surprise and interest, then differences are good candidates for startling the public into reading the article. Headlines showcasing remarkable differences include, "Weird Shrimp Has Astounding Vision" (2008), "Cancer-Sniffing Dog to Be Cloned" (Chan-Kyong 2008), "Cat Plays Furry Grim Reaper at Nursing Home" (2007), "Chihuahua Saves Boy from Rattlesnake" (2007), "Electric Shark Snot" (2007),[6] and "Animals That Glow in the Dark—Bioluminescence Is Not New Technology" (Bourque 2008). This essay is not arguing that such astounding differences and their corresponding headlines do not exist. Rather, I am suggesting that one common media project is that of highlighting (and perhaps exaggerating) surprising similarity. There may well be several other media projects aimed at other aspects of animal behavior and cognition, though I suspect that these too are founded on assumptions about what humans can or cannot do and go on to measure animals by the same yardstick.

Primate Similarities in Cognition

While all animals are candidates for presenting us with some sort of surprising similarity, primates are ideal candidates for bringing us subtle and sophisticated similarities because they are closely related to us in evolutionary terms and bear both physical and sometimes social resemblances to humans.

"Go Ahead, Rationalize, Monkeys Do It Too" is a New York Times science section headline from November 6, 2007. Far beyond mere addition or toolmaking, rationalization is a sophisticated mental feat in self-deception—we gather evidence that our decisions are good while ignoring evidence to the contrary, we paint ourselves in a rosy light to preserve a sense of self or self-efficacy that may not be entirely true. What is a compelling human example? In a decades-old study by Brehm (1956), adult humans were asked to rank the desirability of different household items and then asked to choose one (for example, juicer versus toaster). Upon selecting one—e.g., the juicer—the subjects were once again asked to rank the desirability of the toaster. Without fail, the unchosen item was later ranked as less desirable than it had been ranked originally, suggesting that we unconsciously bolster our choices with rationalizations to place them in the best light. In the partner primate study (Egan, Santos, and Bloom 2007), Capuchin monkeys who preferred red, green, and blue M&Ms equally were asked to select a single color. Once red was chosen over blue, for example, the monkeys would consistently favor other colors over blue, even though before the forced choice, the blue ones were seemingly just as tasty. So, capuchins seem to devalue the item not chosen just as humans do, thus avoiding the cognitive dissonance known as "buyer's remorse."

In a moment, I will point out alternative interpretations for these study results that may be slightly less anthropocentric, but first I note that the primate study is a fairly precise imitation of the human study. That is, this primate research clearly sought to reproduce results found in a human study: The experimental structure is similar, and the understanding of the results appears to have been developed from the standpoint of the human study. While science must bootstrap from one study to the next, this example suggests that we use established facts and interpretations of human behavior to set up our exploration of animal behavior and non-human mental states.

I now turn to a brief complication of the standing interpretation of these experimental results. Far from establishing rationalization, these experiments could easily support alternative interpretations. For example, when forced to choose, one may suddenly find oneself weighing all sorts of attributes that were previously not present to consciousness. Perhaps I liked the toaster and the juicer equally well until I had to select one, at which point I decided that I toast items far more often than I juice, that the juicer takes more counter space, and that I have a supply of bagels in the freezer already, while fresh fruit has become rather expensive. That is, further reflection may reveal reasons that are not rationalizations, but legitimate and thoughtful indicators of preference. To interpret these experimental results as indicative of rationalization that protects us from buyer's remorse is to take a rather cynical (and human) approach to the data.

Likewise, there is more than one interpretation of the capuchin behavior. When allowed only one M&M, perhaps the capuchins do favor the taste of the red dye over the blue, or perhaps they favor a specific experimenter who is offering them the forced red choice. We should not assume so readily that capuchins do not taste the M&M dyes or that they do not have social motivations for their behaviors. The correct placement of Occam's razor here is quite unclear. Should we consider their behaviors to be based on gustatory sensations? On social proclivities? Or on a tendency to rationalize their choices? The fact that the researchers chose to mirror the interpretation and conclusion of the human experiment rather than keep to the probably more conservative interpretation of sensory preference (or even social preference in a social mammal) shows a departure from the Aristotelian and Cartesian tradition that can be explained by a bias toward finding surprising similarity between the species.

Another interpretive problem appears when we explore the structure of the study more closely. In this study baseline preference was measured by how long the capuchin subjects took to retrieve individual M&Ms. Since individual M&Ms were presented, a hungry monkey might find any color equally desirable and reach for it equally quickly. After being forced to choose, the fact that the monkeys will favor the color they previously chose is not necessarily a sign of rationalization. They may simply follow a conservative paradigm of selecting again what has been successful in the past. Well-trained laboratory monkeys who have been subjects of many types of cognitive tests may well believe that remembering what has been presented in the past, and selecting it again, will lead to further rewards. (This is a common cognitive paradigm known as *win-stay*—one continues to choose what one has been rewarded for choosing. Such behaviors are often rewarded, as win-stay behavior is considered to be rational and revelatory of mental processes common to executive function.) But the study was explicitly designed to attempt to measure a common response to cognitive dissonance and was specifically modeled according to the choices the humans made between toasters and juicers in the original cognitive dissonance study. In their article, Egan, Santos, and Bloom (2007) ask, "Are humans unique in their drive to avoid dissonant cognition, or is this process older evolutionarily, perhaps shared with nonhuman primate species?" (p. 978). Clearly, the researchers were motivated by the possibility of finding commonality—surprising similarity in cognitive process—between humans and capuchins. Thus the popular headline emerging from the cognitive dissonance study, "Go Ahead, Rationalize, Monkeys Do It Too," is not so far from the intentions of the researchers, and the goal of finding similar behavior toward cognitive dissonance obscures both other possible similar reactions and potential cognitive differences that can manifest in the same reaction.

Another example will serve to illustrate similar points. In an article with the headline "Young Chimp Beats College Students" (Ritter 2007), researchers report that chimps do better on tests of short-term memory than human adults. The popular article explicitly notes that we generally think that humans are overall better at "executive function" type cognitive tasks than nonhuman primates, so to discover that chimps actually outperform us is surprising. In the partner scientific article, the authors assert this same motivation for their research when they note, "The general assumption is that, as with many other cognitive functions, it [chimpanzee memory] is inferior to that of humans; some data, however, suggest that, in some circumstances, chimpanzee memory may indeed be superior to human memory" (Inoue and Matsuzawa 2008). While one might debate exactly what it means to have a better memory (the chimp performance advantage was not in accuracy but in speed—chimps could press a series of presented numbers faster than we humans could), the explicit research agenda was to explore the similarities in cognition between humans and chimps, and the authors of both articles work on the implicit assumption that a case in which chimps outperform humans would be surprising.

In a third popular article, "Primates Expect Others to Act Rationally" (Lavoie 2007), we find that a variety of primates, ranging from cotton-topped tamarins to chimps to rhesus macaques, seem to respond to the gestures of people using theory of mind, that is, they respond to actions using behaviors consistent with their attributing intentionality to humans. These primates could distinguish between the case of an experimenter intentionally tapping on a container to indicate that it was full of food and unintentionally knocking it with a swinging or flopping hand. The second experiment added to these findings, indicating that these primates distinguished between an experimenter "pointing" to a container with an elbow when her hands were full and an experimenter with empty hands motioning the elbow accidentally toward a food container. Like humans, tamarins, chimps, and macaques assumed that an elbow would be used for pointing when the hands were full, but that hands were more likely to be used for pointing if they were free. While it is not clear that these experiments actually test for "rationality" in any classical sense of the term, they do seem to indicate that these primates perceive and act on goal-directed behavior and treat certain human gestures as intentional.

In the companion scientific article (Wood et al. 2007), the authors explicitly note that their experiments were designed to reveal whether these primates have the same "mindreading" capabilities as humans do, and whether they use them in the same ways. While the researchers make no specific claims about the rationality of their subjects, they do note evidence for both theory of mind and an understanding of intentional, goal-directed behavior and suggest that the ability to use such an "intentional stance" may have evolved over forty million years ago in our shared ancestors.[7] The researchers pose the puzzle early in the article that while "humans are capable of making inferences about other individuals' intentions and goals . . . presently unclear is whether this capacity is uniquely human or is shared with other animals" (Wood et al. 2007:1402).

Such findings in nonhuman primate cognition mirror other recent findings regarding the emotional landscape of nonhuman primates, showing that chimps become more frustrated when an experimenter intentionally teases them by offering food and then taking it away than they do when an experimenter offers food and then accidentally drops it (Call et al. 2004). In popularized studies of surprising emotional sophistication such as "Chimps Get Angry, Not Spiteful" (2007), we find that chimps seek revenge on those who purposefully steal food (that is, they try to destroy the food), but they leave in peace those who are just luckier than they are to get food. Chimps would collapse a table containing food when a thieving chimp was eating, but when food was simply made unavailable to them but available to other chimps, the others were left to eat in peace. This media article suggests that chimps are acting on a rather sophisticated and shared principle of just behavior, leaving punishment for thieves but otherwise accepting fate. The original scientific article, "Chimpanzees Are Vengeful But Not Spiteful," explicitly states the exploration of a human behavioral parallel as a motivation for the study when they explain, "People will willingly suffer a cost to punish others. . . . However, it is not known whether animals other than humans react to harmful actions directed toward them by retaliating against the perpetrator, and whether they react to disproportionate outcomes by behaving spitefully toward the fortunes of others" (Jensen, Call, and Tomasello 2007). The authors are curious about the evolution of a sense of justice and whether humans uniquely will sacrifice themselves in order to achieve punishment of wrongdoers. In a rare moment the researchers conclude that the chimps do not achieve exactly the same responses as humans.[8]

These examples strongly suggest that researchers often participate in actively attempting to uncover surprising similarity between human and nonhuman primate cognition. This is not, in itself, a completely wrongheaded endeavor. Seeking *surprising similarity* might be scientifically valuable, as clearly the development of shared capacities over millennia gives us clues as to the importance of those capacities for survival as well as their environments of origin, their lineage, and their relative success as adaptive capacities. Science ought to commit itself to potentially uprooting biases in traditional thought, and we progress by recognizing when our a priori commitments are not supported by empirical evidence. However, when we search for similarity we place ourselves at the center of the research and explore animals in relation to us. This may interfere with our ability to discover animals' cognitive abilities and propensities on their own terms, in their own context, and in light of their own goals and biases. Thus the hopeful directive of this chapter is that we ought to try to notice when we are being self-centered and strive to do better in order to best serve our processes of discovery.

Anthropocentrism in Measuring Intelligence and Emotionality

All these studies reveal presuppositions we have for defining and measuring *emotionally sophisticated* or *intelligent* behavior. Intelligent creatures, according to these experiments, do such things as use symbols, attribute mental states to other creatures, and have good memories. Likewise, emotionally sophisticated creatures tend to reduce cognitive dissonance and displays of frustration, anger, and spite through rationalization (that is, they control spiteful urges). Clearly, these are limiting and distorting parameters for both *intelligence* and *emotionality*. While intelligence probably does demand symbolic manipulation and memory, the attribution of mental states to other creatures (theory of mind) may well be only a likely rather than an essential component of intelligence and not at all a prerequisite. As humans it is easy for us to conflate intelligence with rationality and both of these with the ability to attribute mental states to others, but other important components of intelligence, such as the ability to problem solve and exhibit flexible behaviors when presented with new stimuli, may be independent of symbol use and intentionality attributions. Our *form of life*, as Wittgenstein might call it, pulls us to navigate the world through intentionality and symbolic thought. Similarly, we humans often take subtle shades of emotion (beyond the simpler feelings of fear, anger, desire, and contentment) to indicate cognitive sophistication, and so we find it easy to look for the more complicated feelings of spite or jealousy (or the ability to control spite and jealousy) in similar animals.

This approach, while important in terms of noting evolutionary points of similarity as well as evolutionary origins of certain traits and abilities, is inadequate in at least three respects. First, for such experiments to reveal similar traits, the subjects must also be cooperative, food motivated, trusting, and familiar with people, as well as implicitly accepting of basic principles of folk physics, object permanence, cause and effect, and the correctness of generalizing from past experience; and they also must generalize in specific ways along salient features of the experiments. That is, relevant similarities are assumed in order for us to prove other relevant similarities. Of course, with Quine (1977), we might simply think that all experiments test multiple, conceptually linked hypotheses and that results confirm or disconfirm complex hypothetical premises as wholes rather than as atomistic postulates. We have no way of knowing, given disconfirming evidence, exactly which of our dozens of assumptions is incorrect or faulty. But we can step back from the entire project and note that it is motivated by the discovery of similarity and that it assumes concomitant similarities in the psychology of nonhuman primates. We can note that other projects that both assume and seek difference are equally possible and perhaps equally valuable.

Second, these assumptions about primate cognition stand in partial contrast to what we explore in humans using standard IQ tests such as the Stanford Binet, making the claims about animal intelligence incomparable to our studies on human intelligence. The Binet focuses on vocabulary, mathematical problem solving, pattern recognition, and visual and verbal memory, and human children are ranked in terms of their general intelligence as a result of their performance on these subsections of the test. Of course, the Binet has been widely criticized as being a measure of enculturation, whiteness, and relative affluence as well as exposure to specific educational opportunities; however, it also is a proven measure of relative successfulness and achievement later in life. Given the importance of Binet scores to children, parents, and scientists studying intelligence, we should perhaps be surprised that ethologists have not in recent years developed any type of cognitive test that is particularly parallel to vocabulary or word definition tasks, or mathematical problem solving, when they have approached the study of primate cognition.

Some of this is because primates are not verbal in the way that we are and so cannot be asked to define terms or solve word problems. Some of this is because training primates to complete far simpler tasks takes years and significant training talent. And I think some of this is because we have no clear idea of what constitutes successful achievement in the later lives of nonhuman primates. The Binet was developed to predict scholastic achievement, which it does to some extent, but it has proven to be a fabulous predictive tool for financial and social success in the Western world. What would a pseudo-Binet predict for primates? We do not have a clear standard against which we can measure the external validity of such a test. We can critique the ability of the Binet to measure intelligence by finding students with low IQs who achieve good grades in college or locating young adults with high IQs who are unsuccessful in the workplace and in providing for themselves in other ways. But what is the parallel to academic and personal success for a nonhuman primate? When we say that chimps are smarter than, as smart as, or equal to their human counterparts, nothing clearly counts as the falsifying case—both because a nonhuman primate Binet does not exist and because, if one did exist, it would be quite unclear what sort of success it would have to measure in order to measure intelligence. So it is unclear what any of our claims about the comparability of human and nonhuman primate intelligence actually mean.

My third point follows from the second; none of these studies is focused on, or designed to measure, any emotions or cognitive processes that are distinctively nonhuman. While mapping the evolutionary progression and development of cognitive abilities is important, some of the most interesting aspects of evolution are anomalies. The duck-billed platypus gives us wonderful information about flexibility, adaptability, and genetic mutation, even though these creatures are far from human. Likewise, creatures that make different sorts of inferences or sport different types of concepts or practice other forms of species-specific cognition (having nonpropositional content, using context-based thought patterns, etc.) can offer us much information for the overall research project of cognitive ethology. Rather than measure the relative lack of spitefulness in chimpanzees, why not develop new notions of emotional response that are particular to chimpanzee life in the wild and then study it in a controlled setting? Humans, after all, may well get jealous because they very often live in a competitive world of private possessions and exclusive, serially monogamous relationships. Why would we expect to find jealousy in species that live communally and polygamously? (Perhaps, but perhaps not.) And more expansively, in a communal, polygamous society of apes, what emotions might develop with which humans in general might be less familiar? Can traditional Mormon society, with its practices of marrying many wives to one husband, help us reconceive our experimental design? *Big Love* might inform the typical Western atheist, Christian or Jewish scientist in a surprising way. What nuances of emotion does the third wife feel toward the first and second wives? To the fourth? Such a model could productively inform experiments in several primate societies, and I call for such nuances of emotion to be explored, defined, and developed for experimental use. Better yet, can polygamous primate societies explain Mormon culture, rules, emotions, and mores?[9]

Moving away from emotions and more toward thought and cognition, I want to raise the question of nonpropositional inferences. Last semester one of my Ph.D. students stated both that primates make inferences and that they do not think. I thought it was a brilliant confusion that reflects the general state of thought in the field. On the one hand, as cognitive ethologists and comparative psychologists, we are intrigued by the potential to discover exactly what types of inferences nonhuman animals make. On the other hand, we make every attempt to stay away from anthopomorphisms such as attributing "thought" or "deduction" to nonhuman animals. We can't have it both ways. If animals make inferences, then they participate in some form of thought, even though we suspect that they do not think in sentences or make word-based deductions. Fodor (1990) may well be right to say that dogs probably cannot distinguish between thoughts such as "the meat in my dish is good tonight" from "the food in that receptacle in the corner that I usually eat from is delicious." There is substantial difference in content at a refined level, and dogs are probably unable to discuss those differences. However, they are well able to make inferences about the location of the food and the familiar eating situation (so as to locate that food behind a screen, to actively beg for the food in order to have a door opened for them, or to perform a sequence of tricks in order to have access to food). Since our basic intuition is that animals do make inferences without propositional thought, and since we are highly propositional thinkers, experiments that seek to highlight nonpropositional inferences could reveal the necessary conditions of thought broadly construed and familiarize us with the structure of nonlogocentric inference. Nonpropositional inference patterns will probably still depend on assumptions of object permanence, linear time, and some form of autobiographical memory in the animals tested. But such experiments can be refined to reveal deductions that support the survival and flourishing of the animal rather than explore what we might expect them to be able to do when we have trained them to use a symbol board, human-constructed syntactical patterns, or human labels for objects.

We could begin our science with fewer assumptions of similarities and aim our experiments at discovering differences. We could think about what we actually mean by intelligence and emotional sophistication and define these terms in such a way that they apply across species, rather than using terms designed to describe humans in animal cognition research. We could reconsider the structures of inference and cast them in a way that is nonpropositional. This sort of science would tell us about minds generally, rather than how animal minds relate to human minds, and would stop placing the human mind at the center with the other minds merely in orbit. If thought has a structure, then the lack of propositionality of that structure may be more interesting than seeking how much we can get nonhuman primates to poke at keyboards, gain vocabularies, and work in our propositional world. And from this type of scientific project, a new series of headlines might emerge: headlines about the nature of thought in the animal kingdom or contributing factors for survival for creatures with specific capabilities in specific contexts or new forms of emotion that humans have yet to experience commonly. Rather than react to the dividing line between humans and animals that has been produced by an insecure Western tradition clinging to the (shrinking) specialness of humanity, we can actively seek to explore the nature of mind and intrigue the public with the findings that will emerge from this new paradigm.

In the meantime, we will be confronted with more headlines like this one: "Smarter Than You Think: The Animal Kingdom Is Home to Much Greater Intelligence Than Has Been Previously Acknowledged, with Scientists Seeing Evidence of Human-like Traits Everywhere" (Leake and Warren 2010).

Notes

1. EthologicalEthics Forum (forum messages) and Google Alerts (Google search results on "animal consciousness and cognition," July 1, 2007–August 10, 2009). Retrieved from http://pets.groups.yahoo.com/group/ethologicalethics & http://www.google.com/alerts
2. Some headlines suggest *amazing differences,* as these too capture the attention and imagination of the media-consuming public. While this chapter will focus on surprising similarity, I would like to suggest that researchers working to uncover and report amazing differences are also taking a basic anthropocentric stance: noting what humans do and then exploring animal behavior and cognition with a fundamentally human framework.
3. Balcombe made this point beautifully. He notes that the scientific community has at least twenty journals discussing experiments and results on animal pain and none on animal pleasure.
4. These headlines all contradict the Heideggerian notion that humans are defined by, and separated from the animals by, their unique use of technology.
5. Robert Lurz at Brooklyn College brought me this objection.
6. The original link (http://blogs.discovery.com/news_animal/2007/11/electric-shark-.html) is defunct, but the equally dramatic "Shark Snot Key to Following Bloody Trails" can be found here: http://dsc.discovery.com/news/2007/11/06/shark-gel-blood.html.
7. Why then did the popular headline feature *rationality* rather than *mindreading* or *intentionality?* Perhaps the popular headline indicated rationality because

rationality is more surprising than mind reading—I'll leave that question to reader speculation.

8. Daniel Povinelli's initial work on similarities between apes and humans in self-recognition was noted in popular media (Achenbach 2004). Since then, he has published quite a bit of scientific research highlighting the difference between humans and animals. I did not see this work highlighted at all in the popular headlines, and I suggest this could be explained by the fact that his conclusions are not surprising, i.e., they fit with the traditional Western belief in a dividing line between human and animal cognition. He reveals neither surprising similarity nor astounding differences.
9. Thanks to Julie Smith for making this point and thus making this paper less anthropocentric.

References

Achenbach, J. 2004. "Monkey See, Monkey Recognize: What Are Animals Thinking?" *National Geographic* 205, no. 1 (January): 3.

Aristotle. 1941 [322 bc]. *The Basic Writings of Aristotle.* Trans. R. McKeon. New York: Random House.

Balcombe, J. 2006. *The Pleasurable Kingdom.* London: Macmillan.

Bourque, H. 2008. "Animals That Glow in the Dark—Bioluminescence Is Not New Technology." Retrieved from http://voices.yahoo.com/animals-glow-dark-bioluminescence-795963.html?cat=32.

Brehm, J. W. 1956. "Post-Decision Changes in the Desirability of Alternatives." *Journal of Abnormal Psychology* 52:384–389.

Call, J., B. Hare, M. Carpenter, and M. Tomasello. 2004. "Unwilling vs. Unable: Chimpanzee's Understanding of Human Intentional Action." *Developmental Science* 7:488–498.

"Cat Plays Furry Grim Reaper at Nursing Home." 2007, July 27. *MSNBC.* Retrieved from http://www.msnbc.msn.com/id/19959718/.

Chan-Kyong, P. 2008. "Cancer Sniffing Dog to Be Cloned." *Discovery,* June 20. Retrieved from http://english.scienceweek.cz/cancer-sniffing-dog-to-be-cloned-iid-41106.

"Chihuahua Saves Boy from Rattlesnake." 2007. *MSNBC,* July 23. Retrieved from http://www.msnbc.msn.com/id/19901704/.

"Chimps Get Angry, Not Spiteful." 2007. *Age,* July 17. Retrieved from http://www.theage.com.au/news/World/Chimps-get-angry-not-spiteful-study/2007/07/17/1184559744220.html.

"Counting, Speaking Parrot Alex Dies." 2007. *ABC News,* September 11. Retrieved from http://www.nanotechclearinghouse.com/module-NewsFeeds-view-option-article-artid-271528.html.

Descartes, R. 1999 [1637]. "Discourse on Method." In *Discourse on Method and Related Writings.* Trans. D. M. Clarke. New York: Penguin.

Dennett, D. 1996. *Kinds of Minds.* New York: HarperCollins.

"Dolphin Whistles Offer Signs of Language Ability." 2000. *New York Times,* September 5. Retrieved from http://www.nytimes.com/2000/09/ 05/science/dolphin-whistles-offer-signs-of-language-ability.html.

Egan, L., L. R. Santos, and P. Bloom. 2007. "The Origins of Cognitive Dissonance: Evidence from Children and Monkeys." *Psychological Science* 18:978–983.

Fodor, J. A. 1990. *A Theory of Content and Other Essays.* Cambridge: MIT Press.

Inoue, S., and T. Matsuzawa. 2008. "Working Memory of Numerals in Chimpanzees." *Current Biology* 17:1004–1005.

Jensen, K., J. Call, and M. Tomasello. 2007. "Chimpanzees Are Vengeful But Not Spiteful." *PNAS* 104:13046–13050.

Kalla, R. 2007. "African Monkeys Sexually Harass Women." *Buzz Media,* August 25. Retrieved from http://www.thebuzzmedia.com/african-monkeys-sexually-harass-women/.

Kant, I. 1981 [1785]. *Groundwork of the Metaphysics of Morals.* Indianapolis, IN: Hackett.

Lavoie, A. 2007. "Primates Expect Others to Act Rationally." *Harvard Gazette,* September 6. Retrieved from http://news.harvard.edu/gazette/story/2007/09/primates-expect-others-to-act-rationally/.

Leake, J., and G. Warren. 2010. "Smarter Than You Think." *Sunday Times,* January 17. Retrieved from http://www.timesonline.co.uk/tol/news/science/biology_evolution/article6991028.ece.

"Mirror Test Reveals Elephants Are No Dumbos." 2006. *SeedMagazine,* November 3. Retrieved from http://seedmagazine.com/content/article/mirror_test_reveals_elephants_are_no_ dumbos.

Neighbour, M. 2007. "Monkeys Ape Sex Harassment." *Scotsman,* August 25. Retrieved from http://news.scotsman.com/latestnews/Monkeys-ape-sex-harassment .3320883.jp.

Njeri, J. 2007. "Monkey Misery for Kenyan Women Villagers." *BBC News,* August 24. Retrieved from http://news.bbc.co.uk/2/hi/africa/6959209.stm.

Pain, S. 2002. "Crow Reveals Talent for Technology." *Science* 297 (August 8). Retrieved from http://www.newscientist.com/article/dn2651-crow-reveals-talent-for-technology.html.

Pickrell, J. 2005. "Parrot Prodigy May Grasp the Concept of Zero." *National Geographic,* July 15. Retrieved from http://news.nationalgeographic.com/news/2005/07/0715_050715_parrotzero.html.

Quine, W. V. O. 1977. *Ontological Relativity and Other Essays.* New York: Columbia University Press.

Randerson, J. 2007. "Crows Match Great Apes in Skilful Tool Use." *Guardian,* August 17. Retrieved from http://www.guardian.co.uk/science/2007/aug/17/1.

Ritter, Malcolm. 2007. "Young Chimp Beats College Students." *Newsvine*, December 3. Retrieved from http://www.newsvine.com/_news/2007/12/03/1138165-young-chimp-beats-college-students.

Rolen, P. H. 2007. "Sexist Monkeys Terrorize Village." *NowPublic*, August 24. Retrieved from http://www.nowpublic.com/sexist-monkeys-terrorize-village.

"'That Damn Bird!'—Alex, Genius Parrot Who Dazzled Scientist Dead at Thirty-one." 2007. *Discovery*, September 11. Retrieved from http://www.dailygalaxy.com/my_weblog/2007/09/obit-alex-geniu.html.

Uexküll, J. von. 1985 [1909]. "Environment [*Umwelt*] and Inner World of Animals." Trans. C. J. Mellor and D. Gove. In G. M. Burghardt, ed., *Foundations of Comparative Ethology*, pp. 222–245. New York: Van Nostrand Reinhold.

"Weird Shrimp Has Astounding Vision." 2008. *ScienceDaily*, May 15. Retrieved from http://www.sciencedaily.com/releases/2008/05/080513210456.htm.

Wong, K. 2001. "Dolphin Self-Recognition Mirrors Our Own." *Scientific American*, May 1. Retrieved from http://www.scientificamerican.com/article.cfm?id=dolphin-self-recognition.

Wood, J. N., D. D. Glynn, B. C. Phillips, and M. D. Hauser. 2007. "The Perception of Rational, Goal-Directed Action in Nonhuman Primates." *Science* 317:1402–1405.

6

Can Animals Make "Art"?

Popular and Scientific Discourses About Expressivity and Cognition in Primates

JANE C. DESMOND

In May of 2005 three paintings by a very special artist were sold by Bonham's Auction House in London for approximately $30,000.00 U.S. dollars (de Vries 2005). Not so unusual perhaps, except that this artist was an ape, and the sum was the highest ever paid for a work of art created by a nonhuman animal. This chapter analyzes how the growing global trade in art by elephants, apes, dolphins and other nonhuman animals functions as a contestation of humanism and asks how the culturally specific category of "art" changes when the species producing it changes. What is at stake in the naming of these objects as art, and for whom? Why do these works command such relatively high prices? Who is the target market? How do such products relate to notions of "the primitive"? And, ultimately, what does this challenge to the humans-only category of art making mean for a posthumanist vision of beings in the world? If the artist is an ape, or the ape is an artist, does that designation have potential political implications for the political status of apes—indeed for their representation, literally and figuratively—as political subjects?

Some of the questions that arise are the following: What does it mean (and for whom?) that nonhuman animals perform actions of drawing, painting, or sound making that produce products perceived by humans as visual art or music? Is the realm of the arts just one more in the ever receding panoply of elusive abilities that supposedly defines the ultimate dividing line between human and nonhuman animals? Toolmaking, language, emotions, souls, art? What aspect of "the human" and of "mind" does art stand for in these debates about animal capacities?

Intentionality, aesthetic pleasure, design capability, cultural knowledge of representational conventions . . . all these are aspects of art making that emerge in discussions and perceptions of animal art. And each references a humanistic value of individual expressivity, intelligence, mind, and aesthetics that have historically been used to separate not only humans from others, but some humans from other humans at specific places and times. Appreciation of the "artistic" has been denoted as a dividing line separating the urbane sophisticate from the rural rube, the upper classes from the lower, the cosmopolitan from the provincial, the socially "advanced" from the "primitive," and the so-called first world from the so-called third. These oppositional categories depend on a concept of "art" and "art making" as a realm of social practice separated from the production of objects for everyday utility, a realm in which only those truly "gifted" can excel, and a realm, although embedded in market economies, supposedly carrying values that ultimately cannot be commodified: the beautiful, the truthful, and the creative.

I want to be careful here, while still taking the opportunity to broadly sketch the contours of art making, to recognize that this constellation of beliefs has a social and geopolitical history. The generalizations offered here are those that I believe have been operational in European and European settler societies at least since the nineteenth century and still hold sway, explicitly or implicitly, in those sites today. And, given the impact of the U.S. and European art worlds on the global art market, the impact of these concepts extends much more broadly too. However, I want to limit these remarks to the current art market in Europe and the U.S.

Like most aspects of the art market, the animal art market operates on a principle of rarity. Only a few members of a few species have produced "artworks" that sell at high prices. Among the painters are Tillamook Cheddar, a fox terrier in New York City; Gambi and Premja, two Lithuanian dolphins; several elephants in Thailand; Cholla the horse; Koko the gorilla; and Alexander the orangutan. Artworks by these artists are for sale on Koko's Web site, in Thailand at an elephant sanctuary, in Brooklyn at the store Tillie Ltd., at the Sea Museum Dolphinarium in Klaipeda, Lithuania, and through major art auction houses like Christie's in New York. The artworks are unique, handmade (sic) objects or live recordings of music "jam sessions," like Kanzi the bonobo's sessions with Peter Gabriel, and thus retain a Benjaminian aura of the "original" (Benjamin 1968).[1] What they stand for—their contestation of "the human"—is the highest part of their value. This is not the case with all products sold as animal "art."

In understanding the stakes at play in discussions of primate artists, we must put their artwork and the market for it in a broader context of "artwork" by all sorts of animals. Today, consumers can purchase art by dogs, dolphins, sea lions, and even rats and turtles. The idea of art making by animals varies depending on the species that created the product. This is not an aesthetic judgment, with turtles making "bad" paintings and apes being "good" artists, rather the calibration has to do with the notion of the animal's *intent* to make an aesthetic object and to make an object through specific selective aesthetic choices about color, line, shape, and composition. The higher up the presumed evolutionary chain of intelligence (understood in human terms), the greater the value of these works, because they represent not only novelties but also "evidence" of subjectivity.

The resultant "art by animals" products, 99 percent of which would be termed "abstract" (i.e., nonrepresentational) art, may be complex or simple, smeared or brushed, large or small, made with a lip, a trunk, a brush or a foot, and brightly multicolored or in a limited palette. A painting by a sea lion is not always immediately distinguishable from a painting by an elephant. Both are likely to be composed of sweeping brushstrokes in layers of colors reaching out toward but not completely filling the canvas. The issue becomes whether or not the animal intended to make aesthetic choices or merely produced physical actions that resulted in aesthetic choices passively being made (either by default—a smearing together of yellow and blue making green), or through the intervention of the human keeper or companion, who may determine which colors are used and in what order as well as when the painting is "done." We will come back to this issue of choice making when we consider primates in particular.

On the "lower" end of the spectrum, we see more tongue-in-cheek examples, like that of the British rat Tony Blair, who produces "sculptures" by gnawing on apple cores, or the works of the Ratistes in the U.S., whose tiny footprints track colorful paint across canvas to produce Scamperart. These paintings actually look like rat footprints, which is part of their appeal—art as the trace of the animal's body on the canvas. But Koopa the turtle produces turtle art that looks a lot like Jackson Pollock's canvases rather than like turtle footprints, thus catapulting his turtle products into another dimension of aesthetic appeal.

Evolutionary Stakes and Political Subjectivity: The Question of Aesthetic Intent

For some scientists of human evolution, one question is whether an "aesthetic sense" is a transspecies phenomenon, detectable in birds and apes, for example, but most highly developed among humans. This group cares about art by animals because they see animals as human pre-history. As Frans de Waal (2001) notes, many biologists regard the New Guinea bowerbird's hutlike nests as evidence of aesthetic expression. Male bowerbirds build elaborate nests, decorating the doorway with colorful objects like berries and flowers which they arrange and rearrange, analyzing the patterns from a distance and then flying in again to move a petal just so until the composition looks right, with an eye to attracting a female mate. For de Waal (pp. 151–152), this may not be art making, but it raises the question of whether human aesthetic urges, or—more accurately—actions and desires that are expressed in the realm that has historically come to be called the artistic, may "go deeper than culture" and relate to basic features of our perceptual systems, like our eyes and ears. As additional evidence he offers the facts that many birds must learn the songs they sing, that they are not born with this knowledge, and that many bird populations have different "dialects" reflecting regional variations, just as landscape paintings of the Rhone and the Rhine vary in style, even during the same historical periods. Some birds are even better singers than others, pioneers in the development of new songs, as de Waal (p. 155) notes.

While these notions of an aesthetic genealogy appear to connect some evolutionary dots, the real stakes lie in debates about art by primates, the group that includes humans too. There are two realms in which debates about art making by animals operate: that of science—including especially the work of comparative psychologists and primatologists working on ape cognition and language capabilities—and that of the lay person.

For comparative psychologists and primatologists, ape paintings have served as data for investigations of eye-hand coordination, tool use, and cognitive studies of symbol making (Beach, Fouts, and Fouts 1984a, 1984b; Boysen, Berntson, and Prentice 1987; de Waal 1999; Tanaka, Tomonaga, and Matsuzawa 2003). For instance Boysen, Berntson, and Prentice, based on a coding of 618 drawings by three chimpanzees, conclude that chimps will engage in drawing activities without training or reinforcement, and "this behavior may reflect their intrinsic interest in exploratory and manipulative play" (1987:82).

Primate specialist Frans de Waal (1999) goes farther: "Apes can deliberately make what looks like art to humans," he says. While apes do not seem to strive to create enduring works of visual art that will "please, inspire, provoke shock, or produce whatever effect it is that the human painter seeks to achieve," they do seem to enjoy the visual and kinesthetic act of making the drawing or painting, he asserts (de Waal 1999:B6). And other experiments, like those conducted in the 1950s by ethologist (and painter himself) Desmond Morris, show that apes make considered choices about where to make marks on the paper in relation to what is already there, seeking apparently a sense of balance, not making marks randomly on a page. Anecdotal evidence indicates they also appear to have a sense of when a painting is "finished." Congo became agitated if Morris tried to remove a painting before Congo wanted him to. Nor could he be implored to paint more on an image once he stopped. While these incidents could certainly be attributed to other factors such as boredom or distraction from the task at hand, the multiplicity of such anecdotes in reports of ape painting indicates they deserve further investigation.

The work by Boysen et al. (1987), de Waal (1999), Tanaka et al. (2003) (who is using touch screens in Japan to trace scribbling patterns by young chimps), and Morris (2005), suggest the possibility of what might be called a protoaesthetic, components of manual and visual choice making that are necessary to but not sufficient for the development of something humans call art making. This interpretation would be important for those concerned with evolutionary issues and human development.

This search for or desire for a protoaesthetic impulse or ability also underlies, I believe, both the public's passion for paintings by apes, and the use of painting as an enrichment activity by primate keepers in zoos and sanctuaries, which is quite widespread. The broad contours about the arts sketched earlier are operative in both the scientific and popular discourse realms, although the scientific realm may parse these concepts more complexly and with greater precision. In both realms there is a huge "WOW" factor—the sense of a frontier being crossed, a limit being broken. If an ape can make art, then . . . what? What is the passion to know what follows that ellipsis?

Although some early isolated case studies of primates showed that some of them like to draw or paint, chimp art really broke through to the popular consciousness in the 1950s. It is not merely coincidental that this postwar period coincided with a widening acceptance of abstract art as "legitimate" artistry. That shift in art history prepared the ground for lay people to see ape art—primarily based on gestural marks, not representational strategies—as Art. This is when ethologist Desmond Morris, trained at Oxford and a surrealist painter himself, began to feature a young chimpanzee named Congo on his popular London television show *Zoo Time.* It was Congo's abstract paintings that were so recently put up for auction at Bonham's in London alongside works by Warhol and Renoir. Although the Warhol and the Renoir failed to sell, Congo's art did, and for far more than the anticipated equivalent of $1,000 to $1,500 U.S. dollars each.

As CBS news reported, American Howard Hong, a self-described contemporary painting enthusiast, paid $26,352 for three brightly colored abstract tempera paintings by Congo (de Vries 2005). This marketability took the Bonham's curator of modern and contemporary art, Howard Rutkowski, by surprise. "We had no idea what these things were worth," he said. "We just put them (on sale) for our own amusement" (de Vries 2005). The surprising sale was reported in both mainstream and arts-specific media, including National Public Radio, CBS News, the London-based *Guardian*, and *Science* magazine online, demonstrating the artworks' status as entertainment news, science, art, and oddity ("Chimp's Art Fetches £14,000" 2005; "No Chump Change for Chimp Art" 2005). Perhaps, like the oeuvre of so many artists, Congo's price climbed because the artist was dead, having succumbed in 1964 to tuberculosis at the age of ten (de Vries 2005). At his most prolific in his youth, Congo produced about four hundred drawings and paintings between the ages of two and four (de Vries 2005). For Desmond Morris, works like these, and not the art of early humans, "represent the birth of art" (see Dodds 2006; Morris 2005).

Reports of the sale lent additional newsworthiness to a recent retrospective of Congo's work titled Ape Artists of the 1950s at the Mayor Gallery in London in July of 2005. The art critic of the *Sunday Times* in London, Waldemar Januszczak (2005a), also writing about the show in the *Australian* (Januszczak 2005b), found his beliefs that only humans can truly paint with intentional rather than accidental aesthetics challenged by the exhibit. He admits that "I like Congo's paintings. A couple of them I love." Calling the London show at the Mayor Gallery "fascinating and slightly worrying," he describes Congo as a "talented" painter who made active color and compositional choices, who threw a tantrum if a human tried to take a picture away from him before he was finished, and who refused to add to any painting he regarded as completed, despite entreaties to do so. Each of these actions serves as evidence of intentional aesthetic production. Qualities of unmuddied color, symmetrical balance, and, at his best, a "mood [that is] pure Kandinsky," makes Congo's works demonstrations of profound achievement, in Januszczak's words.

But even in this article we see the smirk, the long-time trace of the artist as monkey, as fop, and self-important boor traced by scholar Thierry Lenain in his 1997 book on aesthetics called *Monkey Painting*. The title given to Januszak's article is "Monkey Master" and the resounding accent of the missing question mark is impossible to miss. Even the cavalier rhyming *m*'s of monkey master depend on a disregard for the particularity of the painter—for Congo is a chimpanzee, not a monkey. But the word *monkey*, aside from its alliterative use, also conjures up images of an organ grinder's monkey, a trickster, and a miniature and comic humanoid. The fear that we might be monkeys after all erupts through the tongue-in-cheek titling.

But there is another side to this coin: Morris is concerned with tracing the origins of human abilities to their nonhuman primate past (thus assuming, of course, that apes do not have cultural history but rather live in the present as mere exemplars of our long-distant evolutionary cousins). But these discussions rarely recognize what sociologists and theorists of art know—that representational systems are not inherent results of human eye-hand coordination and perceptual abilities, but are historically distinctive symbolic systems and are learned both actively and passively by members of specific human communities. Just think of the difference between the visually flat medieval paintings of saints and the lush three-dimensional images of Michelangelo, just to draw one example from well-known Western European traditions. When some lay persons and even some scientists refer to ape drawings as a mode of protoaesthetics, or representational art, they ignore the fact that "art" is a category of social activity that has a specific history and a different history in specific times and places among various communities.

To better understand these practices and their linkages to what human primates do, we must frame art making as a cultural activity, and this doesn't necessarily mean that apes don't paint, but that maybe some do. This is especially resonant when we consider the cases of individual apes like Washoe, Koko, Michael, Kanzi, and Panbanisha, all stars in long-term communication research who have been trained to "speak" with humans through sign language or the use of lexigrams. After all, Koko and her now deceased companion Michael, for example, have been raised as members of a bispecies community that is full of symbolic, visual images (they look through store catalogues, look at books, and watch videos). And they have learned to perceive and to name at least some objects, and perhaps even concepts, which is a culturally specific representational act in itself. Perhaps this juncture is the origin of their art making if, in fact, we are to accept that characterization. At the very least they complicate the question of meaning and suggest the need to consider seriously what might be happening when the artist is an ape.

But there are other reasons why a documentable artistic ability among nonhuman primates, and even among other animals, might be discomfiting. If animals do produce works of art, might they not be more like us—expressive, self-aware, reflective—than we would like to admit? And, if so, might their already contested status—as property, as commodity, as "animal," and hence without rights, and with few legal protections—be harder and harder to maintain?

How Do You Teach an Ape to Paint?

For this part of the discussion, I want to leave aside painting by Washoe, Kanzi, and Koko. Washoe, for example, was involved in studies on representation and schemata. My understanding from a report on that research by the Fouts is that Washoe was not "taught" to paint in any formal way (see Beach, Fout, and Fout 1984a, 1984b). Most of the paintings I will describe here, by contrast, are produced through operant conditioning and target training, by apes who do not have access to sign or symbolic language.

In the sessions I observed at the Oklahoma City Zoo, painting is usually taught through operant conditioning, just as any other activity might be. For example, the training to facilitate medical procedures such that the primate will present her chest to the bars of the cage so her heartbeat can be checked by a stethoscope. I had the opportunity of watching primate keeper Jennifer Davis on two different days as she painted with Toba, a forty-year-old orangutan, and with Gracie, an eight-year-old female gorilla, and her father Tatu, a silverback. Toba, a self-taught painter, is unusual, while Gracie and Tatu are more typical, learning to paint through specific instructional techniques.

Toba, the orangutan, sports a reddish gold comb-over look on her head and long tangles of fur. She has been painting for a couple of years. Although Jennifer was not the one to introduce her to painting, it appears that upon being presented with the materials she took to it right away without formal instruction. The first day I spend with her, she is not in the mood to paint, despite Jennifer and my painting on the opposite side of Toba's bars, hoping to entice her to join us. But on the second day she is ready to go. She loads the brush with paint herself, choosing from among several colors, and then paints on the paper we've put inside her cage. She does two things while painting that she's never done before. First, she shakes the brush when loaded with paint, getting dribs and drabs to spray on the paper instead of just strokes; second, she holds the paper upright in one hand while making strokes with the paintbrush in the other. This variety indicates that she is actively problem solving and making choices in how to conduct this activity on this particular day. As a reward at the end of the ten-minute session, I get to feed her some yummy Yoplait yogurt, a treat for both of us.

But Toba is unusual, and most primates have to be taught to paint; this is done through operant conditioning. In operant conditioning a stimulus, like a verbal command, is paired with a reward when the proper response is performed by the primate. "Targeting" or touching a specific area is a commonly taught response. Painting involves a set of multiple steps—holding a brush, touching the brush to paper or canvas, and returning the brush to the keeper. Commands like "take" and "give" can prepare the animal to take a brush loaded with paint and then to give the brush back to the keeper. The animal is trained to touch the brush to the paper when the keeper holds the canvas up to the bars of the cage and asks the animal to "target" the brush onto the spot—in this case, to hit the paper.

"Targeting" is the stage at which Jennifer is with Gracie, who has been training for approximately a couple of months. Gracie received both verbal praise and a few tasty grapes as reinforcement for touching the paint brush to the canvas. As for her father, on his very first painting lesson, the goal was to get him to hand back a loaded paint brush to the keeper through the bars of his cage, so that the keeper could interrupt this action by sticking the paper in the way, so that the brush left a mark. The question of creativity is not being explored here, although keepers I've talked to often try to give the animal a choice so that part of the enrichment activity is the opportunity to exert control over one's environment by choice making. Which brush? Which color? Nevertheless, keepers intervene in all sorts of ways to heighten the odds that a painting will be salable—aesthetically pleasing—so that the colors won't be muddy and the paint will cover more than one spot on the paper. They turn the paper, remove it at a certain point, offer a limited pallet of colors that "go together," and so on.

But, for our purposes here, my concern is with this process as an enrichment exercise. One of the primary aspects of the enrichment is the one-on-one activity that gives the ape attention from the keeper, most often undivided attention. Erica Thiele, chimp enrichment coordinator at the M. D. Anderson Research Center in Texas, calls this shared activity "an intimate behavior" (interview by author in Bastrop, Texas, February 7, 2008). Thiele makes painting available to her captive charge Joey the chimpanzee as an enrichment activity. Joey, she says, apparently loves painting, as evidenced by his vocalizations "uh uh uh," in response to praise at the end of a session. And while the painting exercise itself is a positive experience for him in her estimation, he also likes the tasty treats he receives afterward. Joey gets a coke at the end of his painting sessions, says his keeper Thiele. Apes also like to paint other things, including their cages, so the act of mark making can be enriching/engaging/entertaining in itself.

But while these products may be sold as "paintings" and may engage to some degree with the issues of "expressivity" and aesthetic choice making, the emphasis here is on "enrichment" for the animals. The development of money-making items is merely a byproduct that can help support the primate program. (Joey's canvases, for example, sell for $250 for a 5" by 7" minimasterpiece, and twice that for an 8" by 11" canvas.) A different level of intent seems to emerge among those few apes who are "bicultural" in that they have been raised in human-generated visual worlds replete with conventions of human aesthetic design. Let me turn now to the small but crucial category of those apes who are "language enabled."

Koko the gorilla, Panbanisha and Kanzi the bonobos, and Chantek the orangutan each participates on a regular basis in art making (see figure 6.1). Koko has her own Web site with paintings on sale (www.Koko.org), and many of her pieces are self-titled, like the poetic *Pink Pink Stink Nice Drink*—an acrylic of sweeping blues, greens, and pinks all rushing upward from the bottom right to upper left part of the canvas. Hovering on the brink of representation, this piece, according to the Web site of the sponsoring organization, the Gorilla Foundation, is "inspired by a nearby flowering meadow with a stream running through it." The title is explained thus: "Koko's word for flower is 'stink' even though she admits that she loves their smell" ("Koko's World" n.d.). So the title references a very pink flowering area by a stream that is nice to drink from—a representation that translates a sense of the vision, smell, and taste of the three-dimensional world onto a two-dimensional canvas.

Koko and Michael have also produced images of other beings, often from memory. These portraits of animals, like that of Michael's dog Apple, or Koko's picture of her pet fledgling blue jay, while rarely unambiguously representational, do give a new meaning to the category of portraiture.

FIGURE 6.1. Michael, *Stink Gorilla More.* Color painting. Photo courtesy of Dr. Ron Cohn/The Gorilla Foundation/koko.org.

For example, Michael's painting *Apple Chase* consists of whites and grays sweeping across the paper. Although he had a large selection of colors to choose from, Michael used the black and whites that match the colors of Apple's coat. His title, *Apple Chase,* seems to combine a memory of a being with that of an event, recalling his favorite game of chase with Apple. (See www.kokomart.org for images of paintings by Koko and by Michael for sale.)

In addition, upon request, both gorillas have produced paintings expressing their interpretations of specific emotions—the meanings of which they have come to understand through sign language—including *love, hate,* and *anger.* This level of interspecies communication was unavailable with the 1950s ape artists represented in the recent retrospective of Congo's work. None possessed the linguistic knowledge to communicate in a way that humans could understand, and so they could not be asked to paint certain things or ideas or emotions. Granted, in the case of Koko and Michael, the paintings and their titles, and the interpretations of the titles and their referential meanings, are all products of bispecies collaboration. Gorilla knowing is filtered through the medium of human concept-based communication, in the English language, as transposed into American Sign Language. In other words, the gorillas speak and understand a form of English, but the humans don't speak Gorilla.

The whole concept of artwork and art making is always already, in this instance, constructed through human categories of meaning. But even allowing for that, these paintings come perilously close to the status of "artwork" as that which is a visual representation produced for the pleasure of looking at it or of making it, but not for a utilitarian reason. These works seem to combine a sense of mark making with imagination resulting in a product, which is then perceived as "art" by someone else, thus completing the hermeneutic circle uniting perception with interpretation.

If apes *are* artists, what does that imply about humans' obligations to them? Already there are moves afoot in the European Union to grant special status to great apes—a sort of in-between animal and human status garnering legal protections (Glendinning 2008). And even dogs and cats will soon have legally mandated "freedoms" in some European counties—freedom from hunger, from the elements, from isolation. If more studies and more popular reports describe animals, and especially apes, as artists, it becomes harder and harder to deny their sentient and intelligent status.

In a liberal humanist social orientation, where individual rights, the rule of law, and a belief in the importance of individual expressivity are crucial underpinnings of social formations, the line between human and nonhuman primate becomes ever more indistinct. Especially in a poststructuralist, posthumanist vision, with decentered subjectivity and an emphasis on a socially constructed "I" as a position to be occupied, not an essence to be expressed, the social construction of the category "animal" as that which is not human is increasingly exposed as an epistemology with a specific history, not as a "fact" naming an already extant reality. In either case, animal artists subvert the presumed privilege of the human. And this, perhaps, is the utopian ideal (or fantasy?) that people purchase when they buy a painting by Koko to put on their wall.

Notes

1. Walter Benjamin argues that in the age of mechanical reproduction, hand-made unique objects retain a special value, an "aura."
2. Of course, not all communities value enduring material artistic representations. The implicit reference here is to a European-derived value system for painting, sculpture, and so on.

References

Beach, K., R. Fouts, and D. Fouts. 1984a. "Representational Art in Chimpanzees." *Friends of Washoe Newsletter* 3, no. 4: 2–4.

——. 1984b. "Representational Art in Chimpanzees, part 2." *Friends of Washoe Newsletter* 4, no. 1: 1–4.

Benjamin, W. 1968. "The Work of Art in the Age of Mechanical Reproduction." In *Illuminations: Essays and Reflections,* pp. 217–252. New York: Schocken.

Boysen, S. T., G. G. Berntson, and J. Prentice. 1987. "Simian Scribbles: A Reappraisal of Drawing in the Chimpanzee (*Pan troglodytes*)." *Journal of Comparative Psychology* 101:82–89.

"Chimp's Art Fetches £14,000." 2005. *Guardian,* June 21. Retrieved from http://www.guardian.co.uk/uk/2005/jun/21/arts.artsnews.

de Vries, L. 2005. "Dead Chimp's Art Sells Big: Three Works by the Late Chimpanzee 'Congo' Sell for $25,620." CBS News, June 20. Retrieved from www.cbsnews.com/stories/2005/06/20/entertainment/main703057.shtml.

de Waal, F. B. M. 1999. "Apes with an Oeuvre." *Chronicle for Higher Education,* November 19, B6.

——. 2001. *The Ape and the Sushi Master: Cultural Reflections of a Primatologist.* New York: Basic Books.

Dodds, A. 2006. "Ape Artists of the 1950s." *Frieze* 99 (May). Retrieved from www.frieze.com/issue/review/ape_artists_of_the_1950s/.

Glendinning, L. 2008. "Spanish Parliament Approves 'Human Rights' for Apes." *Guardian,* June 26. Retrieved from http://www.guardian.co.uk/world/2008/jun/26/humanrights.animalwelfare.

Januszczak, W. 2005a. "Even Picasso Was a Fan: What Makes the Paintings of Congo the Chimpanzee So Beguiling, Asks Waldemar Januszczak." *Sunday Times,* September 25. Retrieved from entertainment.timesonline.co.uk/tol/arts_and_entertainment/article569970.ece.

——. 2005b. "This Is a Masterpiece . . . and This Is the Artist—Monkey Master." *Australian,* October 8, edition 5, NSW Review, p. 1. Retrieved from http://www.theaustralian.com.au/; content retrievable from http://www.artistsezine.com/WhyChimp.htm.

"Koko's World." N.d. The Gorilla Foundation. Retrieved from http://www.koko.org/world/art_still.html.

Lenain, T. 1997. *Monkey Painting.* London: Reaktion.

Morris, D. 2005. "Ape Artists of the 1950s." Retrieved from www.artnet.com/galleries/exhibitions.asp?gid=725&cid=80738.

"No Chump Change for Chimp Art." 2005. National Public Radio, June 21. Retrieved from http://www.npr.org/templates/story/story.php?storyId=4712948.

Tanaka, M., M. Tomonaga, and T. Matsuzawa. 2003. "Finger Drawing by Infant Chimpanzees (*Pan troglodytes*)." *Animal Cognition* 6:245–251.

PART III

Embodiments and Interembodiments

7

Toward a Privileging of the Nonverbal

Communication, Corporeal Synchrony, and Transcendence in Humans and Horses

GALA ARGENT

> There seem to be instinctive tendencies on the part of [social animals] to move in the direction [in] which other animals are moving, such as is found in any group of cattle drifting across the prairie together as they graze . . . [but] they do not enter into the life of the individual so as to determine that life throughout.
>
> (MEAD 1967 [1934]:239)

> Armies do not march in step for exercise.
>
> (BYERS 1977:137)

The relationship between the human and the horse—particularly the ridden horse—has fascinated both riders and writers at least since Xenophon in the first millennium BCE penned the first treatise exploring the psychology of the horse. Since then, many societies have applied mystical, occult, or religious connotations to the human ability to work well with horses (Dierendonck and Goodwin 2006:35). In the United Kingdom, for instance, a secret society of horsemen was entered through an initiation ceremony that included communication of the "horseman's word," which was said to bestow powers over horses when whispered in the horse's ear (p. 35). This is probably the source of our contemporary understanding of the term *horse whisperer* as someone with an ability to work with horses in a nonviolent and seemingly invisible fashion; human and horse appear in sync in ways that communication is indiscernible to others who are not so attuned.

Extensive scholarly attention from widely disparate academic realms—much from authors who are themselves "horse people"—has focused on explaining the relationship and manner of interaction between horses and people. Sociologists (Brandt 2004; Brown 2007), ethologists (Mills and McDonnell 2005), anthropologists and archaeologists (Argent 2010; Lawrence 1985; Olsen et al. 2006; Oma 2007b), critics and cultural historians

(LeGuin 2005; Weil 1999), feminists (Birke, Bryld, and Lykke 2004; Birke and Parisi 1999), and philosophers (Hearne 2007; Sharpe 2005) have all addressed this relationship from various perspectives.

In many of these discussions the relationship between humans and the ridden horse is described in rather esoteric terms.[1] In one study, many informants "mentioned having an intense, non-verbal communication with their animals which gave them a feeling of oneness. For example, many participants who discussed horses talked about feeling so connected to their horses when riding that all they had to do was think about a command (such as turning) and the horse would do it" (Brown 2007:336). Moreover, informants in sociological studies describe an association that is both physical and psychological, that reaches into the telepathic, spiritual, and metaphysical (p. 336; also Sharpe 2005:212), and that often allows the person a connection with something larger than herself (Brown 2007:336). Primary authors, as well, speak of "mutual becomings" (Oma 2007a). They make statements such as "to ride a horse well, in the sense of creating a harmonious partnership, we must 'become horse'" (Birke and Parisi 1999:64), and "I have come to appreciate just how important a forgetting of our separate human self is if we are to ride well" (Game 2001:8–9). What emerges is a general trend describing human-horse interactions as seemingly both extrasensory and transcendent.

Why do these people say such mysterious and cryptic things? Are these informants and scholars all a part of the same New Age tribe, somehow bewitched by horses to chuck rationality out the window? Further, why do horse people—myself included—nod with enthusiastic understanding at these statements, while nonhorse people cant their heads quizzically and skeptically?

Conceptualizations that privilege the verbal elements of language are deeply rooted in Western notions of mind and self; from the Greeks forward, philosophers and scholars have assumed verbal language was a necessary precursor to both reflection and intelligence (Lingis 2007:45; Sanders 2007; Sharpe 2005:169–189). This long-standing correlation between human verbal language processes and mind is problematic for (at least) two reasons. First, it discounts the vital importance of nonverbal communication in *human* lives. Second, it discounts the incredible complexity and effectiveness of nonverbal communication systems, as used by both human and nonhuman animals, to create meaning.

Accessing the nature and depth of nonhuman animal minds is thus contingent upon moving away from the focus on the verbal toward adequate models with which to appreciate, comprehend, and explain nonhuman embodied "language" (e.g., Sanders 2007; Shapiro 1990; Smuts 2008). In what follows I propose another means of exploring elements of horse-horse and human-horse interactions, one based upon models and theories of human nonverbal communication. I then explore the capability for moving together in corporeal synchrony in humans and in horses—intra- and interspecifically. I also discuss the human need for, and pleasurable and transformative effects of, such embodied synchronous movement. Finally, I raise questions that might allow us to reconsider notions of agency and subjectivity in horses.

Nonverbal Communication in Humans and Horses

Before discussing human-horse interactions in terms of a human communication model, I will note that what allows this interspecies communication at all is that both are social animals. Wild horses, like humans, live in long-term, hierarchical family and social units with complex statuses and roles, where young are taught appropriate norms of social interactions (Boyd and Keiper 2005:55–56; Fey 2005:83; Morris 1988:49; Sigurjónsdóttir, Dierendock, and Thórhallsdóttir 2002). Like humans (Schutz 1966), horses are driven to participate within groups by inherent needs for control, inclusion, and affection (Godfrey 1979:4–8; Morris 1988:54). Social hierarchy is an overestimated factor in equine social relations because within horse society, "interaction with one another is based, not on domination or even confrontation, but on cooperation and approval-seeking" (Sharpe 2005:197). Horses bring this to human-horse interactions. They are inherently cooperative and they "can sense what a person wants them to do and will try to understand a person's intent" (Dorrance and Desmond 1999:1). Furthermore, horse sociality is "based on friendships and context dominance, rather than rigid formal dominance" (Morris 1988:52). Horses form "long-term, cooperative alliances between unrelated individuals" (Fey 2005:83) and "tend to be faithful to their playing partners" (Sigurjónsdóttir, Dierendock, and Thórhallsdóttir 2002:4–5). "Preferred attachment [appears] not only between dam and foal, but also among peers of all ages, genders and *between species*" (Dierendonck and Goodwin 2006:30, my emphasis). In their lives horses will have only one or two such close preferred partners (Fey 2005:86): They choose individuals as "best friends," and those friends can be human.

As social animals, both humans and horses have the need to convey information within the social contexts of their respective communities. While humans communicate using verbal and nonverbal messages, horses use nonverbal means. There are two central considerations to offer about nonverbal communication, as it is understood by human communication scholars. First, in the field of human communication, "language" is most often understood as encompassing *both* verbal *and* nonverbal messages. Nonverbal communication can be broadly defined as those messages sent through other than linguistic means, the goal of which—as with verbal language—is to "create shared meaning between a sender and a receiver" (Guerrero, DeVito, and Hecht 1999:6). The second vital point is the incredible importance, for humans, of messages conveyed through nonverbal means. As much as 93 percent of meaning is transmitted nonverbally (Ferguson 2008:176); 60–65 percent of social meaning is derived nonverbally (Guerrero, DeVito, and Hecht 1999:4); and nonverbal messages are so important in conveying meaning that when they conflict with verbal messages, we tend to believe the nonverbal (Knapp and Hall 2002:15–16). Nonverbal communication is primarily used to convey affective and relational messages, and elements of intentionality. In other words, it is through nonverbal channels that humans "engage in corporeal-kinetic sense-makings, forging a sense of themselves and of the world, and, in a social sense, forging an intercorporeal world, a world of common understandings" (Sheets-Johnstone 2002:104). Might the same be said about horses?

The manner in which horses communicate with each other might be understood through models of human nonverbal communication. Here "structural studies" of human nonverbal communication look at how the messages are sent—what movements, bodily parts, sounds, objects, or distances are manipulated to convey meaning. Although there are other subcategories, three are most important to this discussion: movement, touch, and space.

Kinesics is the study of body movement and position, including facial expressions. Smaller-scale kinesics addresses facial expression, which deals with *which* emotion is experienced (Knapp and Hall 2002:8–9). As with people, the horse's face is particularly expressive and conveys myriad emotions and intentions. The ears contain thirteen pairs of muscles to control movement (Kiley-Worthington 2005:56) and are particularly expressive. Pinned ears, for instance, mean "Don't push me further," but one ear cocked, "I'm listening." Nostrils wrinkle in disgust and flare with excitement, and eyes convey pain, exhaustion, fear, anxiety, apprehension, relaxation, submission, and anger (Morris 1988:33–34). At its largest scale, kinesics deals with posture and gestures. Horses convey a variety of messages through movement of specific parts of the body. A slight swish of the tail, for instance, can convey "Stop, you are annoying me," while a large swish might mean "You're in for it now!" As with humans, horses filter their interpretations of kinesically sent meanings through their understanding of the individual sending the message and the context of the interaction.

The study of tactile communication behaviors, *haptics*, is viewed as "one of the most potent forms of nonverbal communication" (Knapp and Hall 2002:9). For humans, touch can be used to express power, sexual interest, positive or negative affect, play, responsiveness, comfort and healing, or symbolic meanings (such as greetings and farewells) (Knapp and Hall 2002:284–288). For the horse, touch also factors into sexual activity, fighting and play, but it is most often seen in positive contactual behaviors. Horses engage with others in a variety of nurturing behaviors: Mares lick and nuzzle foals to comfort them, young and adults co-groom, and cohorts stand head to tail and swish flies off each other (Godfrey 1979:4; Morris 1988:54). In addition, horses signal their own desire for caring attention (Godfrey 1979:8). In other words, horses are gregarious, seek warmth and protection from other horses, and communicate their desire for closeness and contact.

Coupled with touch, the use space determines how humans physically make contact with others. The study of the perception and use of social and personal space is termed *proxemics* and deals with the spacing of interactional distances to convey both dominance and intimacy (Guerrero, DeVito, and Hecht 1999:174; Knapp and Hall 2002:155–163). Within the study of proxemics, it is understood that "the boundaries of the self extend beyond the body" (Hall 1966:11). Such boundaries can be visualized as invisible bubbles or three-dimensional spheres of actual and psychological space, extending out from the body to first intimate, then personal and social space (Hall 1966:13–14, 119–122), with closer interactional distances associated with closer relationships (Knapp and Hall 2002:161). Like humans, horses' use of space plays into both dominance and affiliative relations (Dierendonck and Goodwin 2006:30–31). Like humans, horses stay in close proximity, rest, co-groom, approach, and follow their preferred partners (Fey 2005:83–86).

Horses share social characteristics and needs with humans, use similar nonverbal communication modalities to meet these needs, and can choose members of other species as friends. Because of this, they are able to come together with humans through a co-created and understood embodied language, potent in its ability to create relational meaning, and compelling in affective force. This comparison of "continuities" between humans and horses is not meant to discredit or disrespect the "discontinuities" between the two (cf. Noske 1997:126). In all but vision, for instance, horses' sensory abilities far exceed those of humans (Saslow 2002), and it is likely that their skills at reading nonverbal messages and intentionality also exceed ours (see Keeling, Jonare, and Lanneborn 2009). I now turn to how humans and horses use bodily movement, touch, and space in moving together—intra- and interspecifically.

Moving Together in Synchrony and Rhythm

The ability of humans to synchronize nonverbal behaviors is both pan-human and innate (Hall 1976:72–73). When talking, humans synchronize with each other the tempo and rhythms of their speech patterns and other movements such as eye blinking (Hall 1976:72–75, 1983:177). People in interactions move together "in a kind of dance, but they are not aware of their synchronous movement and they do it without music or conscious orchestration. Being 'in synch' is itself a form of communication" (Hall 1976:71). Synchronization extends beyond dyadic to group interactions. "Church ceremonies, cheerleading at sport events, rock concerts, [and] dances" are examples of moving together in synchrony (Byers 1977:137). When in sync, "the players [constitute] a single, living, breathing body" (Hall 1983:163); they "function, in part, as a single organism" (Byers 1977:138).

Horses, too, have ability for intraspecific corporeal synchrony. Horses run flat out, over rough terrain, seemingly of one mind, distances between individuals maintained; no one is jostled or trampled. Various models have been proposed for the type of collective cognition necessary for synchrony of movement in fish schools, insect swarms, and bird flocks (Couzin 2008; Couzin et al. 2005; Reynolds 1987; Sumpter 2006; Sumpter et al. 2008), and vertebrates (Couzin and Krause 2003). Rifa (1990:167) examined how groups of horses synchronize behaviors—such as "grazing, walking, standing, standing up, lying up, lying down, suckling and mutual grooming"—and found "10% complete synchrony, 81% partial synchrony and 64% materno-filial synchrony." Thus these studies are concerned with coordinated transitioning from place to place or with the synchronization of behavioral activities. What concerns me here, however, are smaller-scale dyadic synchronies—those of individuals with each other—such as those noted in dolphins (Fellner, Bauer, and Harley 2006; Smuts 2008). I am interested in how horses use kinesic, haptic, and proxemic communication modes to create meanings and whether those meanings might be accessible to us.

Over the years I have noticed in horses the ability to move in exact foot-for-foot synchrony, and I asked a group of breeders if they had photographs of this. The figures here convey what they too have observed. Figure 7.1 show foals moving accurately in synchronization with their dams.

The ability extends to horses working in tandem, where horses often fall into synchronous movement (figure 7.2). Horses also synchronize their movements with humans. To get a horse to trot when lunging (working from the ground with a long line connected to the horse's halter or bridle), one stomps her feet in the cadence one wants the horse to take on. When walking or trotting together, horses will coordinate their movements with people in the same way they do with other horses, as shown in figure 7.3, of an "in-hand" judging classes. This ability translates to riding. Even though they are quadrupeds, the movement of our hips and seat when we ride mirrors the horse's leg movements: Their walk is like our walk, their trot like our jog, and their canter like our canter on two feet. Within this mutual cross-species embodiment of movement, rider and horse each, at times, follow and lead, as in a dance (cf. Game 2001).

We might now return to the questions presented at the beginning of this essay with some partial answers. Sharpe (2005:212) suggests "that it is in fact the horses' remarkable ability to read *bodies* that enables it to predict human behaviour so accurately." Members of both species can, as with humans and dogs using deception in play, assess the project of the other (Mitchell and Thompson 1986). Rather than reading deception, however, horses can choose to read human intentions in order to synchronize their movements with them, to participate as actors in the joint action (Sanders 2007) of riding. I suggest, then, that it is not only horses' highly developed nonverbal skills but also their capacity to—and choice to—move together in synchrony with us, combined with a superior ability to assess intentionality, that allow riders at times to wonder if they merely *thought* a request, to which the horse telepathically responded, or actually *asked for it* nonverbally. But does this fully answer the questions posed? Is this explanation enough to make clear why informants describe their relationships with their horses as spiritual and metaphysical?

FIGURE 7.1. Foundation Appaloosa mares and foals moving in synchrony. *Top,* Shavano Domino and CTR Shavana Reign; *middle,* DREA Comanche Redeagl and DREA Fires Sugareagl; *bottom,* Pratt Toby Girl and DREA Ochoco El Diablo. Photos by Cheryl Woods (*top*) and Milton Decker (*middle, bottom*).

FIGURE 7.2. Breton draft horses working in tandem. Photo by Yvon Le Berre.

FIGURE 7.3. Humans and horses moving in synchrony in in-hand judging classes. *Above,* Thunderstryk; *below,* Breton draft horse. Photos by Wayne Lecky (*above*) and Yvon Le Berre (*below*).

Entrainment and Transcendence in Humans and Horses

In a cogent exploration of human-horse interaction, Anne Game (2001) discusses the concept of entrainment as the manner in which horse and rider become in sync with each other's rhythms and co-create the act of riding. Having thus far discussed *that* people and horses come together—both intra- and interspecifically—in rhythmic conjoint movement, I will now expand upon Game's notion of entrainment, the process *underlying* such synchrony. The difference between the two concepts, as outlined by anthropologist Edward T. Hall (1983), is that "while synchrony and entrainment appear to mean the same thing, they focus on different aspects of the same process. Synchrony is the manifest observable phenomena; entrainment refers to the internal processes that make this possible, i.e., the two nervous systems 'drive each other'" (p. 225, n. 1).

Although entrainment certainly can yield destructive and nefarious results (e.g., Canetti 1960), my concern here is with how it might be seen to function positively, in the horse-rider context. Entrainment occurs when we automatically begin tapping our fingers and toes or swaying our bodies while listening to music (Mithen 2006:25). Most religious rituals use music or dance, and thus the process of rhythmic synchronous entrainment, to enhance the experience (McNeill 1995:67). In other words, we become in sync with other beings *through* entraining with some action of rhythmic, synchronous movement such as music, dance, sports activities, and religious and public rituals.

Not only is there an innate urge for humans to entrain with others (Byers 1977:138; Hall 1983:178); it is pleasurable. There is something about moving together that powerfully calls out to us. Military leaders well know that moving together in rhythm encourages emotional bonding and shared identity, facilitating a sense of boundary loss and feelings of oneness with something larger than oneself (McNeill 1995:8). As communication scholar Paul Byers (1977:137) noted, "Armies do not march in step for exercise." Describing his basic training experience in the U.S. Army, historian William McNeill—who explored the effects of "muscular bonding" when humans move together purposefully (1995:151)—reported that "marching aimlessly" produced in him: "a sense of pervasive well-being . . . more specifically, a strange sense of personal enlargement; a sort of swelling out, becoming bigger than life, thanks to participation in collective ritual. . . . Moving briskly together and keeping in time was enough to make us feel good about ourselves, satisfied to be moving together, and vaguely pleased with the world at large" (p. 2). Through the process of entrainment, manifested in synchronous movement, we transcend our senses of individual selves; we are still "there," but our identities expand outside of our boundaries. When we entrain, our bubbles of space and our experience encompass more than our senses alone provide; we are both giving and receiving identity to something larger—and in such transcendence lies ecstasy. Such "ecstasy happens to our *selves*. It is a momentary transformation of the knower, not merely a transformation of the knower's experience" (Jourdain 1997:328). The experience of boundary loss in social synchrony is "the deeply felt, yet often unspoken experiences of being *of* a group . . . [which moves] *beyond felt resemblances to experienced fact of social connections and unity*" (Turino 1999:241). It is no longer the doing, but the being, that becomes important. We *become* a part of the larger event. Here the medium is indeed the message: moving together in embodied synchrony produces its own results.

It is perhaps this entrainment—this corporeal synchrony-induced sensation of boundary loss—that more fully answers the question as to why both informants and primary writers describe their interactions with horses as generally transcendent, spiritual, and metaphysical and as allowing connection with something larger than themselves. Might something of the same be said about horses? Or are humans, through the process of riding, simply exploiting horses as a means to attain such ecstasy themselves?

Happiness, Work, and Play

I have discussed how the manner in which horses use movement, touch, and space can be understood through human models of communication. I have further outlined how horses have the ability to move in synchrony with each other and with humans. Given these abilities, and given the fact that people find transcendence in moving together with each other, and in moving together with horses, is it possible that horses find something of the same in moving together with other horses and with us?

In much of the broader academy, human relations with domesticates have been heavily influenced by an overemphasis on frameworks of exploitation and dominance, fostering a "domination model" of horse-human relations, tracking a Marxist approach to human-human interactions (Cassidy 2007:7). Here horses are coerced through "the whip, spur, harness and hobble, all of them designed either to restrict or to induce movement through the infliction of physical force, and sometimes acute pain" (Ingold

2000:73; also Dietz 2003; Tuan 1984). Those who subscribe to this domination paradigm would answer my questions regarding horses' finding transcendence in corporeal synchrony in the negative. Like George Herbert Mead in the epigraph to this chapter, McNeill (1995) also posits that animals moving together in herds, flocks, or schools does not constitute the same level of meaning as does human dance, because the animals do not keep a regular beat (p. 183, n. 22; but see Kroll 1981; Lawrence 1985; LeGuin 2005). I disagree with both Mead and McNeil. Movement, for the horse, does "determine that life throughout" (Mead 1967:239). As prey animals who are also social, moving together with others is vital not only to survival but also to every aspect of their social lives, throughout their lives.

Based upon similarities in humans' and many animals' chemical and neurobiological systems, it is now recognized that both share at least the primary emotions of enjoyment, anger, fear, sadness, disgust, and surprise (Bekoff 2007:10). If, as I have shown, horses have both the *desire* and the *ability* to move in synchrony, then might we not suggest that they experience something of the same *result* of such entrained activity as we do—happiness, joy, perhaps even transcendence? (see Balcombe 2009; Hearne 1994). The problem is, we cannot even ask this question while subscribing to the dominance model of human-horse relations, for the simple reason that this paradigm strips horses of their agency. Here horses are packages of instincts, acted upon, not co-actors within human-nonhuman relations. This does not jibe with the actual experience of riding with a large, powerful, and potentially dangerous animal. One does not "tell" a horse, one "asks." Philosopher and horse trainer Vicki Hearne (2007) notes that "because we ride them, because they *carry* us, it is particularly hard to avoid noticing not only that horses know us but that they know us without yielding their own volition, which continues to belong to the horse" (p. 115).

In order for riding to occur at all, it is not only we humans who must entrain to the experience of riding, it is a choreography of two. As aptly put by Game (2001): "once we think of riding in relational terms, it ceases to be an action carried out by a rider and becomes a situation of rider and horse simultaneously carrying and being carried by each other and something more—the spirit of riding" (p. 9).

If we allow for social agency in horses, we can conceive that entraining with each other—and perhaps with us—through shared purpose and joint action might constitute for horses "being a part of something larger than themselves." Equine researcher Kiley-Worthington (2005:211) notes that horses seem to "take pleasure in various group movements that they do with other equines, but also with humans." Moving together is *the way* horses belong, and horses who have been well and gently socialized to participate in joint action with a rider seem to want to, and try very hard to, get the synchronous movements "right" (cf. Hearne 2007:162). When they do, "riders report that their horses enjoy those moments too, that they also seem 'proud' or pleased with their performance. . . . [It seems] that experiencing this floating harmony, which is so difficult to achieve, addresses a capability and pleasure-reward in both human and non-human partners" (Evans and Franklin 2010:176).

This raises the question of "work." If it is neither physically nor psychologically painful, do horses differentiate between intraspecific, conjoint synchronous movement, which humans *construct* as either "work" or "play," in a similar manner? Indeed the distinction is often fuzzy for humans. Is performance in a marching band, choral society or dance troupe work, or play? Does the answer depend upon choice or payback? Regardless, for humans, synchronizing with others through rhythm "relieves the tedium of repetitive tasks" (McNeill 1995:48), and fosters rapport through entrainment (Tickle-Degnen and Rosenthal 1990). With this in mind, we might ask if, at an individual level, it is not only the horse and rider who entrain to a mutual project but also the plow horse and plowman, the elephant and the mahout, the sheepdog and the shepherd (and perhaps the sheep as well).

This is not to discount that horses have been exploited, for they most certainly have been and continue to be. But when the domination model is taken to the extreme, as it is in some animal rights discussions (e.g., Bunting 1997:64), the agency of horses is negated and we "[see] only animal slaves" (Weil 2010:723). When we focus solely upon issues of domination and hierarchy, we miss entirely the part of the social picture in which horses are cooperative and effective communicators, with each other and with humans, who can choose to move synchronously with humans (or not). I suggest, rather, that the relationship between individual horses and humans is not *necessarily* exploitive (Lawrence 1984:39), and can be dynamic, complex, mutual, co-created, bidirectionally cooperative, and perhaps transcendently pleasurable for both parties. It is worth noting that the secret "horseman's word," said to convey powers over horses when whispered in their ears, was "both in one" (Evans 2008:246), reflecting the harmonious partnership that can occur between the two when both approach the interaction cooperatively. Perhaps indeed there is a power to the words, although it is not mystical at all. It is not unreasonable to suggest—given horses' acute ability to read intentionality and innate desire to cooperate—that, if communicated to horses honestly, some of them might choose to answer the humans who approach them in this manner with like intent.

I have argued that the manner in which horses communicate nonverbally can be accommodated in large degree by models of human nonverbal communication, that this mode of interaction is at least as potent as human verbal "language," and that horses are more highly attuned at nonverbally reading intentionality than are humans. Rather than considerations of what we do not share with horses—the perceived deficits in verbal, cognitive and choice-making capabilities—my view has been toward exploring the social and communicative aspects we do share with them and how these manifest when the two species come together. I have discussed notions of corporeal synchrony, entrainment, boundary loss, bonding, joy and ecstasy that might be seen to function similarly for both humans and horses as social actors. Through this I hope I have shown that an intra- and interspecific model of assessing nonverbal communication and embodied synchronous movement, when coupled with an allowance for the different abilities and types of intelligences afforded by these modes of interacting, can open doors for new appreciations of animal minds.

Notes

I owe a debt of gratitude to Bob Mitchell and Julie Smith for their assiduousness in bringing this project to fruition and to Amy Nelson and Jane Costlow for talks which spawned the ideas that prompted this work. I also wish to thank the many members of the Foundation Appaloosa Breeders listserve who answered my call for photographs of horses moving in synchrony.

1. *Ride* is a loaded and inadequate term for the process by which human and horse traverse the landscape together because it does not acknowledge the agency of the horse. A motorcycle or train, which we "ride," does not decide to jump or not to jump, to stop or not to stop, to put itself in this situation but not that one, or even to allow or disallow itself to be mounted. Rather, I conceive of riding as a "joint project" (Shapiro 2008:14) that might better be described as *riding with.*

References

Argent, G. 2010. "Do the Clothes Make the Horse? Relationality, Roles, and Statuses in Iron Age Inner Asia." *World Archaeology* 42, no. 2: 157–174.

Balcombe, J. 2009. "Animal Pleasure and Its Moral Significance." *Applied Animal Behaviour Science* 118:208–216.

Bekoff, M. 2007. *The Emotional Lives of Animals.* Novato, CA: New World Library.

Birke, L., M. Bryld, and N. Lykke. 2004. "Animal Performances: An Exploration of Intersections Between Feminist Science Studies and Studies of Human/Animal Relationships." *Feminist Theory* 5, no. 2: 167–183.

Birke, L., and L. Parisi. 1999. "Animals, Becoming." In H. P. Steeves, ed., *Animal Others: On Ethics, Ontology, and Animal Life,* pp. 55–73. Albany: SUNY Press.

Boyd, L., and R. Keiper. 2005. "Behavioural Ecology of Feral Horses." In D. S. Mills and S. M. McDonnell, eds., *The Domestic Horse: The Evolution, Development, and Management of Its Behaviour,* pp. 55–82. Cambridge: Cambridge University Press.

Brandt, K. 2004. "A Language of Their Own: An Interactionist Approach to Human-Horse Communication." *Society and Animals* 12, no. 4: 299–316.

Brown, S.-E. 2007. "Companion Animals as Selfobjects." *Anthrozoös* 20, no. 4: 329–343.

Bunting, G. 1997. *The Horse: The Most Abused Domestic Animal.* Toronto: University of Toronto Press.

Byers, P. 1977. "A Personal View of Nonverbal Communication." *Theory Into Practice* 16, no. 3: 134–140.

Canetti, E. 1960. *Crowds and Power.* New York: Farrar, Straus and Giroux.

Cassidy, R. 2007. "Introduction: Domestication Reconsidered." In R. Cassidy and M. Mullin, eds., *Where the Wild Things Are Now,* pp. 1–25. Oxford: Berg.

Couzin, I. D. 2008. "Collective Cognition in Animal Groups." *Trends in Cognitive Science* 13, no. 1: 36–43.

Couzin, I. D., and J. Krause. 2003. "Self-Organization and Collective Behavior in Vertebrates." *Advances in the Study of Behavior* 32:1–75.

Couzin, I. D., J. Krause, N. Franks, and S. Levin. 2005. "Effective Leadership and Decision Making in Animal Groups on the Move." *Nature* 433:513–516.

Dierendonck, M. C. van, and D. Goodwin, 2006. "Social Contact in Horses: Implications for Human-Horse Interactions." In M. C. van Dierendonck, ed., *The Importance of Social Relationships in Horses,* pp. 28–44. Utrecht: Proefschrift Universitat. Also online at http://igitur-archive.library.uu.nl/dissertations/2006–0419–200436/full.pdf.

Dietz, U. L. 2003. "Horseback Riding: Man's Access to Speed?" In M. A. Levine, C. Renfrew, and K. Boyle, eds., *Prehistoric Steppe Adaptation and the Horse,* pp. 189–199. Cambridge: McDonald Institute for Archaeological Research.

Dorrance, B., and L. Desmond. 1999. *True Horsemanship Through Feel.* Bath, NH: Diamond Lu.

Evans, G. E. 2008. *The Horse in the Furrow.* London: Faber and Faber.

Evans, R., and A. Franklin. 2010. "Equine Beats: Unique Rhythms (and Floating Harmony) of Horses and Their Riders." In T. Edensor, ed., *Geographies of Rhythm: Nature, Place, Mobility, and Bodies,* pp. 173–186. Aldershot: Ashgate.

Fellner, W., B. Bauer, and H. E. Harley. 2006. "Cognitive Implications of Synchrony in Dolphins: A Review." *Aquatic Mammals* 32, no. 4: 511–516.

Ferguson, S. D. 2008. *Public Speaking: Building Competency in Stages.* Oxford: Oxford University Press.

Fey, C. 2005. "Relationship and Communication in Socially Natural Horse Herds." In D. S. Mills and S. M. McDonnell, eds., *The Domestic Horse: The Evolution Development and Management of Its Behaviour,* pp. 83–93. Cambridge: Cambridge University Press.

Game, A. 2001. "Riding: Embodying the Centaur." *Body and Society* 7, no. 4: 1–12.

Godfrey, J. F. 1979. *How Horses Learn: Equine Psychology Applied to Training.* Lincoln, NE: Authors Guild.

Guerrero, L. K., J. A. DeVito, and M. L. Hecht, eds. 1999. *The Nonverbal Communication Reader.* Long Grove, IL: Waveland.

Hall, E. T. 1966. *The Hidden Dimension.* New York: Anchor.

——. 1976. *Beyond Culture.* New York: Anchor.

——. 1983. *The Dance of Life.* New York: Anchor.

Hearne, V. 1994. *Animal Happiness: A Moving Exploration of Animals and Their Emotions.* New York: Skyhorse.

——. 2007. *Adam's Task: Calling Animals by Name.* New York: Skyhorse.

Ingold, T. 2000. *The Perception of the Environment: Essays in Livelihood, Dwelling, and Skill.* New York: Routledge.

Jourdain, R. 1997. *Music, the Brain, and Ecstasy.* New York: Harper.

Keeling, L. J., L. Jonare, and L. Lanneborn. 2009. "Investigating Horse-Human Interactions: The Effect of a Nervous Human." *Veterinary Journal* 181:70–71.

Kiley-Worthington, M. 2005. *Horse Watch: What It Is to Be Equine.* London: Allen.

Knapp, M. L., and J. A. Hall. 2002. *Nonverbal Communication in Human Interaction.* 5th ed. Australia: Wadsworth/Thompson Learning.

Kroll, P. W. 1981. "The Dancing Horses of T'ang." *T'oung Pao* 67, no. 3–5: 240–269.

Lawrence, E. A. 1984. "Human Relationships with Horses." In R. K. Anderson, B. L. Hart, and L. A. Hart, eds., *The Pet Connection,* pp. 38–43. Minneapolis: Center to Study Human-Animal Relationships and Environments, University of Minnesota.

——. 1985. *Hoofbeats and Society: Studies of Human-Horse Interactions.* Bloomington: Indiana University Press.

LeGuin, E. 2005. "Man and Horse in Harmony." In K. Raber and T. J. Tucker, eds., *The Culture of the Horse: Status, Discipline and Identity in the Early Modern World,* pp. 175–196. New York: Palgrave MacMillan.

Lingis, A. 2007. "Understanding Avian Intelligence." In L. Simmons and P. Armstrong, eds., *Knowing Animals,* pp. 43–56. Leiden: Brill.

McNeill, W. H. 1995. *Keeping Together in Time: Dance and Drill in Human History.* Cambridge: Harvard University Press.

Mead, G. H. 1967 [1934]. *Mind, Self, and Society: From the Standpoint of a Social Behaviorist.* Chicago: University of Chicago Press.

Mills, D. S. and S. M. McDonnell, eds. 2005. *The Domestic Horse: The Evolution, Development, and Management of Its Behaviour.* Cambridge: Cambridge University Press.

Mills, D. S., and J. McNicholas. 2005. "The Rider-Horse Relationship." In D. S. Mills and S. M. McDonnell, eds., *The Domestic Horse: The Evolution, Development, and Management of Its Behaviour,* pp. 161–168. Cambridge: Cambridge University Press.

Mitchell, R. W., and N. S. Thompson. 1986. "Deception in Play Between Dogs and People." In R. W. Mitchell and N. S. Thompson, eds., *Deception: Perspectives on Human and Nonhuman Deceit,* pp. 193–204. Albany: SUNY Press.

Mithen, S. 2006. *The Singing Neanderthals: The Origins of Music, Language, Mind, and Body.* Cambridge: Harvard University Press.

Morris, D. 1988. *Horsewatching.* New York: Crown.

Noske, B. 1997. *Beyond Boundaries: Humans and Animals.* Rev. ed. Montreal: Black Rose.

Olsen, S. L., S. Grant, A. M. Choyke, and L. Bartosiewicz, eds. 2006. *Horses and Humans: The Evolution of Human-Equine Relationships.* BAR International Series 1560. Oxford: Archaeopress.

Oma, K. A. 2007a. *Human-Animal Relationships: Mutual Becomings in Scandinavian and Sicilian Households, 900–500 b.c.* Oslo Arkeologiske Series No. 9. Oslo: Unipub.

——. 2007b. *Horses in Scandinavia, 500 bc –1000 ad: Enduring Structures, Symbols of Transformation, and Mutual Becomings.* Paper presented at the Nordic Theoretical Archaeology Group conference, Aarhus, Denmark.

Reynolds, C. W. 1987. "Flocks, Herds, and Schools: A Distributed Behavioral Model." *Computer Graphics* 21, no. 4: 25–34.

Rifa, H. 1990. "Social Facilitation in the Horse (*Equus caballus*)." *Applied Animal Behaviour Science* 25:167–176.

Sanders, C. 2007. "Mind, Self, and Human-Animal Joint Action." *Sociological Focus* 40, no. 3: 320–336.

Saslow, C. A. 2002. "Understanding the Perceptual World of Horses." *Applied Animal Behaviour Science* 78:209–224.

Schutz, W. C. 1966. *Interpersonal Underworld: A Reprint Edition of FIRO, a Three-Dimensional Theory of Interpersonal Behavior.* Palo Alto: Science and Behavior.

Shapiro, K. 1990. "Understanding Dogs Through Kinesthetic Empathy, Social Construction, and History." *Anthrozoös* 3:184–195.

——. 2008. *Human-Animal Studies: Growing the Field, Applying the Field.* Ann Arbor: Animals and Society Institute.

Sharpe, L. 2005. *Creatures Like Us? A Relational Approach to the Moral Status of Animals.* Exeter: Imprint Academic.

Sheets-Johnstone, M. 2002. "Introduction to the Special Topic: Epistemology and Movement." *Journal of the Philosophy of Sport* 29:103–105.

Sigurjónsdóttir, H., M. C. van Dierendock, and A. G. Thórhallsdóttir. 2002. "Friendship Among Horses—Rank and Kinship Matter." Havemeyer Foundation workshop on horse behavior. Retrieved from http://www3.vet.upenn.edu/labs/equinebehavior//hvnwkshp/hv02/hv02auth.htm.

Smuts, B. 2008. "Embodied Communication in Nonhuman Animals." In A. Fogel, B. King, and S. Shanker, eds., *Human Development in the Twenty-first Century: Visionary Ideas from Systems Scientists*, pp. 136–146. Toronto: Council on Human Development.

Sumpter, D. J. T. 2006. "Principles of Animal Collective Behaviour." *Philosophical Transactions of the Royal Society B* 361:5–22.

Sumpter, D., J. Buhl, D. Biro, and I. Couzin. 2008. "Information Transfer in Moving Animal Groups." *Theory in Biosciences* 127, no. 2: 177–186.

Tickle-Degnen, L., and R. Rosenthal. 1990. "The Nature of Rapport and Its Nonverbal Correlates." *Psychological Inquiry* 1, no. 4: 285–293.

Tuan, Y.-F. 1984. *Dominance and Affection: The Making of Pets.* New Haven: Yale University Press.

Turino, T. 1999. "Signs of Imagination, Identity, and Experience: A Peircian Theory for Music." *Ethnomusicology* 43, no. 2: 221–255.

Weil, K. 1999. "Purebreds and Amazons: Saying Things with Horses in Late-Nineteenth-Century France." *Differences: A Journal of Feminist Cultural Studies* 11, no. 1: 1–37.

——. 2010. "Shameless Freedom." *Journal of Advanced Composition* 30, no. 3–4: 713–726.

8

Thinking Like a Whale

Interdisciplinary Methods for the Study of Human-Animal Interactions

TRACI WARKENTIN

In this essay I explore the possibilities for humans to imagine what some experiences are like for other animals—whales in particular. My work involves developing methods for being attentive to the worlds of whales in a way that informs and inspires an appreciation of how different species encounter and engage with each other's worlds. I use practices of careful attentiveness and embodied imagination to approximate what is experientially significant to an individual of a different species. I believe, like Midgley (1995), that we grasp the meaning of each other's behavior through a corporeal understanding of what it means to be us.[1]

I present in this chapter interdisciplinary methods derived from a theoretical framework that I developed for investigating and approximating the subjective experiences of orcas, belugas, and bottlenose dolphins in human-animal encounter programs in captivity at public aquariums in the United States and Canada. While there seems to be a plethora of methods for studying human-human interactions and whale-whale interactions, there is a distinct lack of methods specific to studying *human-whale* interactions, both in terms of data collection and analysis. By casting a wide net across disciplines, however, I have found promising leads in the work of psychologist Kenneth Shapiro (1997), who has devised a method of "kinesthetic empathy" with his dog, Sabaka, of practical phenomenologist Elizabeth Behnke (1997), who developed practices of "interkinaesthetic comportment" with cats, and of anthropologist Thomas Csordas (1993, 1999), who uses "somatic modes of attention" to approach meaning in the embodiments of human others. Each has profoundly informed my development and application of interdisciplinary methods for investigating human-whale interactions and whale experience. Informed by my fieldwork with bottlenose dolphins, belugas, and orcas at Sea World, Orlando and Marineland, Canada, this discussion highlights the potential for my methods to contribute to practices and theories related to the study of animal minds. One particular interaction at the viewing area of the orca pool at Marineland, Canada will illustrate my attempts to create field research protocols that examine interaction from the perspective of embodiment.

Embodied Minds, Embodied Methods

In my study I assumed that humans' and other animals' perceptual and cognitive experiences are embodied, i.e., that having a body is necessary for having experiences, and that the body as an organism, including its neural and sensory (perceptual) systems, influences how those experiences are created, interpreted, and expressed. I believe that for humans (whether using linguistic or nonlinguistic forms of communication), as well as for other animals, conceptual thought has a visceral, sensory basis (Lakoff and Johnson 1980) and that nonlinguistic behaviors (particularly visual and acoustic gestures) can express embodied experience (Midgley 1995). I use forms of bodily movements, especially patterns in such movements, as well as the physical and social environment of an organism, to interpret what it is doing and thus, to some degree, its experience. A variety of theorists (Code 1993; Grosz 1993; Haraway 1989, 1991; Harding 1993; Nagel 1974) have influenced my methods and assumptions. Rather than working from explicit hypotheses, I allowed my findings to emerge (as much as possible) from what I observed. To further this end, I kept a fieldwork journal in which I created richly phenomenological descriptions in anecdotal, narrative accounts of my observations in order to reflect on them. This journal allowed me to recognize assumptions, biases, preconceptions, and judgments in my observations. I viewed both humans and whales as participants in my study and, to allow them to be equal participants, focused on nonlinguistic behavior only.

At the outset of my studies, I searched at length for methods of recording movements. I looked to dance and choreography and found Labanotation, but it turned out to be very complicated and not quite suitable. It was not designed to record interactive movements but rather to document the movements of each dancer separately. I also looked to whale ethograms, which are used extensively in ethology to describe and record whale behaviors, but again found it difficult to accommodate interactive movements, particularly ones between whales and humans. In addition, it was not dynamic enough, as ethograms are typically expressed as two-dimensional, static sketches or brief written descriptions. Ultimately, I found that the best way for documenting dynamic interactions was to video record them. With my digital video camera, I was able to record gestures and actions as they happened, to review the interactions later, and, as many times as necessary, to allow patterns to emerge as well as to recognize specific moments that were not very common.

Adding to these logistical benefits, the video maintained my focus on the gestural interactions between whales and humans. In a small but significant way, this treated the whales and humans with some measure of equality as they were simultaneously recorded as research subjects and active agents. To these ends, I also chose not to conduct formal interviews with human participants, which would have privileged human speech to talk *about* the experience, distancing the data from the immediate interactions between humans and whales and excluding the whales from this form of contribution to the research. Such an emphasis on dynamic, situated expression led me to look to theories of embodiment and interpretations of gesture. I found unique examples of methods that could help me from three scholars in completely different disciplines, but all working with embodiment in applied ways.

Meaning in Gesture

All three begin with an assumption that bodies, however physically and physiologically different, can be grounds for interpreting nonlinguistic behavior. Cultural anthropologist Thomas Csordas (1993) purposefully adopts such an assumption in his research on charismatic healing in Catholic and Navajo rituals, describing it as an engagement in "somatic modes of attention." Csordas's term refers to multiple modes of sensory engagement rather than just visual attention, which is predominant in conventional methods of observation. He insists that even the act of imagining is multisensory and not exclusively visual. He also stresses that one must engage one's own body not only as a way of sensing the embodiments of others but also as a way of grasping the meaning expressed in those embodiments. Attentiveness to one's own body in the process of attending to others "allows us more immediately to grasp or recognize a set of socially salient bodily dispositions of posture, bearing, and physique" (Csordas 1999:148).

Similarly, psychologist Kenneth Shapiro (1997:294) has devised an "empirical phenomenological method" of "kinesthetic empathy," specifically developed to be exercised between beings of different species. He describes the method as "an investigatory posture of bodily sensibility adopted to promote empathic access to the meaning implicit in an animal's postures, gestures, and behavior" (p. 292). Shapiro developed and used this method in phenomenological studies with his dog, Sabaka. He asserts that this method has applications beyond just himself and Sabaka, explaining that there is a visceral basis for his ability to empathize with Sabaka and that this visceral basis is one all embodied beings share to some extent. Because of this, Shapiro suggests, we humans have the capacity to bodily approximate another being's experience, including that of other species.

Attending with one's body goes far beyond just looking toward or gazing at another body; it means turning full bodily engagement toward others; it means that our very own bodies can actually tell us something about others and about their worlds if we are attentive to them (Csordas 1993:138–139). Taking up a similar perspective on embodiment in yet another disciplinary field, philosopher and practical phenomenologist Elizabeth Behnke (1997) has explored "interkinaesthetic comportment," the ongoing adjustments of postures, gestures, actions made in relation to others in human-human and human-animal interactions. Like Csordas and Shapiro, she is additionally interested in how external factors, both social and physical, influence situated exchanges. The contention is that in particular places and social situations we tend to adopt certain patterns of movement and habits of posture and gesture, some of which may, over time and through repetition, become sedimented into a repertoire of micromovements that Behnke calls "ghost gestures." She finds that the "tacit choreography of everyday life [and] micromovements [give] witness to our sociality insofar as they are not only socially shaped, but perpetuate certain styles of intercorporeal interaction and sustain certain modes of responsivity" (p. 181). In other words, there is a tendency to embody, in the form of recognizable postures and gestures, certain social and cultural expectations. Behnke gives the example of a child who, when repeatedly urged to concentrate harder on her homework, merely displayed "a kinaesthetic pattern that is visibly expressive of 'trying'" (p. 190). That is, she "hunched over her work, frowning and staring at the page, clenching her teeth and gripping her pencil tightly" (p. 190).

What Is It Like to Be a Whale?

Overall, Behnke, Shapiro, and Csordas describe visceral practices that I was able to transpose into practical methods for generating data and for making *sense* of that data during analysis. I paid close attention to how the humans and whales were moving, gesturing, and acting in response to each other and in relation to the social and material contexts they were in. One small problem remained, however; I am not a whale. And so I needed some understanding of how whales perceive and interact with their surroundings and express their agency. I found this in complementary studies of phenomenological biology and ecological psychology (reported in Warkentin 2009).

How animals perceive phenomena in their environments is the focus of phenomenological biology.[2] By understanding their perceptual abilities, we can gain an understanding of how their environments appear to them and, combined with observations of animal behavior, we can better understand which phenomena might be particularly significant to them. For example, whales are thought to have excellent eyesight; but an awareness of their typical marine environments, which can be dark and murky, suggests that they also rely greatly on their ability to echolocate to find food, to navigate, and to communicate. Thus information about whales provides insight into whale agency, as we can discern possible opportunities for action, also known as "affordances" in ecological psychology (Gibson 1979; Reed 1996), and pay attention to which opportunities are taken advantage of and which are avoided or resisted.[3]

Keen attention must also be paid to specifics of environments if we are to get a sense of how another being experiences it. In doing so, if I may simplify things, we must take note of at least two dimensions of environments: their physical qualities and their social milieu. To illustrate the first, an aquarium's most distinguishing physical characteristic is its enclosure. Captive whales are not free to leave or explore beyond the boundaries of the pool walls. The pools themselves are typically round and made of concrete, creating an acoustically charged environment for the whales, which likely affects their use of echolocation to sense their surroundings. For example, according to psychologist Howard C. Hughes (2001), "the sides of the aquarium tank are excellent reflectors of sound energy. As a result, sonar signals reverberate off the sides and bottom of the tank, and the echoes are very intense. To the dolphin, it might seem a little like sitting inside a big bass drum while the band plays a rousing Sousa march" (p. 99). On the other hand, remarking on affordances for playful activities, Anthony Weston (personal communication, September 14, 2007) has suggested that whales echolocating and vocalizing in captivity experiment with their modulations in a sense similar to human experiences of singing in the shower. The second dimension, social milieu, can also influence how whales behave in the space and with others there. Social hierarchy can arise among whales in the same pool, even if they have had no relationship prior to their capture or arrival at the facility. Orcas form matriarchal societies, meaning they are led by the oldest female (or, in some cases, the only female) of the group, and she can greatly influence the behavior of the other whales. Unique relationships also develop between trainers and the whales they work most closely with, and these can involve trust and affection, or coercion and fear, depending upon the individuals and the specific situation. The social dynamics within the pool thus likely influence the behavior of the whales in various ways, as do the kinds of interactions with human visitors that take place at its borders. Before we can investigate interspecies dynamics, however, we must explore how the social milieu of aquarium spaces might also influence human behavior.

Social Milieu and Ghost Gestures

By contrast to the relatively obvious effects the physical features of the public aquarium can have on whale and human behavior and interactions, the effects of social milieu can be far more subtle and complex, necessitating greater detail in explanation. For this part of the discussion, I return to Elizabeth Behnke's notions of kinesthetic comportment and ghost gestures. Ghost gestures are kinesthetic patterns that have become sedimented in our bodies, not in a rigidly fixed sense but rather as a repertoire of micromovements and interkinesthic comportments (Behnke 1997). For instance, ghost gestures related to controlling the body, like "bracing" and "holding back," are some of the most commonly shared among humans (p. 192). Many people will be able to identify with holding back "in, for instance, an educational system that expects people to sit still and shut up, producing a 'schooled body'" (p. 192).

As Behnke reminds us, embodiments of ghost gestures are not merely hangovers from past experiences, "but may be an all too appropriate response to networks of power relations operating here and now" (p. 193). Similarly, ghost gestures are not "merely the manifestation of individual 'psychological problems' that have been 'somatized' in 'holding patterns.' Rather, they may be clues to the way(s) bodies and bodily movements in general are ongoingly shaped in a particular social milieu" (p. 193). Behnke reminds us that careful attention must be brought to our own "ghost gestures," to our involuntary, interkinaesthetic comportments. She warns that we tend to embody ghost gestures and a "toxic intercorporeity" (p. 196), which, if left unattended, can be particularly counterproductive to interspecies interactions, as those depend entirely on nonlinguistic communication and body language. She explains that toxic intercorporeity often takes the forms of "bracing and numbing oneself" and then "vacating the area" (p. 195) and calls this a "set/vacate" pattern. Iterations of such set/vacate patterns were evident in my videos of humans at underwater viewing windows of the aquariums. I experienced the patterns repeatedly as I observed them while I myself attempted to adopt Csordas's somatic modes of attention. I did not fully grasp them, however, until I witnessed what I consider to be the corporeal opposite of a set/vacate pattern in a strikingly rare encounter between two children and a very young orca at Marineland, Canada in Niagara Falls.

Interkinesthetic Comportments at Marineland

Just before dusk on a warm day in August, I wandered down to the darkness of the underwater viewing area of the orca pool at Marineland, Canada, known as Friendship Cove. Three adults stood quietly a few feet away from the large rectangular window. A woman was filming her two young children, a boy of about seven years of age and a girl of about five. They had crawled under the bar in front of the window and were pressed against the glass waving and calling to attract the attention of the mother and calf orcas in the pool. To the children's delight, Athena, the one-year-old female calf, broke away from her repetitive circling of the pool alongside her mother and made a beeline for the children. She floated down so that they were level with each other and made eye contact. Then Athena nodded her head up and down, and the children nodded their heads up and down; the children shook their heads from side to side, and Athena shook her head from side to side. The children spoke directly to Athena, asking her questions, and she in turn vocalized by squeezing air out of her blowhole. Transfixed, I watched them play and respond to each other for approximately twenty minutes. Throughout, the children were highly attentive to Athena's body and her gestures, just as she was to theirs; and they all used their bodies to express themselves and engage with each other.

It is significant that this interaction arose between young children and a young orca. It raises important questions about potential differences between youth and adult behavior in human-whale encounters and about the effects of captive spaces on such human behaviors. The children were not content to stand behind the barrier or accept the restrictive layout of the space, and they were very expressive in their movements and more open in their bearing than any of the adult humans who were present. The human children did not appear self-conscious in their actions and spoke directly to Athena as though she could hear and understand them. They addressed her by name, as a subject and unique individual, creating an intersubjective space of interaction. By contrast, their parents and the other adults all appeared to obey the rules, standing back behind the railing and quietly observing from that distance. I saw a difference also between the young orca, Athena, and her mother, Kiska, an adult orca. Athena did not behave in the same way as Kiska, who repeatedly swam by the window with her eyes closed, rarely, if ever, stopping to look out or make eye contact with the humans on the other side of the glass. Several times Athena broke out of the routine circling of the pool and swam directly over to where the children were. She obviously took notice of the children and initiated engagements with them.

Is it possible that the children were able to engage Athena because they had not yet adopted ghost gestures of inhibited postures or internalized social expectations of adult behavior in this place? Was it because they were allowed and encouraged by their parents to be playful, free to encounter Athena as a potentially communicative subject, that they were able to actively and imaginatively create space for possibilities in an otherwise restrictive setting? Another striking contrast in this interaction was between the playful spontaneity of the children and Athena and the attempt by the human mother to impose a structure related to previous experiences with authority figures. She tried several times to choreograph her children's movements and to do so by adopting the gestures of the trainers they had seen: she instructed the children to raise one arm and turn around in circles as they had seen the trainer do to get the whale to roll over. She also interrupted the interactions to compose the situation. She asked the children to position themselves to be in the frame of her video camera as she recorded them with Athena. These interjections by the human mother as well as the postures of the other adults who were in the viewing area made me wonder if their actions and gestures might be expressions of "deeply sedimented kinaesthetic patterns," those ghost gestures Behnke was referring to that express self-consciousness and the need to protect oneself and conform to expectations of polite, socially appropriate and mature behavior in the milieu of public spaces.

Behnke (1997) explains that ghost gestures may be kinds of "inadvertent isometrics" (p. 191). Inspired by the work of Thomas Hanna, Behnke expanded on the idea of isometric exercises (e.g., press palms together hard) to explain that involuntary "kinds of postural distortions and chronic tensions are not merely 'states' of a thing called 'body,' but can be understood instead as ongoing kinaesthetic 'holding patterns' in which movement is simultaneously produced and arrested, executed and countered" (1997:191). These holding patterns typically involve embodiments of restraint, such as holding still or getting a grip on oneself, that are associated with what is appropriate for polite behavior in public places.

The dominant processes of socialization in Western societies involve the internalization of at least some culturally appropriate modes of bodily expression in public situations. As Csordas (1993) reminds us with his practice of somatic modes of attention in cultural phenomenology, "the ways we attend [to other bodies] with our bodies, and even the possibility of attending, are neither arbitrary nor biologically determined, but are culturally constituted" (p. 140). Embodiment for Csordas is both an existential condition or "mode of presence and engagement in the world" (p. 135) and a methodological tool, in that "embodied experience is the starting point for analyzing human participation in a cultural world" (p. 135).

The embodiments of the adults at the underwater viewing window at Marineland, Canada, as well as at other captive sites I examined, certainly seem to illustrate Behnke's isometrics of the "set/vacate" pattern. Indeed, throughout my observations of participants at the underwater viewing windows of aquariums, I typically saw patterns of adult human behavior involving a stance of distancing, protection, and power: leaning back, arms folded, and silently watching or observing from behind the camera lens, taking photographs and video. Similarly, the common behavior of adult whales typically involved them swimming slowly in circles with their eyes closed, in their own way vacating the space by holding back from engaging in any way with the humans on the other side of the glass.

The physical and the social dimensions of captive spaces work in concert to shape behavior, postures, and gestures within captive settings. The lighting, typically bright within the pool, dark on the dry side, tended to hinder opportunities for the whales to look out and see the humans, particularly when they were standing back from the window in the semidarkness. On the other side of the glass, the window itself has the appearance of a giant television screen, which may have encouraged an objectification of the whales as passive, watchable, aesthetic objects and enabled a sense of voyeurism (Desmond 2001:182).[4] In fact, Behnke suggests that ghost gestures can reinforce subject-object relations; she states that "the ghost gesture of 'set/vacate' is, in short, part of the kinaesthetic means whereby the Cartesian mind/body dualism familiar to us from philosophical reflection can be concretely lived out at the level of individual moving bodies" (1997:194). Applying Behnke's theory, I wonder if the adult humans were expressing a psychologically embedded anthropocentric dualism of "human subject and animal object."

Wittingly or not, adult humans may have been performing ghost gestures to express assumptions that whales are relatively generic objects for human pleasure, meant to be looked at (framed in the viewing window and through the camera lens) and to be entertaining (as they are advertised to be on television and Web sites). The bulk of footage from my fieldwork video certainly suggests this to be the typical case. Rare exceptions arose, however, showing that the social, material, phenomenal space of the underwater viewing area could provide mutual affordances, like the opportunities shared by both Athena and the children at Marineland, Canada. In this case the window provided opportunities for some forms of reciprocity as well. They could make eye contact, as they could be level with each other. The window allowed them to see each other's whole bodies and to respond with gestures and verbal communication, as they could also hear each other through the glass.[5] Such responsiveness is not due to the window alone, though. The orca and children themselves facilitated reciprocity in the interaction through an embodied openness and intersubjective engagement.

Conclusion: Embodiment and Imaginative Generosity in Research

For most adults encountering other animals in this way—as individuals capable of reciprocity who learn and make things happen, that is, as subjects with agency—demands an exercise of imaginative generosity. It involves an attempt at imagining what the worlds of others *might* look like from their own embodied standpoints and a keen attentiveness to expressions of their agency as both enhanced and limited by social others and by material surroundings. It requires an effort to grasp continuities and negotiate boundaries of otherness, while at the same time appreciating uniqueness and even radical difference. Moreover, it involves a humble acceptance of what we do not know and perhaps cannot know about the perceptual abilities of other animals because of the relative limitations of human understanding. We should remember, though, that even perceptual challenges should not cause us to conclude that interspecies understanding is completely impossible or irrelevant. As phenomenological biologist Adolf Portmann insists, "The inwardness of these forms, widely different from our own organization, speaks to us through its appearance. That we cannot translate this language into human words is no reason not to see the appearance itself. If, in a distant country, I attend a dramatic performance, of which I understand not a single word, I shall not on that account assert that nothing at all is being presented, that nothing is happening but a random noise" (as translated in Grene 1965:28).[6] Although imagining transspecies perceptual experience can be difficult, even as a hypothetical exercise, it may help us to appreciate the embodied agency of other animal beings and it could be invaluable to the way we conduct research about animal minds. Referring to the renowned work of phenomenological biologist Jacob von Uexküll (1957), environmental philosopher Neil Evernden (1993) asks us to enter the lifeworld bubble, or *umwelt,* of the female tick. He explains that "when we ourselves step into one of these bubbles, the familiar meadow is transformed. Many of its colourful features disappear, others no longer belong together but appear in new relationships. In short, we step into a completely new world, but a world unimaginable to the mechanist with his belief in animals as automatons responding to stimuli rather than as subjects who help create their own worlds" (Evernden 1993:79). At the very least, then, the interdisciplinary methods I have developed and the theoretical framework they are based upon may encourage research that foregrounds the agency and subjectivity of the animal participants. My focus on situated behavior as a method for attending to other animals can serve as a humbling approach to the unique ways animal and human individuals outwardly express their embodied agency in complex social environments.

Notes

1. Robert Mitchell (this volume) calls this kind of process "kinesthetic-visual matching," or KVM, and employs it as a pre- or nonlinguistic bridging structure between ascribing states of consciousness to others and to the self.
2. See, for example, the work of Jacob von Uexküll (1957), Adolf Portmann (e.g., 1964), and Marjorie Grene (1965).
3. For a fuller discussion of agency, *umwelten,* and affordances, see Warkentin (2009).
4. In *Staging Tourism,* Jane Desmond says of the experience of aquarium exhibits, "it is clear who is viewing and what is being viewed. The exhibits are lit, but we are in darkened rooms. Like peeping toms staring in through a lighted window, we observe unobserved" (2001:182)
5. There is even a glimpse here of the potential for ethical affordances in whale-human interactions, a theory and practice I developed through research reported here and in earlier publications (Warkentin 2010, 2011).
6. Whether or not Portmann is putting forth an argument for the psychological inner world of animals with this passage is open for interpretation, as another translation of his work (Portmann 1964:156) suggests otherwise, at least for some species. My interest in Grene's translation is in the way she expresses the idea that, although we lack knowledge about the inner life of other animals, we should not deny its existence.

References

Behnke, E. 1997. "Ghost Gestures: Phenomenological Investigations of Bodily Micromovements and Their Intercorporeal Implications." *Human Studies* 20:181–201.

Code, L. 1993. "Taking Subjectivity Into Account." In L. Alcoff and E. Potter, eds., *Feminist Epistemologies,* pp. 15–48. New York: Routledge.

Csordas, T. J. 1993. "Somatic Modes of Attention." *Cultural Anthropology* 8, no. 2: 135–156.

——. 1999. "Embodiment and Cultural Phenomenology." In G. Weiss and H. F. Haber, eds., *Perspectives on Embodiment: The Intersection of Nature and Culture,* pp. 143–162. New York: Routledge.

Desmond, J. 2001. *Staging Tourism: Bodies on Display from Waikiki to Sea World.* Chicago: University of Chicago Press.

Evernden, N. 1993. *The Natural Alien: Humankind and Environment.* 2d ed. Toronto: University of Toronto Press.

Gibson, J. J. 1979. *The Ecological Approach to Visual Perception.* Boston: Houghton Mifflin.

Grene, M. 1965. *Approaches to a Philosophical Biology.* New York: Basic Books.

Grosz, E. 1993. "Bodies and Knowledges: Feminism and the Crisis of Reason." In L. Alcoff and E. Potter, eds. *Feminist Epistemologies,* pp. 188–216. New York: Routledge.

Haraway, D. 1989. *Primate Visions: Gender, Race, and Nature in the World of Modern Science.* New York: Routledge.

——. 1991. *Simians, Cyborgs, and Women: The Reinvention of Nature.* New York: Routledge.

Harding, S. 1993. "Rethinking Standpoint Epistemology: 'What Is Strong Objectivity?'" In L. Alcoff and E. Potter, eds., *Feminist Epistemologies,* pp. 49–82. New York: Routledge.

Hughes, H. 2001. *Sensory Exotica: A World Beyond Human Experience.* Cambridge: MIT Press.

Lakoff, G., and M. Johnson. 1980. *Metaphors We Live By.* Chicago: University of Chicago Press.

Midgley, M. 1995. *Beast and Man: The Roots of Human Nature.* New York: Routledge.

Nagel, T. 1974. "What Is It Like to Be a Bat? *Philosophical Review* 83:435–450.

Portmann, A. 1964. *New Paths in Biology.* Trans. A. J. Pomerans. New York: Harper and Row.

Reed, E. S. 1996. *Encountering the World: Toward an Ecological Psychology.* Oxford: Oxford University Press.

Shapiro, K. 1997. "A Phenomenological Approach to the Study of Nonhuman Animals." In R. W. Mitchell, N. S. Thompson, and H. L. Miles, eds., *Anthropomorphism, Anecdotes, and Animals,* pp. 277–295. Albany: SUNY Press.

Uexküll, J. von. 1957. "A Stroll Through the Worlds of Animals and Men: A Picture Book of Invisible Worlds." In C. H. Schiller, ed., *Instinctive Behavior: The Development of a Modern Concept,* pp. 5–80. New York: International Universities Press.

Warkentin, T. 2009. "Whale Agency: Affordances and Acts of Resistance in Captive Environments." In S. McFarland and R. Hediger, eds., *Animals and Agency: An Interdisciplinary Exploration,* pp. 23–43. Leiden: Brill.

——. 2010. "Interspecies Etiquette: An Ethics of Paying Attention to Animals." *Ethics and the Environment* 15, no. 1: 101–121.

——. 2011. "Interspecies Etiquette in Place: Ethical Affordances in Swim-with-Dolphin Programs." *Ethics and the Environment* 16, no. 1: 99–122.

9

The Meaning of "Energy" in Cesar Millan's Discourse on Dogs

JULIE A. SMITH

Contemporary scholar Zhang Longxi (2005) tells the following story about two ancient Chinese philosophers:[1]

> Zhuangzi and his rival, Huizi, are strolling on the bridge over the Hao River. Zhuangzi says, "Out there a shoal of white minnows are swimming freely and leisurely. That's what the fish's happiness is." Huizi replies, "Well, you are not a fish, how do you know about fish's happiness?" Zhuangzi says, "You are not me, how do you know that I do not know about fish's happiness?" Huizi replies, "I am not you, so I certainly do not know about you. But you are certainly not a fish, and that makes the case complete that you do not know what fish's happiness is."
>
> (p. 3)

Zhang explicates the anecdote as follows:

> The crucial point Zhuangzi makes in this passage is to pursue Huizi's dry logic vigorously. . . . Huizi never has a moment of doubt about what he knows, namely, that Zhuangzi is not a fish, ergo he does not know fish's happiness. Throughout the conversation, Huizi's negative knowledge, his

> conviction that there is a difference between Zhuangzi and a fish . . . is stated most positively and assuredly. His skeptical attitude toward knowledge thus rests on his unreflective confidence in his own negative knowledge of the difference of things. For Zhuangzi, however . . . the difference between man and fish is by no means a fact established a priori.
>
> (pp. 4–5)

I cite this anecdote as my introduction to Cesar Millan because Millan also claims to know the mind of an animal and to locate his knowledge in his experience of its movement. Zhuangzi sees the minnows moving free and easily, and he concludes that they are happy. Millan claims to know dogs' minds by means of their bodily movements. Those minds he calls their *energies*, a term that refers to what we might call frames of mind.

Millan's capacity to read energies of dogs is related to, but not identical with, a classic ability that all humans have—the ability to recognize a match between the look of movement and the feeling of movement. I raise my arm, and you know how that feels even though you have not raised yours; you relate the look of what you see to a particular feeling you have had of that movement. This is called kinesthetic-visual matching or KVM. Descriptions of classic KVM do not specify whether the knowledge of movement one has while watching the sight of movement entails reexperiencing imaginatively the feeling of movement or instead the grasping intellectually of the connection between the two modalities. Whichever it is, I will argue here that Millan's exceptional gift to interact with dogs arises from something similar to the first interpretation: He is able to feel in himself the movements he sees dogs make as they make them. To distinguish this from KVM, I am calling this KVT, or kinesthetic-visual transfer. I will also examine whether Millan's discourse of energy presumes that dogs themselves experience KVT. I think that it does.

Cesar Millan: Background

But first a synopsis of the story Millan tells about his life with dogs. Millan was born in the town of Culiacan, Mexico, and spent much of his early childhood on his grandfather's farm about an hour from his hometown. There he lived among free-roaming dogs who trekked about in loosely formed packs of five to seven animals. In 1990, at the age of twenty-one, he entered the United States illegally with the hope of becoming a professional dog trainer like those he imagined behind the scenes in the television shows *Lassie* and *Rin Tin Tin*. When he arrived in San Diego he worked as a dog groomer and then moved to Hollywood, where he was employed as a kennel boy at an upscale dog-training establishment. Recognized for an ability to manage difficult dogs, he was nonetheless unhappy with conventional training methods and left his job to accept employment washing limousines. From his employer at the limousine rental company, Millan learned the basics of running a business, and he soon started his own school for dogs, the Pacific Point Canine Academy. Also from his boss, who would say to influential friends, "I've got a Mexican who's good with dogs," Millan acquired a prestigious clientele; eventually he was sought out by celebrities and rescue groups, and he founded the Dog Psychology Center in the warehouse district of south Los Angeles, where he kept a pack of rehabilitated dogs. In 2002 the *Los Angeles Times* published a feature article about him, which led to offers of a television show. *The Dog Whisperer with Cesar Millan* began on the National Geographic Channel in 2004 and at this writing is in its seventh season with an estimated ten to fifteen million viewers.

In the television show and in his two books, Millan rejects the label *dog trainer* and calls himself a dog rehabilitator. He dismisses dog training as human psychology rather than dog psychology: "Dog training is created by humans; everything you do from the dog training world is human psychology: sit, stay, down, good boy. Show me a dog that teaches another dog that way." For Millan, training does not give access to a dog's mind: "Just because a dog knows how to obey doesn't mean she's balanced. When you train a dog, . . . you get access only to conditioning. And conditioning doesn't mean anything in the dog world" (Millan and Peltier 2006:225). For Millan, training does nothing to address the psychological states that produce problem behavior in dogs, such as fear, anxiety, nervousness, dominance, aggression. Millan's goal is to return dogs to a state of mind he calls "balance," which he describes as the mental state toward which a dog's natural instinct tends if its needs are met. Calling balance variously "the deepest form of resting mode," or what humans would call a meditative mode (p. 213), mental stability, a mind free of issues, Millan defines it as a calm-submissive state appropriate to a dog who is a follower in a dog pack. That state can be nurtured by humans, according to Millan, by giving the dog exercise, specifically a walk that recapitulates the ancestral migrations of dog packs; by enforcing rules, boundaries, and limitations through a human version of pack leadership; and by providing affection as a reward only for calm-submissive pack behavior.

Millan has attracted fierce criticism. Where his supporters see leadership, his detractors see heavy-handed dominance. Where his admirers see intervention for psychologically troubled dogs, his critics see old-style aversion training. What Millan calls calm submission, his opponents label learned helplessness. Academics especially have located his methods within the paradigm articulated by Yi-Fu Tuan in his seminal 1984 book, *Dominance and Affection: The Making of Pets*.[2] Lisa Jackson-Schebetta (2009) writes that Millan's dog training is based on a slave-master dichotomy that mythologizes the relationship between humans and dogs "into a narrowly conceptualized dominance paradigm through which the non-human animals are presented as commodities that conform to the human animal's desires" (p. 137). She represents Millan himself as a willing participant in a neocolonialist agenda offering bourgeois America a performance of the American dream in tandem with a new colonized other: "The dominance paradigm Millan sells includes himself: he can acceptably be the alpha male of the dog pack because he is like a dog, and therefore, is himself subject to culture's dominion over nature, the developed and colonial United States' dominion over the backward and developing Mexico" (p. 155). While some people may have this disparaging view, many more are awed by Millan's gift at communicating across the human-animal divide.

How he does that has been the subject of investigation and speculation. In his 2006 essay in the *New Yorker,* Malcolm Gladwell examined what he calls Millan's "movements of mastery."[3] He began by suggesting that "everything we know about dogs suggests that in a way that is true of almost no other animals, dogs are students of human movement" (p. 52). Gladwell watched tapes of Millan with two movement specialists and reported on their findings: "Combinations of posture and gesture are called phrasing, and the great communicators are those who match their phrasing with their communicative intentions—who understand, for instance, that emphasis requires them to be bound and explosive. To Bradley [Karen Bradley, graduate director of dance at the University of Maryland] Cesar had beautiful phrasing. . . . 'He's dancing,' Bradley said. 'Look at that. It's gorgeous. It's such a gorgeous little dance'" (p. 53). Suzi Tortora, a New York dance-movement psychotherapist who uses dance to communicate with autistic children agreed: "The phrasing is so lovely. It's predictable. To a dog that is all over the place, he's bringing a rhythm" (as cited in Gladwell 2006:55). Gladwell concluded that these analysts confirmed his own impressions that Millan used "extraordinary energy and intelligence and personal force . . . on behalf of the helpless" (p. 56). The gift of movement these dance consultants see in

Millan relates well to Millan's idea of energy in movement. As I suggested earlier, his idea of energy can be explained as a version of what is called kinesthetic-visual matching.

What Millan Means by "Energy"

Millan explains all his interactions with dogs in terms of *energy* and attributes his successes at rehabilitating them to understanding their energy and taking on the kind of energy necessary for a pack leader. When Millan talks about energy, he closely relates mind and body. Sometimes he uses the term to name the physiological state of a dog or a human at any particular moment; sometimes he talks about energy by compounding words, as when he refers to "emotional-mood-energy signals" (Millan and Peltier 2007:226). More than different levels of exertion, energy makes internal states manifest in movement: emotions, but also attitudes, bodily chemistry, even an entire mental demeanor or personality. Energy is the look of another's inner life at that particular moment made visible and communicable through complex movement.

Millan describes himself as having searched for an explanation for the communicative energy he sees in dogs. He says of his experience at the Hollywood kennel, "Unconsciously, I was beginning to apply the dog psychology I had learned from years of observing dogs on my grandfather's farm. I was interacting with the dogs the way they interacted with one another. . . . I couldn't have explained in words what I was doing at the time. . . . Everything just came instinctually to me" (Millan and Peltier 2006:47). Millan claims he found the language for what he knew intuitively in books: "I began a program of self-education, reading everything I could get my hands on about dog psychology and animal behavior. . . . I was finding ways to articulate the things I intuitively understood" (p. 53).

Although Millan traces his idea of energy back to popular books of dog psychology and human self-help literature, he describes something quite different from what one finds there. The books of dog behavior that he cites compartmentalize dogs' mental faculties in very conventional ways. Millan does seem to borrow his term *energy* from motivational speaker Wayne Dyer, as he admits. For Dyer (1989), energy is a spiritual system, a universal life force: Human beings "are part of the life force that is onesong. . . . We fit into and harmonize with the entire energy system called the universe" (p. 162). Dyer (1992) coaches humans to activate their energy: "What is needed is a shift from the inert energy of wanting to the active energy of doing and intention" (p. 76).

Dyer's spiritualism is missing from Millan's discourse of energy, at least with respect to dogs, although Dyer's motivational rhetoric does sometimes play out in Millan's interaction with human clients. One important idea in both writers is the link between energy and mental states: "Virtually everything about the physical you is a result of . . . how your thoughts become energized into action" (Dyer 1992:76). When Millan parodies dogs' manner for humans, he always does so to convey states of mind, and he clearly believes that different mental states translate into different kinds of energy, itself translated into macro and micro movement. One cannot effectively convince a dog that one is the pack leader without the right mental states, says Millan, which transpose into a manner of being in the body. Thus, for example, Millan successfully instructed a young girl with cerebral palsy to walk her dog in the manner of a pack leader, suggesting a distinction between gross bodily movements and a manner of translating mental states into body energy.

Both Millan and Dyer also say that energy enables universal communication between beings. Millan claims that energy is a language that everyone has: "Isn't there a language we can learn that means the same thing to every creature? . . . I'm happy to report that the universal language . . . already exists. . . . It's a language all animals speak without even knowing it, including the human animal. . . . Even human beings are born fluent in this universal tongue, but we tend to forget. . . . This truly universal, interspecies language is called *energy*" (Millan and Peltier 2006:61). As described, Millan's approach seems to securlarize Dyer's spiritualistic energy. However, as it plays out in his relations with dogs, it brings to mind a capacity cognitive theorists call kinesthetic-visual matching. Like KVM, Millan's idea of energy seems an intuitive capacity to relate the sight of movement and the feeling of movement.

As stated earlier, KVM is a capacity to associate the feelings of movement of one's own body with the look of movements made by someone at a distance, "something visually over there," such as one's image in a mirror or another being engaged in imitative movements."[4] The "matching" that occurs in KVM entails a cognitive understanding of a relationship between movement and vision, the power to put two very different modalities together. I am able to imitate your movements not just because I can copy the look of them, but because I can connect that look with my knowledge of how it feels to make those movements. KVM does not, as

I understand it, *necessarily* entail a capacity to feel in oneself the movements one observes as one observes them. Rather, KVM is a cerebral understanding of a relationship between vision and movement, a grasp that what one sees is isomorphic with the experiences of movement one has had in one's own body. What I will describe links sight and movement in a different way than this interpretation of KVM.

I am suggesting that while Millan obviously possess the basic human ability to connect the sight of bodily movement with an understanding of similar movements in himself ("classic" KVM), he also brings something else into play. That something else is a capacity that enhances or operates as a version of KVM. Like KVM, it may be explained by intuition or instinct and may be developed through experience within a particular kind of sight-movement context. I argue that that "something else" is a capacity I would call KVT or kinesthetic-visual transfer. I suggest that KVT is not just an *understanding* of a vision-kinesthetic relationship but an *experience* of it—reaction, sensation in oneself of the feeling of movement while observing movement in another. When Zhuangzi looks at the fish, he seems to feel what it is like to swim freely and leisurely, although not executing those movements himself. In fact, many words in English confound the difference between three components of a visual-kinesthetic experience: the look of movement to an observer, the feeling of it in the one executing the movement, and the feeling of it reproduced in the observer. If I describe fish as "moving freely and leisurely," am I describing how they look to me? how I think they must feel when they move in that way? how they make me feel? All three, I would say.

To my mind, a special or enhanced capacity for visual-kinesthetic transfer with respect to dogs explains Millan's exceptional abilities at imitating them, an ability not explainable by traditional KVM, however much experience one has with a particular kind of look-movement interaction. When Millan removes food from a dangerous dog who guards its bowl obsessively (as in Fincke 2004), he does so by slowly moving in and then placing himself between the bowl and the possessive dog, the way that dominant dogs do. While this seems simple, and while his goal is to teach other people how to do it, I expect that very few people could reproduce the movements as he does, that is, as if by second (or even first) nature. That ability seems to depend on remembering how it felt within himself when he watched those movements in dogs. Also, when Millan imitates dogs' manner for a human audience, his face and upper body become startlingly and comically doglike. As anyone who watches comedians regularly will attest, effective parody of others' movements is extraordinarily rare. I am arguing that such exceptional abilities require a capacity to feel in oneself the movements being made by others—in Millan's case, the sensation of movements that dogs have as they make them. This visual-kinesthetic transfer is what can explain Millan's idea of the flow of energy from dogs to himself and from himself to dogs. This is not something made possible solely by an intellectual grasp of a sight-movement connection, or classic KVM, no matter how much experience one has had with it.

Millan's discourse of energy, just as Zhuangzi's insistence that "the difference between a man and fish is by no means a fact established a priori," is about connectedness between beings. If I feel how it feels to move as you do, I am likely to feel myself as not particularly separate from you. In spite of the fact that Millan insists that humans must never confuse dogs with humans, his connectedness to dogs is the compelling subtext of every episode of his television show, *The Dog Whisperer.* Although this is performed rather than stated, Millan does once say, "To say that I 'love' dogs doesn't even come close to describing my feelings and affinity for them" (Millan and Peltier 2006:22). The transfer of feelings of movement through sight might well be a powerful bonding mechanism within a social group; Milan's unique ability is to experience that with another species.

Do Dogs Have KVT?

Millan repeatedly says that dogs read the energy of others all of the time and do so more accurately than do humans. No doubt he believes that dogs read each other and read him. But the question is, "how do they do it." Most people assume that dogs interpret gestures and behaviors that they see in others, that is, that they learn that certain movements in others have associated behavioral effects. Using the term *kinesthetic empathy,* Shapiro (1997) argues that dogs are adept at reading movements in terms of intentions. Others believe that dogs interpret not just bodily gestures but expressions, even smells. These capabilities seem not the same thing as what I am calling kinesthetic-visual transfer because they are cognitive acts, not experiences.

In Millan's discourse of energy, dogs understand others in a way less clumsy than cognitive interpretation; their understanding seems more holistic, dramatic, and immediate. Millan states, "All animals communicate using energy, constantly. Energy is beingness. Energy is who you are. . . .

That's how your dog sees you. Your energy at that present moment defines who you are" for them (Millan and Peltier 2006:97). He goes on: "In terms of your energy, dogs know you thoroughly at every moment they are in your presence: I can't emphasize enough that dogs pick up every energy signal we send them. They are reading our emotions every minute of the day" (p. 123). If true, this is different from interpreting body language, from understanding external signs as indicative of internal states. More visceral and comprehensive, it entails a kind of reexperiencing of others' system-states, a "picking up" of others' mental conditions in a way more literal than metaphorical.

Dogs' ability to respond to energy has evolutionary importance for the stability of the dog pack, says Millan. Dogs comprehend energy because they value stable energy within a pack. And such stable energy is only possible through the pack leader. Millan says, "all animals can evaluate and discern *what balanced energy feels like*" (Millan and Pelteir 2007:217, my emphasis). Thus when he describes dogs "sensing" the energy of others, he seems to mean receiving an impression that impacts their own sense experience:

> If you watch the news clips from Hurricane Katrina, when the abandoned dogs of New Orleans started coming out of their homes, they automatically began to take up with one another and form packs,. . . . In one photo of such a pack, I noticed a big old Rottweiler, a German shepherd, and some other big dogs. But they were being led by a beagle! Why did they choose to follow the beagle? Because the beagle had a better sense of direction . . . and she obviously had *leadership energy*. Animals know that if another animal shows the determination and takes the leadership role firmly, they should go with her. They don't say, "Look, You're a beagle. I'm a Rottweiler. I don't follow beagles. . . . The Rottweiler *sensed that the beagle was in a calm-assertive state*, and that's all she was looking for in a leader.
>
> (p. 166, my emphasis)

Millan never explicitly describes the mechanism for how dogs understand the energy of others beyond the idea of "sensing." But his accounts are associated with processing complex movement in a holistic way:

> In the wild, different animal species intermingle effortlessly. Take the African savannah or a jungle, for instance. At a watering hole in a jungle, you might see . . . many different species sharing the same space. How do they all get along so smoothly? . . . All these animals are communicating with the same

> relaxed, balanced, non-confrontational energy. Every animal knows that all the other animals are just hanging out, doing their own thing—drinking water, foraging for food, relaxing, grooming one another. Everybody's feeling mellow and no one's attacking anyone else. Unlike us, they don't have to ask one another how they're feeling. The energy they are projecting tells them everything they need to know.
>
> (Millan and Peltier 2006:62)

Millan's scene at the waterhole describes animal actions that require specific and intricate kinds of motions—drinking, foraging, grooming. Movement is the look of interiority: "everyone's feeling mellow." And how does that look of movement transfer to others? Surely Millan does not mean that the animals are calculating intentions in terms of visible exertion levels or behaviors. Millan says that each animal projects its energy to others; by this he seems to mean that animals feel or *experience* the intertwined mental-physical states of others as transmitted by their movements.

Millan often describes energy as something that is instantly transferred from one animal to another, and one can almost see that transfer while watching the interactions between Millan's dogs on many episodes of his television show. Millan maintains a dog pack of over forty rehabilitated dogs. He often brings a new dog into the pack temporarily as part of its rehabilitation. It is a heart-stopping sight to see forty dogs converge on a new dog, and Millan states repeatedly that the new dog must be in a calm-submissive state or its undesirable energy will have a disastrous ripple effect through the pack. That such energy is transferred so quickly from animal to animal, and that the dogs experience it so intensely, makes one think not of a process of "understanding" but rather of an electric current of feeling passing from one dog to another.

The most compelling evidence for me of the dogs' capacity for visual-kinesthetic transfer is that they can read Millan as they would another dog. The foundation of Millan's rehabilitation program is to convince dogs that humans are the pack leader. This means more than bossing the dogs around. It has to do with an ability to use his body to transfer to the dogs the feeling of authority as coming from him, to give them that sight-kinesthetic experience of a state called I-am-in-charge. It also entails conveying and maintaining within followers a feeling of calmness and security transferred from the movement-manner of the leader of the pack to the followers.

There is great resistance to giving dogs anything like a capacity for KVM. If dogs are afforded KVM, this would mean that they have an intellectual understanding that the movements they see in others are isomorphic with the experiences of movement in their own bodies. For cognitive theorists, dogs and most other nonhuman animals do not possess KVM in relationship to anything but the feel and look of their own bodies. The reason that they are not afforded KVM is that the capacity is closely associated with mirror self-recognition and imitation, and dogs show little or no ability for either. The argument goes that if they had KVM they would engage in more imitative behavior and give more explicit evidence of mirror self-recognition. Further, KVM is considered the "gateway" capacity to many sophisticated cognitive abilities. This includes perspective taking, that is, "recognition that a perspective different from one's own exists and is experienced by others" (Mitchell 1997:33). And perspective taking is thought to be the path to empathy.

However, if Millan's discourse of energy gives to dogs something like what I have called kinesthetic-visual transfer, then it also gives them not exactly empathy but a kind of identification with others. It does not necessarily afford dogs the same perspective that humans have, that is, that others are separate from themselves yet operating from invisible internal states. However, it does give them a particular kind of relating. If dogs have a capacity to feel in themselves the movements they see in others, then they likely can experience identification, one that blurs the boundary between self and other. This kind of collapse might entail a more intense empathy than an intellectual understanding that others are not oneself but have comparable mental states. Additionally, this blurring may explain why dogs might not be interested in imitation. Imitation depends on experiencing the salience of "difference"; imitative actions bridge perceived gaps between self and other. If dogs have KVT, they may experience no such gaps.

When I reach for the leash to take my dogs on a walk, they jump up excitedly, tails wagging furiously, bodies dancing in circles. These movements express their eager anticipation of a walk, as every human knows. Millan would add that my dogs are also reading *my* energy, perhaps within the context of a complex network of my inner states. If so, they probably know that I have decidedly lower levels of enthusiasm for this walk than they do. Although they appear to care only about their own feelings, who knows? If my demeanor sends to them feelings of "I'd rather be reading a book," maybe they are trying to transfer to me their feelings of "won't this be great." How interesting to think dogs might not only experience the transfer of feelings through movement but be able to send back movement-messages.

Notes

1. I am indebted to my colleague Jian Guo for my knowledge of this work.
2. For an overview of the issue of human dominance over pets, see Fudge.
3. I am indebted to Karla Armbruster for calling my attention to this essay.
4. On KVM, see Mitchell's several essays. He has been the theorist who has developed most fully the implications of KVM. I am grateful to him for reading and commenting on this essay. He does not necessarily share my views.

References

Dyer, W. W. 1989. *You'll See It When You Believe It.* New York: Morrow.

——. 1992. *Real Magic: Creating Miracles in Everyday Life.* New York: HarperCollins.

Fincke, S. A., dir. 2004. "Lucy and Lizzie." Written by M. J. Peltier. In S. A. Fincke, producer, *The Dog Whisperer with Cesar Millan,* October 18. Los Angeles: MPH Entertainment.

Fudge, E. 2008. *Pets.* Stocksfield, UK: Acumen.

Gladwell, M. 2006. "What the Dog Saw: Cesar Millan and the Movements of Mastery." *New Yorker* 82, no. 14 (May 26): 47–57.

Jackson-Schebetta, L. 2009. "Mythologies and Commodifications of Dominion in *The Dog Whisperer with Cesar Millan.*" *Journal for Critical Animal Studies* 7, no. 1: 137–159.

Millan, C., and M. J. Peltier. 2006. *Cesar's Way: The Natural, Everyday Guide to Understanding and Correcting Common Dog Problems.* New York: Harmony.

——. 2007. *Be the Pack Leader.* New York: Harmony.

Mitchell, R. W. 1997. "Kinesthetic-Visual Matching and the Self-Concept as Explanations of Mirror-Self-Recognition." *Journal for the Theory of Social Behavior* 27, no. 1: 17–39.

——. 2002. "Kinesthetic-Visual Matching, Imitation, and Self-Recognition." In M. Bekoff, C. Allen, and G. M. Burghardt, eds., *The Cognitive Animal: Empirical and Theoretical Perspectives on Animal Cognition,* pp. 345–351. Cambridge: MIT Press.

Shapiro, K. J. 1997. "A Phenomenological Approach to the Study of Nonhuman Animals." In R. W. Mitchell, N. S. Thompson, and H. L. Miles, eds., *Anthropomorphism, Anecdotes, and Animals,* pp. 277–295. Albany: SUNY Press.

Tuan, Y-F. 1984. *Dominance and Affection: The Making of Pets.* New Haven: Yale University Press.

Zhang, L. 2005. *Allegoresis: Reading Canonical Literature East and West.* Ithaca: Cornell University Press.

10

Inner Experience as Perception(like) with Attitude

ROBERT W. MITCHELL

That animals move themselves provides most people with the idea that they are psychological, and most people have little difficulty coming up with psychological descriptions of animal activities from the animal's perspective. People recognize that Rover wants to go out or is beginning to feel tired or presume that Rover feels guilty about destroying the couch. Often in discussions of an animal's mind, however, people want to or try to describe what is termed the animal's "inner private" (or conscious) experience in a manner similar to the way we describe our own such experience (e.g., Griffin 2001). Some such experiences are perceptions—experiences like feelings of movement or pain, seeing a dog, hearing a tune. Other such experiences seem based (in some way) on perception—things like auditory or visual mental images. Sometimes the desire is to describe the conceptualization behind the experience (see Mitchell 2001)—does Rover see a ball, a round object, or something he can play with? In such descriptions some visual and auditory perceptions seem "less" inner or private than others, because (if I see or hear) I can discern the thing or sound perceived by Rover and me, even if I cannot tell *how* the things or sounds are experienced or interpreted by Rover. People at times assume that vision and audition are fully intersubjective (identically shared) perceptions, in part because people can agree generally (by using or accepting the same words) about *what* they see and hear and in part because both animals and people act as if they are experiencing what they see and hear in the same ways. People make aspects of shared experiences intersubjective through language, but often do not attend to the fact that their own inner private experiences are not actually shared (Meyers and Waller 2009).

Often people end up saying that they can know what other people are experiencing (because of intersubjective meanings), but cannot know what animals are experiencing. For example, recently I heard a biologist argue that we couldn't know if a calf, held immobile while having a hot iron attached to a segment of its head for several seconds to remove its developing horns, experienced "the same" pain as "a person" would under similar circumstances, which was part of an extended argument about whether or not it was important to alleviate the pain of calves during dehorning. But unless you assume that calves have *no* pain experience during the administration of painful stimuli, lack of knowledge of exactly "how pain feels" to the calf, or whether it is like that of humans, is irrelevant (Singer 1975). I want to argue that we often have just as much access to animals' experiences as we have to other people's experiences. For example, a person writhing in pain tells us as much about their pain as an animal doing so, whether or not the person is screaming "It hurts" or "I'm in pain." I have no more access to the pain experience of the person than I do to that of the animal; in both cases I have to make judgments using knowledge from their behavior and context. Humans *can* specify through language that their experience feels like something that is or appears intersubjectively understandable—e.g., "it feels like a hammer keeps hitting me"; "it oscillates in intensity" (more on intersubjectivity later in this chapter).

People often want to argue, at this point, that they know what human pain is like, so they assume another person's pain is or would be the same as theirs in a similar situation. But there are great varieties of human pain, if connoisseurs and victims of pain are a guide, and there is no knowing what exactly someone is experiencing. So although we recognize the type of psychological experience of both animals and people, we just don't know exactly how it feels or is experienced for either. We know how to describe another's experience, but we don't know all aspects of it that are known directly to the experiencer. An example makes this clearer: an acquaintance of mine once informed me that as a child he did not wear glasses and consequently saw double. When he first experienced corrective lenses, suddenly everything was integrated. He was surprised that everything was a singular whole, that everything could *not* be doubled, because he thought that everyone saw double. It never occurred to him that people were in any way different from himself in their perceptual experience. The problem of other people's minds was just not a problem for him—he assumed that everyone saw the world as he did, and everyone (except his eye doctor) assumed that he saw the world as they did. And to some degree, everyone was right—he did see (to some degree) what others saw, he just saw it twice. Almost everyone would have described accurately (to some degree) his psychological experience (he had visual experiences), but they would have missed aspects of that experience that were private (his visual experiences were double). Of course if he told people he saw double, he would have made his inner private experiences known, but most of us don't talk about what we take to be common ground. We make the same mistake my acquaintance did when we think that other humans' experiences are exactly like our own. My acquaintance was right that people have experiences like his own, but he was wrong in believing that they were *exactly* like his own. We can only know so much about other people's experiences, just as we can only know so much about animals' experiences. Interestingly, the fact that his eye doctor knew that my acquaintance saw double indicates that tests of some mental experiences can provide knowledge of normally private aspects of one's inner mental states.

The chapter concerns how much language can effect intersubjectivity, or how much intersubjectivity language can effect, and why we expect that others have inner private experiences. My chapter is intended to elucidate two aspects of this "other minds" problem: why we believe that there are other minds (other experiencers) and what inner private mental states are like. My ideas derive in part from a fascinating debate between the philosophers Ludwig Wittgenstein and Peter Frederick Strawson in the 1950s that examined how English-speaking humans linguistically describe their own inner private experiences (see Mitchell 2000). It is surprising to me that this debate has not had more of an impact on the problem of other minds, especially as it relates to animals. What is interesting to me about the debate is that it leads readers to recognize that understanding other minds is all about analogies of one sort or another, but not the analogy that is usually posited to explain our belief that there are other minds. In my view the debate leads readers to understand what kinds of things inner private mental states are as well as why we believe that other entities (people as well as animals) have them.

Wittgenstein and the Problem of Other Minds

Wittgenstein (1953) found the existence of other minds a significant problem for philosophers, but recognized that this existence was not a problem for most people. He drew the reader into the problem by accepting the commonsense view. In answer to the questions "What gives us *so much as the idea* that living beings, things, can feel?" (#283) and "How could one so much as get the idea of ascribing a *sensation* to a *thing*?" (#284), he responded, "And now look at a wriggling fly and at once these difficulties vanish and pain seems able to get a foothold here, where before everything was, so to speak, too smooth for it" (#284). He (#275) wrote, "Look at the blue of the sky and say to yourself 'How blue the sky is!'—When you do it spontaneously—without philosophical intentions—the idea never crosses your mind that this impression of colour belongs only to *you*." But, for philosophers, the question is how we could possibly know that your blue and my blue (your experience of the sky and mine) are the same experience. Perception includes inner private experiences (e.g., of color) that cannot be shared directly, even if we agree that the term (*blue*) is appropriate to tie together our experiences and the color of the sky.

Wittgenstein (1953) raised questions about whether there can be evidence for inner private mental states (one's own or another's) and about the importance of language in understanding mental states. He argued that, in determining that something is the case, one commonly uses either criteria (a principle) or symptoms (empirical correlations). For example, the criterion for a team's earning points in basketball is that a team member shoots a basketball through that team's hoop during a basketball game, whereas the symptom is intense and extreme jubilation from fans for that team. But in the case of experiences, e.g., perceptions, we have neither criteria nor symptoms by which to discern them. This analysis led Wittgenstein (#293) to the idea that language about sensation need not have a consistent referent (your blue and my blue may differ): "If we construe the grammar of the expression of sensation on the model of 'object and name' the object drops out of consideration as irrelevant" because a sensation (the "object") has no independent means of recognition. For Wittgenstein, then, naming a sensation does not indicate that the sensation is an object. This idea does not mean that no experience or sensation exists; rather, "the conclusion was only that a nothing would serve just as well as a something about which nothing can be said. We have only rejected the grammar which tries to force itself on us here" (#304).

Wittgenstein questioned the use of language to *describe* inner private states and be consistent in our naming them. Suppose that a man wanting to keep track of a particular sensation marked "E" in his diary whenever he experienced that sensation. Wittgenstein (#258) believes that this person can have no criterion of correctness, no way of knowing if the sensation marked by "E" is the same every time. Even though a sensation may feel the same as a previous sensation, there is no independent method of verification (presumably one compares a current perception to a corrigible memory of that perception). In writing or saying that one experiences the same sensation, "What I do is not . . . to identify my sensation by criteria: but to repeat an expression" (#290). Wittgenstein continues, playing questioner and critic: "But isn't the beginning the sensation—which I describe?—Perhaps this word 'describe' tricks us here. I say 'I describe my state of mind' and 'I describe my room.' You need to call to mind the differences between the language-games." Words may not be descriptions of *experiences*, though our language suggests that they are; rather, they are often descriptions of the external objects of experiences, which can be cross-validated (Glover 1981). Visual sensations are about external objects and are not objects themselves. For Wittgenstein, "An 'inner process' stands in need of outward criteria" (#580), an experience requires criteria in order to be experienced (#509).

Given that one cannot directly compare experience across individuals, a second problem about the analogy to other minds concerns extrapolating from one's own experience to that of another. Without verification through direct comparison, how can one extrapolate from only one instance? Wittgenstein's concern arises from the classic solution to the problem of other minds: the argument from analogy. As it is classically presented, the problem of other minds is potentially solved by analogical argument. I have a body, which produces experiences (feelings and other inner private mental states), which cause my behaviors; other people have a body very like mine in many ways and produce behaviors very like mine in many ways. I know my own experiences, and can correlate these experiences with my behaviors. Therefore, when I perceive another person behaving as I do in a situation similar to one in which I have a particular experience, I expect they have similar experiences. That is, I extrapolate from similarity in body, behavior, and context to similarity in conscious experience and thereby recognize *that* they have a mind (see, for example, Shorter 1952).

There are, predictably, several problems with the argument from analogy as an explanation for belief in other minds. Wittgenstein focused on a few. He wrote:

> The essential thing about private experience is really not that each person possesses his own exemplar, but that nobody knows whether other people also have *this* or something else. The assumption would thus be possible—though unverifiable—that one section of mankind had one sensation of red and another section another. . . . If I say of myself that it is only from my own case that I know what the word "pain" means—must I not say the same of other people too? And how can I generalize the *one* case so irresponsibly?
> (#272, #293)

Wittgenstein does not intend to deny here that the other's experience is obvious to us: "Just try—in a real case—to doubt someone else's fear or pain" (#303). This obviousness does not, however, mean that we know how the fear or pain feels to another, i.e., know directly or exactly his experience of it, even though we use the same words to describe an inner experience.

To make the problem of cross-person comparisons of consciousness salient, Wittgenstein imagines a scenario (#293) in which everyone had her own box containing something called a "beetle," and everyone could know what a beetle is only by looking in her own box. Everyone might have something different in her box or might have nothing at all (presumably the "beetle" could be the air in the box or nothing—a sleight of hand by Wittgenstein, in that he is using a term to name nothing). So the term *beetle* would not refer to anything commensurate across individuals. Similarly, one's name for a particular inner state one experiences can be used across individuals without being a name for the same experience across individuals (or, as the sleight of hand suggests, for any experience at all). However much we may use the same term to describe our inner experience, there is no evidence available indicating that human experiences under the same name must be identical experiences: There can be differences in the phenomenally accessed subjective aspects (qualia) of our experiences that come under the same name.

Thus language (or at least naming) holds a special place in our understanding of mental states. Although one can have concepts without language, in a way language creates mental states by creating particular concepts: "You learned the *concept* 'pain' when you learned language" (Wittgenstein, #384). Even when we conceive of mental states as inaccessible to others, we still talk about them, and can teach others to talk about them.

> What is it like to say something to oneself; what happens here?—How am I to explain it? Well, only as you might teach someone the meaning of the

> expression "to say something to oneself." And certainly we learn the meaning of that as children.—Only no one is going to say that the person who teaches it to us tells us "what takes place." . . . Rather it seems to us as though in this case the instructor *imparted* the meaning to the pupil—without telling him directly; but in the end the pupil is brought to the point of giving himself the correct ostensive definition. And this is where our illusion is.
>
> (#361–362)

The illusion is, apparently, in the idea that one can point to (i.e., provide an ostensive definition for) one's talk to oneself or describe it to another (#370); the language implies that one can, but the language is (for Wittgenstein) all that there is: "The mental picture is the picture which is described when someone describes what he imagines" (#367), i.e., the mental picture is not of something inside.

Strawson's Response to Wittgenstein

One response to Wittgenstein's concern about criteria for one's own experienced mental states is to state that no independent method of verification is needed, in that inner perceptions (or mental pictures or other experiences) simply are given as such and may be just as corrigible as any externally oriented perception, though memory can serve as a check of sorts (Strawson 1954:84–85, 91). Another response is to wonder if we can in fact identify the same internal state again and again. In some cases—persistent pain—one feels correct in identifying it; in others—an image of one's own or another's face—one feels less inclined to believe that the internal image is exactly or essentially the same each time one imagines it.

Yet another response is that, although one can be skeptical about the idea that language functions to describe mental experience, this skepticism should not lead one to claim that if language does not describe something, there is nothing there. "It is hard to *describe* what it feels like to have a headache or a toothache, but these *occur*" (Penelhum 1957:501). Just because something is unobservable to another does not make it inappropriate to describe that something in terms of things that *are* observable (Strawson 1954:85–86, 90–91), and something's being unobservable to someone does not indicate its nonexistence. (See Strawson [1954:86] on the nonnecessity of having criteria to describe taste experiences.) True, many experiences may not be amenable to linguistic description, which is based on intersubjectively acceptable categories. But "only a prejudice against 'the inner' would lead anyone . . . to deny that I can sometimes say something by way of description of my experiences of having imagery, as well as describing to people what I am imagining" (Strawson 1954:91).

Strawson provided two responses that are relevant here to Wittgenstein's ideas. One is that, in describing mental states linguistically, we always compare them to something we can intersubjectively agree upon—perceptual experiences. (Strawson is accepting those aspects of perception that can be intersubjectively agreed upon as relevant here and is not concerned with the identity of all aspects of experience.) Many of our internal states are analogically comparable to external things presumed to be mutually known or understandable:

> It is . . . true that when we describe "private" or "inner" or "hidden" experiences, our descriptions of them (like our descriptions of their status) are often *analogical*; and the analogies are provided by what we *do* observe (*i.e.*, hear, see, touch, etc.). This is in itself an important fact. It throws light once more on the conditions necessary for a common language. . . . [A] description is none the worse for being analogical, especially if it couldn't be anything else. Moreover, some of these analogies are *very good ones*. In particular the analogy between saying certain words to oneself and saying them out loud is very good. (One can even be unsure whether one has said them out loud or to oneself.) The analogy between mental pictures and pictures is, in familiar ways, less good.
>
> (Strawson 1954:91)

So Strawson's response means that at least some of our inner private experiences are like perceptions in some ways—at least close enough to draw the comparison. In fact, many theorists believe that mental representations develop from perceptual modalities available to the organism (e.g., Finke 1980, 1986; Mitchell 1994; Piaget 1954, 1962 [1945]; Reisberg and Heuer 2005; Ryland 1909). Although experienced mental representations (images) are somewhat perceptionlike and somewhat not perceptionlike, it appears that imagery relies to some degree upon areas of the brain used to experience corresponding perceptions (Thomas 1999). So when we agree about the uses of our common language to describe perceptions, we can also agree about the uses to describe inner private states. But our language does not tell us exactly what others experience.

Strawson's second response is that, without a bridging structure between the experiences of self and others, you cannot know that either you *or* others have mental states, as you would not ascribe mental states to anyone. For Strawson, the fact that one *has* experiences at all—that one ascribes experiences *to oneself*—has implications about our understanding of other minds that become clearer when one asks, "Why are one's states of consciousness *ascribed* at all, to *any* subject?" (Strawson 1959:89). Strawson goes on as follows (and it requires careful attention to understand what he says): "If, in identifying the things to which states of consciousness are to be ascribed, private experiences are to be all one has to go on, then, just for the very same reason as that for which there is, from one's own point of view, no question of telling that a private experience is one's own, there is no telling that a private experience is another's. All private experiences, all states of consciousness, will be mine, i.e., no one's" (Strawson 1958:339; 1959:96). The traditional solution to the problem of other minds by way of analogy argues that I know that other subjects exist and have experiences because "the subject of those experiences . . . stand[s] in the same unique causal relation to body N [the other's body] as *my* experiences stand to body M [my body]," but this analysis "requires me to have noted that *my* experiences stand in a special relation to the [my] body M, when it is just the right to speak of *my* experiences at all that is in question" (Strawson 1958:339).

> There is no sense in the idea of ascribing states of consciousness to oneself, or at all, unless the ascriber already knows how to ascribe at least some states of consciousness to others. So he cannot (or cannot generally) argue "from his own case" to conclusions about how to do this; for unless he already knows how to do this, he has no conception of *his own case*, or any *case* (i.e., any subject of experiences). Instead, he just has evidence that pain, etc., may be expected when a certain body is affected in certain ways and not when others are.
>
> (p. 344)

Strawson is saying here that if your own experiences are the only ones knowable to you, there is no need to identify them as your own, as there's no thought of anyone else and no need to identify them as anyone's. Designating your experiences *as* yours entails a recognition that there are other potential experiences, that is, someone else's, which is only possible if you recognize that there are other subjects of experience. One must have a concept of mental experiences as occurring to others as well as oneself to understand that one's own experiences *are* one's own, otherwise there are just experiences without the need or possibility of explicating whose they are. The argument goes on with various elaborations and is summarized: "To put it briefly: one can ascribe states of consciousness to oneself only if one can ascribe them to others; one can ascribe them to others only if one can identify other subjects of experience; and one cannot identify others if one can identify them *only* as subjects of experience, possessors of states of consciousness" (Strawson 1958:339; 1959:96). Because self-ascription requires attributing states of consciousness to oneself, "a necessary condition of states of consciousness being ascribed at all is that they should be ascribed to the *very same things* as certain corporeal characteristics, a certain physical situation, etc." (Strawson 1958:340). For me to ascribe states of consciousness to myself, I must be able to ascribe them to others and I must have a means of doing so. Strawson describes the means: states of consciousness are "unambiguously and adequately ascribable *both* on the basis of observation of the subject of the predicate *and* not on this basis (independently of observation of the subject): the second case is the case where the ascriber is also the subject" (p. 346). In the observation "The dog is angry," the dog is the subject and "is angry" is the predicate, and I ascribe anger to the dog based on the dog's snarling, its context, etc. In the observation "I am angry," I am the subject directly experiencing anger and ascribing anger to myself on the basis of that experience; yet the mental state of anger is, in Strawson's view, appropriately applied in both situations.

Strawson's account neatly solves the problem of why we believe that there are other minds by making it essential for the ascription of a mind to oneself. If Strawson is correct, there must be a way of recognizing the fusion between behavior in context and inner private experience (in my view; for a different view, see Danto 1967). For Strawson, this indicates that the concept of the person, an entity that incorporates physical and mental predicates, is a primitive concept.

Whereas Strawson seems to believe that language is the main method for recognizing this conceptual fusion between others' behavior and my own experience, Maurice Merleau-Ponty (1964) presents an alternative structure in his phenomenological solution to the problem of other minds. Merleau-Ponty argues that recognition of one's own experience requires a structure or schema which is "relatively transferrable from one sensory domain to the other in the case of my own body, just as it could be transferred to the domain of the other" (p. 118). Specifically, "I can perceive, across the visual image of the other, that the other is an organism, that that organism is inhabited by a 'psyche,' because the visual image of the other is interpreted by the notion I myself have of my own body and thus appears as the visible envelopment of another 'corporeal schema'" (p. 118). Merleau-Ponty offers imitation of others and recognition of self in the mirror (p. 119) as evidence of the "corporeal schema." Both imitation and self-recognition exhibit an ability to match kinesthetic experience within the self to visual experiences of self (in the mirror) and of others (in imitation; see Mitchell 1993, 2010). I call this kinesthetic-visual matching. Merleau-Ponty's ideas about the corporeal schema, and my own about kinesthetic-visual matching, are taken from the French comparative-developmental psychologist Paul Guillaume, whose wonderful study of imitation in his own children (Guillaume 1971 [1926]) led him to posit how children come to understand other minds through kinesthetic-visual matching.

Kinesthetic-visual matching allows one's bodily experience to reach out to another's bodily experience. You can understand this by asking a friend to make a blow up of a Xerox copy of either a left or right hand (not telling you which), placing this picture about ten to twenty feet away, and then asking yourself whether it is a left or a right hand. It is important that you do not visually compare your own hand and the image while deciding which hand it is and also that you not decide the answer before a distance is maintained between you and the pictured hand. Most people, when asked to solve the problem, feel the kinesthesis of their own right or left hand move to the image to compare. Generally, if people are shown an image of a disembodied hand in diverse orientations and are asked to decide if this hand "over there" is a right or left hand, they usually imagine moving their appendage and/or body (as if kinesthetically) until it matches the hand in order to make or confirm their decision (Parsons and Fox 1998; Sekiyama 1982). This imagined appendage or imagined body movement is called "motor imagery." When shown images of objects that are not body parts, people usually visually imagine transforming the image of the object to align—or not—with another image, rather than imagining their body moving. (They sometimes do the same thing when shown images of body parts.) Just as visual imagery seems to operate using brain processes peculiar to vision, and auditory imagery similarly depends on audition, motor imagery depends on motor movements and their kinesthetic accompaniments, in that the time required to make the imagined and real movements are correlated (Parsons 1994).

Recapitulations and Extensions

I reiterate Strawson's two responses to Wittgenstein:

- In describing mental states, we always compare them to something intersubjectively agreed upon—that is, perceptual experiences. It is the aspects of perceptual experiences that *are* intersubjectively available that are compared, not qualia.
- Without a structure to bind behavior and experience for self and other, you cannot know that either you or others have mental states, as all mental states will be your own—i.e., no one's.

One implication of Strawson's responses to Wittgenstein is that inner experiences are perceptions or are analogically like perceptions, so we compare them to perceptions. Like regular perceptions, the same inner perception or perceptual analog can have different attitudes or perspectives toward it. A visual image of a dog can be experienced as my dog, a collie, an animal, a dog I saw on television that time when I was at Janet's house during the party, or any number of things. Which interpretation of the image we understand it to be—our "attitude" toward the image (Rollins 1989)—is usually simultaneously present with the image, so that many of Wittgenstein's (1953:193–210) and others' (e.g., Goodman 1968) concerns about the difficulties of determining what an image represents are already taken care of when we have an image. In essence, images come with their perspective—the agent's "attitude" toward the image—already attached (Rollins 1989:91). In another view, images and perceptions are viewed as more like action-based schemas—rather than fully formed representations, perceptual re-presentations, or pictorial depictions—from which diverse pieces of relevant information can be obtained (O'Regan and Noë 2001; Thomas 1999). The "attitude" toward the image, in humans, includes the reference frame (what is the bottom and what is the top) and typically a name of some sort, often a verbal description, as in "a rabbit" or "Harvey" or "the rabbit I had as a child," though the descriptor need not be verbalized or labeled explicitly (Reisberg 1996; Reisberg and Heuer 2005).

Strawson's ideas have further implications:

- Living beings may have inner experiences without ascribing these experiences to themselves or anyone.

- Ascription of experiences to self and other requires a bridging structure, usually kinesthetic-visual matching, but also language.
- Without a bridging structure, only first-person (unascribed) experiences are possible.
- But, even with a bridging structure, understanding of another's experience is always incomplete with regard to qualia. We can know what experiences are like (they are like my perceptions), but not necessarily exactly how they are experienced.

So the end result of the argument between Wittgenstein and Strawson results in two points for each. Consistent with Strawson's ideas,

- An organism's behavior lets you decide it has private inner experiences. Without the connection between behavior and experience for self and others, there would be no "other minds" problem because there would be no minds knowable as one's own or another's.
- Once you know the perceptual abilities and concerns (attitudes) of such an organism, all of which are knowable from its behavior, you have some understanding of the *kinds* of private inner experiences it can have.

Consistent with Wittgenstein's ideas,

- You cannot know all aspects of an organism's qualia, even if it uses language to describe its qualia.
- Even with human language users, you can't know everything about another's experience—or even your own. Pains can be indescribable and can fade in memory. Mental images can be transformed through thinking about them. Memories can be unreliable.

So assumptions that we can know other human beings' experiences, but not animals' experiences, are problematic, as so often when we say we know other people's experiences we mean that we are making fairly strategic decisions about the meaning of their behavior in this context, rather than knowing their qualia, and the same is true of our understanding of animals' experiences (Mitchell and Hamm 1997). Just as knowing about other people's experiences requires spending time with them, the same is true of knowing about animals' experiences (Mitchell 1997). My elaboration of Strawson's ideas delineates what I expect are the possibilities for inner experiences, that is, that they are perceptions, or are perceptionlike with attitude, and suggests that, in our attempts to understand an animal's inner experience, we may be asking for more information than we can obtain even about other humans who speak the same language.

Note

I appreciate the thoughtfully critical comments of Julie Smith, Robert Lurz, Matthew Pianalto, and Sara Waller about the ideas in this chapter. This chapter builds on part of a previously published work (Mitchell 2000). I appreciate the permission of Psychology Press to use copyrighted material from this previously published work that appeared in P. Mitchell and K. Riggs, eds., Children's Reasoning and the Mind (2000; ISBN: 9780863778551).

References

Danto, A. C. 1967. "Persons." In P. Edwards, ed., *The Encyclopedia of Philosophy*, pp. 110–114. New York: Macmillan and Free Press.

Finke, R. A. 1980. "Levels of Equivalence in Imagery and Perception." *Psychological Review* 87:113–132.

——. 1986. "Mental Imagery and the Visual System." *Scientific American* 254, no. 3 (March): 88–95.

Glover, J. 1981. "Critical Notice [review of *Mortal Questions* by T. Nagel]." *Mind* 90:292–301.

Goodman, N. 1968. *Languages of Art: An Approach to a Theory of Symbols*. Indianapolis: Bobbs-Merrill.

Griffin, D. R. 2001. *Animal Minds: From Cognition to Consciousness*. Chicago: University of Chicago Press.

Guillaume, P. 1971 [1926]. *Imitation in Children*. Trans. E. P. Halperin. 2d ed. Chicago: University of Chicago Press.

Merleau-Ponty, M. 1964. "The Child's Relations with Others." Trans. W. Cobb. In J. M. Edie, ed., *The Primacy of Perception*, pp. 96–155. Evanston, IL: Northwestern University Press.

Meyers, C. D., and S. Waller. 2009. "Psychological Investigations: The Private Language Argument and Inferences in Contemporary Cognitive Science." *Synthese* 171:135–156.

Mitchell, R. W. 1993. "Mental Models of Mirror-Self-Recognition: Two Theories." *New Ideas in Psychology* 11:295–325.

——. 1994. "The Evolution of Primate Cognition: Simulation, Self-Knowledge, and Knowledge of Other Minds." In D. Quiatt and J. Itani, eds., *Hominid Culture in Primate Perspective*, pp. 177–232. Boulder: University Press of Colorado.

——. 1997. "Anthropomorphism and Anecdotes: A Guide for the Perplexed." In R. W. Mitchell, N. S. Thompson, and H. L. Miles, eds., *Anthropomorphism, Anecdotes, and Animals*, pp. 407–427. Albany: SUNY Press.

——. 2000. "A Proposal for the Development of a Mental Vocabulary, with Special Reference to Pretense and False Belief." In P. Mitchell and K. Riggs, eds., *Children's Reasoning and the Mind,* pp. 37–65. Hove: Psychology.

——. 2001. "Book Review [*Animal Minds* by D. R. Griffin]." *Animal Behaviour* 62:1225–1227.

——. 2010. "Understanding the Body: Spatial Perception and Spatial Cognition." In F. L. Dolins and R. W. Mitchell, eds., *Spatial Cognition, Spatial Perception: Mapping the Self and Space,* pp. 341–364. Cambridge: Cambridge University Press.

Mitchell, R. W., and M. Hamm. 1997. "The Interpretation of Animal Psychology: Anthropomorphism or Behavior Reading?" *Behaviour* 134:173–204.

O'Regan, J. K., and A. Noë. 2001. "A Sensorimotor Account of Vision and Visual Consciousness." *Behavioral and Brain Sciences* 24:939–1031.

Parsons, L. M. 1994. "Temporal and Kinematic Properties of Motor Behavior Reflected in Mentally Simulated Action." *Journal of Experimental Psychology: Human Perception and Performance* 20:709–730.

Parsons, L. M. and P. T. Fox. 1998. "The Neural Basis of Implicit Movements Used in Recognising Hand Shape." *Cognitive Neuropsychology* 15:583–615.

Penelhum, T. 1957. "The Logic of Pleasure." *Philosophy and Phenomenological Research* 17:488-503.

Piaget, J. 1954. *The Construction of Reality in the Child.* New York: Basic Books.

——. 1962 [1945]. *Play, Dreams, and Imitation in Childhood.* Trans. C. Gattegno and F. M. Hodgson. New York: Norton.

Reisberg, D. 1996. "The Nonambiguity of Mental Images." In C. Cornoldi, R. H. Logie, M. A. Brandimonte, G. Kaufmann, and D. Reisberg, *Stretching the Imagination: Representation and Transformation in Mental Imagery,* pp. 119–171. New York: Oxford University Press.

Reisberg, D., and F. Heuer. 2005. "Visuospatial Images." In P. Shah and A. Miyake, eds., *The Cambridge Handbook of Visuospatial Thinking,* pp. 35–80. New York: Cambridge University Press.

Rollins, M. 1989. *Mental Imagery: On the Limits of Cognitive Science.* New Haven: Yale University Press.

Ryland, F. 1909. *Thought and Feeling.* London: Hodder and Stoughton.

Sekiyama, K. 1982. "Kinesthetic Aspects of Mental Representations in the Identification of Left and Right Hands." *Perception and Psychophysics* 32:89–95.

Shorter, J. M. 1952. "Imagination." *Mind* 61:528-542.

Singer, P. 1975. *Animal Liberation: A New Ethics for Our Treatment of Animals.* New York: Random House.

Strawson, 1954. "Critical Notice [review of *Philosophical Investigations* by Ludwig Wittgenstein]." *Mind* 63:70-99.

——. 1958. "Persons." *Minnesota Studies in the Philosophy of Science* 2:330-353.

——. 1959. *Individuals.* Garden City, NJ: Anchor.

Thomas, N. J. T. 1999. "Are Theories of Imagery Theories of Imagination? An Active Perception Approach to Conscious Mental Content." *Cognitive Science* 23:207–245.

Wittgenstein, L. 1953. *The Philosophical Investigations.* Trans. G. E. M. Anscombe. New York: Macmillan.

11

The Voice of the Living

Becoming-Artistic and the Creaturely Refrain in D. H. Lawrence's "Tortoise Shout"

CARRIE ROHMAN

Because poetry participates in the rhythmic, the musical, and the incantatory, the poetic representation of animal being is particularly salient to a discussion of cosmic and aesthetic forces. Moreover, as Jorie Graham explained during the 2006 Geraldine R. Dodge Poetry Festival, poetry must be recognized as bodily experience. During her festival presentations Graham reiterated that reading and hearing poetry are not primarily cognitive but rather somatic processes. Such a claim forces us to rethink some of our more conventional notions about literature, the body, and even the creaturely. I want to suggest that one of D. H. Lawrence's often-anthologized animal poems, "Tortoise Shout," reveals an aesthetics of poetry that is rooted in bodily experience, in which the nonhuman voice functions within the context of the Deleuzian refrain. That voice—the shout of the tortoise—carries a rhythmic force that connects human to animal and both to broader cosmological powers in a posthumanist becoming-artistic of the living.

Elizabeth Grosz's work on aesthetics and inhuman forces is useful as we investigate Lawrence's poetic achievement in this text. Working among theories ranging from Deleuze to French feminism and her own rereading of Darwin's evolutionary theory, Grosz (2005) asserts, in an interview with

Julie Copeland, that we need to understand art as "the revelry in the excess of nature, but also a revelry in the excess of the energy in our bodies" (p. 2). Grosz makes the provocative claim that "we're not the first artists and we're perhaps not even the greatest artists, we humans; we take our cue from the animal world. So what appeals to us? It's the striking beauty of flowers, it's the amazing colour of birds, it's the songs of birds" (p. 2). Grosz maintains that art is not primarily conceptual or linked to representation, but rather that art's "fundamental goal is to produce sensations"; "it's about feeling something intensely [while] there may be the by-product of a kind of understanding" (p. 3).

In what may first seem a counterintuitive locating of the artistic outside of human praxis, Grosz (2008) asserts that the intersection of life itself with earthly or even cosmic forces serves as the occasion for what is fundamentally an aesthetic emergence. Grosz describes the "productive explosion of the arts from the provocations posed by the forces of the earth . . . with the forces of living bodies, by no means exclusively human, which . . . slow down chaos enough to extract from it something not so much useful as intensifying, a performance, a refrain, an organization of color or movement that eventually, transformed, enables and induces art" (p. 3). In the aforementioned interview, Grosz (2005) goes on to emphasize the way that her ideas decenter the traditional attribution of art to an elevated, human function: "I think what's radical about what I'm saying is that art isn't primarily or solely conceptual, that what it represents is the most animal part of us rather than the most human part of us. Frankly, I find it really refreshing, in a way, that it's not man's nobility that produces art, it's man's animality that produces art, and that's what makes it of potential interest everywhere" (p. 3).

Before we turn to the particular work of this essay and its examination of Lawrence's poetics, it is important to locate Grosz's claims within the Deleuzian framework that she outlines in her own discussion of the artistic. Deleuze rejects the notion that art is primarily to be understood in terms of intention or representation. Rather, as Grosz (2008) explains, Deleuze suggests that "the arts produce and generate intensity, that which directly impacts the nervous system and intensifies sensation. Art is the art of affect more than representation, a system of dynamized and impacting forces rather than a system of unique images that function under the regime of signs" (p. 3). Readers will recognize the Deleuzian emphasis on intensities here, and Grosz reminds us that the idea of the affective in Deleuze involves a linkage between bodily forces and "cosmological forces," a linkage that emphasizes the human participation in the nonhuman (p. 3).

It is somewhat remarkable to note in this context what Fiona Becket pointed out in her 1997 study, *D. H. Lawrence: The Thinker as Poet*. Lawrence, she reminds us, "confirms the inseparability of the aesthetic and the ontological," especially through his concept of "art-speech" as that which "will play fruitfully across the centers and plexuses of the body of the reader" (p. 29). One segment from Lawrence's writings is particularly salient to our discussion. It comes from his work *The Symbolic Meaning* and explains this concept of art-speech:

> Art-speech is also a language of pure symbols. But whereas the authorized symbol stands always for a thought or an idea, some mental *concept*, the art-symbol or art-term stands for a pure experience, emotional and passional, spiritual and perceptual, all at once. The intellectual idea remains implicit, latent and nascent. Art communicates a state of being—whereas the symbol at best only communicates a whole thought, an emotional idea. Art-speech is a use of symbols which are *pulsations on the blood and seizures upon the nerves*, and at the same time pure percepts of the mind and pure terms of spiritual aspiration.
>
> (Cited in Becket 1997:29–30, second emphasis mine)

Given these claims about art impacting the nervous system, when Grosz (2008) refines her discussion about artistic production as that which "merges with, intensifies and eternalizes or monumentalizes, sensation" (p. 4), we realize that Lawrence's body of work, his life's work, really—with all its investment in the concept of "blood consciousness"—is especially resonant for our purposes.

I want to examine this posthumanist concept of art in D. H. Lawrence's poem, "Tortoise Shout" (see the complete poem in the appendix of this chapter). This text provides an astonishingly rich and provocative example of the kind of transspecies participation in becoming-artistic toward which Grosz's work points. Indeed, the way in which the poem situates a becoming-artistic of the creaturely, specifically through the medium of sexual difference and sexual behavior, marks it as an extraordinary example of these particular concepts in literary discourse.

It is useful, at the outset of our discussion, to note both the factual or scientific realities of the tortoise and also the mythic or symbolic resonance it has tended to carry across cultures. Peter Young's book, *Tortoise*, in the recent Reaktion Books Animal series, helps us with some preliminary situating in this regard. Young (2003) notes in the opening pages of his discussion: "Tortoises look and are old, almost mythical creatures. They are primeval, the oldest of the living land reptiles, their age confirmed by fossil remains. Tortoises are the surviving link between animal life in water and on land. . . . Tortoises . . . have survived for some 225 million years. They are living fossils. Hardy, self-contained creatures, they have endured aeons of major changes, and on a world scale survived geological upheaval, volcanic activity and climatic swings" (pp. 7–8). Amidst a good deal of further biological information, such as the advanced age to which tortoises often live, Young makes note of one fact that our poem takes as its focus: the coitus cry of the male. "Tortoises," he explains, "are basically mute, except for males squealing with delight, sometimes with open mouth, at the climax of mating" (p. 22). In fact, my further investigation suggests that the mating call, or cry, is often very pronounced during the mounting phase in some tortoises, and may continue during segments of the often protracted coitus. Thus it may be more accurate to say that the tortoise is basically mute, except for the male crying during mating activities.

We might also take notice of Young's chapter "Myths and Symbols." Like most animals, tortoises have been subject to varying and sometimes contradictory symbolic use by human cultures. We are familiar with the recurring association of tortoises with qualities such as longevity and perseverance. However, Young reminds us that an astonishingly broad array of cultures have symmetrical creation myths that see the tortoise as supporting, literally, the world itself on its back, sometimes simultaneously as a "model of the world itself" (p. 42). He provides an interesting theory on the historical "route" of these myths, but also makes the Jungian claim that the tortoise seems to have an archetypal presence in "the collective unconscious of the human race" (p. 47). For our purposes, I want to note that the centrality of the tortoise in myths of creation—its association with a truly rudimentary force of generation—gives us an especially poignant purchase on Lawrence's use of the tortoise.

The poem, which ultimately runs to about one hundred twenty lines, begins with three short ones: "I thought he was dumb, / I said he was dumb, / Yet I've heard him cry" (lines 1–3). These opening observations emphasize several important elements of the text to come. Like some of Lawrence's other poems, perhaps most famously "Snake" and "Fish," as I have argued elsewhere, the ineptitude, misapprehension, or limit of human knowledge is made immediately evident (see Rohman 2009). These three lines signal that the human speaker was wrong about the animal's "dumbness" or muteness. What is more, Lawrence emphasizes human reason and language specifically. The narrator *thought and then said* the animal was dumb. Typically, reason and linguistic capacities are hailed as the hallmark of human superiority and dominion over the animal, yet here Lawrence acknowledges that his rational mind and speaking abilities were just plain mistaken. What is it that tells the truth about this animal or this event? It is what the speaker hears: the auditory sense. Not thinking or speaking in symbols, not representation or high-order analysis, but the body's sensorium. And apparently this alternative knowledge, which in Lawrence's terms would be blood-conscious knowledge rather than nerve-conscious knowledge, renders an immediate truth. I thought he was dumb, yet I heard him cry. At the instant when the body registers the tortoise's voice, human assumptions about the animal are revealed as naive.

Lawrence's choice of words here is not to be overlooked. He does not say that he thought the animal was silent; he says "dumb" instead. This word's historical weight is apparent: The OED lists a reference to animal muteness as the word's second primary meaning: "Applied to the lower animals (and, by extension, to inanimate nature) as naturally incapable of articulate speech." Moreover, *dumb*'s polyvalence calls into question our assumptions that animals are stupid, "bestial," unintelligent. He was wrong about animals being *dumb.* Animals are not dumb, and indeed their apparent muteness for the human does not indicate a mental vacuity. Similarly, the choice of *cry* in the opening stanza seems to emphasize something particular about animal emotion. Lawrence does use the word *scream* in the very next line, a term we would associate more with instinct or aggression or even simple physical distress. But the choice of *cry* suggests weeping, which clearly locates this creature in an emotional realm more akin to humans than alien from them.

Thus, in the second stanza, while Lawrence seems to suggest that the tortoise cry is distant, or obscure to us, we already know that this primal scream is in fact deeply shared in some way by the human. The stanza reads: "First faint scream, / Out of life's unfathomable dawn, / Far off, so far, like a madness, under the horizon's dawning rim, / Far, far off, far scream" (lines 4–8). Rather than intimate that the scream is "far" ontologically from the human—foreign, that is, to human being—this stanza sets up the poem's broad ideological claim that the tortoise brings us back to the fundamental, the primeval scream or vibration of the living in general. The scream may be far off to the hapless, unattending, or overly rationalized human passerby, caught unawares. But, as the text unfolds, we recognize how it brings the human speaker and the tortoise irrevocably into the same ontological space. This making coextensive of the human and tortoise takes place in large part through a discursive meditation on extremity.

I will come back to the specific image of the tortoise *in extremis* shortly. For now I want to focus on the poem's subsequent question: "Why were we crucified into sex? / Why were we not left rounded off, and finished in ourselves [?]" (10–12). Lawrence often focuses on a kind of suffering involved in the problem of individual sovereignty versus communal experience and intersubjectivity. These themes are well-outlined by critics of his writing and frequently play themselves out through the problematics of sex. Indeed, Richard Ellmann (1990 [1953]) noted early on the connections between images of crucifixion and sexuality in Lawrence's work: "The metaphor of the cross," he reminds us, "is one of his most dramatic and successful images, for it implies the sacredness, terror, and pain which were for him essential parts of the sexual experience" (p. 194). What interests me here, however, is the way that Lawrence's question evokes the most fundamental biological realities about sexual difference. Grosz (2008) helps us recognize those realities when she discusses the role of sexual selection in the elaboration and creativity inherent in organic life forms:

> The evolution of life can be seen not only in the increasing specialization and bifurcation or differentiation of life forms from each other, the elaboration and development of profoundly variable morphologies and bodily forms, but, above all, in their becoming-artistic, in their self-transformations, which exceed the bare requirements of existence. Sexual selection, the consequence of sexual difference or morphological bifurcation—one of the earliest upheavals in the evolution of life on earth and undoubtedly the most momentous invention that life has brought forth, the very machinery for guaranteeing the endless generation of morphological and genetic variation, the very mechanism of biological difference itself—is also, by this fact, the opening up of life to the indeterminacy of taste, pleasure, and sensation. Life comes to elaborate itself through making its bodily forms and its archaic territories, pleasing (or annoying), performative, which is to say, intensified through their integration into form and their impact on bodies.
>
> There is much "art" in the natural world, from the moment there is sexual selection, from the moment there are two sexes that attract each other's interest and taste through visual, auditory, olfactory, tactile, and gustatory sensations.
>
> (pp. 6–7)

There's a kind of delicious irony embedded in this angle on sexuality, for readers and critics of Lawrence, anyway. Grosz explains that sexual difference is perhaps the prime generator of the excesses and transformations that ultimately establish the becoming-artistic of life itself. In other words, the origins of art at least partly reside in the ways that bodies overcome themselves, create something nonutilitarian, in conjunction with the forces of the earth and the cosmos in the great dance of seduction and mating (not necessarily driven only by reproduction). I think that Lawrence might hate this idea if confronted by it, with its profound yoking of sexual difference and artistic capacity, since Lawrence was often so distraught by the necessities of sex. But perhaps, in a psychoanalytic register, Lawrence's ambivalence toward sexuality reveals a subterranean recognition on his part of this very linkage that requires a moving beyond the self and a deep connection to the body and animality in order to become-artistic.

When Lawrence asks his next question in the poem, we are ushered into the world of vibration, a world of forces that is central to Grosz's analysis, particularly via the concerns of Darwin and Deleuze. Lawrence writes: "A far, was-it-audible scream, / Or did it sound on the plasm direct?" (lines 15–16). A scream that is potentially not audible, a sound that bypasses the ear and "sounds" directly on and throughout the body. This is vibration, the vibratory, and emphasizes the primal connection between bodies, rhythm, and cosmic energies.

That Richard Ellmann in 1953 recognized this element of Lawrence's animal and plant poetics is noteworthy. While Ellmann's overall sense of the tortoise sequence remains insistently anchored in an unquestionable humanism, he nonetheless clearly senses the importance of the vibratory for Lawrence. Ellmann's general claim begins to acknowledge Lawrence's deep sensitivity to animal ontology, though its quick return to human subjectivity as something implicitly nonanimal disappoints: "No poet has a more uncanny sense of what it is like to be, for instance, a copulating tortoise. At their best the poems about tortoises, about elephants, about plants, reveal Lawrence's attitudes toward men, but without relinquishing their hold on the actual object" (Ellmann 1990 [1953]:195–196). My own readings of Lawrence's poetry in *Stalking the Subject* (Rohman 2009) depict a writer much more attentive to animal ontology. Indeed, I would argue that when the animal poems become mere metaphors for human action, they often fail. However, the "passion" with which Lawrence writes about other creatures leads Ellmann to conclude that "nature for Lawrence *is pullulating*; his landscape, his flowers, his animals have *radiant nodes of energy within them*, and he sets up an electric circuit between them and himself" (pp. 196–197, my emphasis).

Grosz (2008) notes that for Darwin, the use of rhythm and cadence, sometimes located in the voice, is an essential element of sexual selection. These powers not only seduce, intensify, and excite, but they can also endanger the creature who excels at them. "Nevertheless," Grosz explains, "it is the erotic, indeed perhaps vibratory, force in all organisms, even those without auditory systems, that seduces, entices, mesmerizes, that sexualizes the body, metabolizes organs, and prepares and solicits it for courtship" (p. 32). Darwin, she explains, insists that "rhythm, vibration, resonance, is enjoyable and intensifying" (p. 32).

It is important to clarify here that Grosz qualifies the emphasis on sexual selection by suggesting that reproduction does not necessarily have to be viewed as the primary telos of these processes. Rather, Grosz (2008) speculates that

> [perhaps] sexuality is not so much to be explained in terms of its ends or goals (which in sociobiological terms are assumed to be the [competitive] reproduction of maximum numbers of [surviving] offspring, where sexual selection is ultimately reduced to natural selection) as in terms of its forces, its effects . . . which are forms of bodily intensification. Vibrations, waves, oscillations, resonances affect living bodies, not for any higher purpose but for pleasure alone.
>
> (p. 33)

Tellingly, Lawrence's subsequent stanza in the poem elaborates the tortoise vibration in ways that suggest an excessive polyvalence associated with this rhythm: "Worse than the cry of the new-born, / A scream, / A yell, / A shout, / A paean, / A death-agony, / A birth cry, / A submission, / All tiny, tiny, far away, reptile under the first dawn" (lines 17–26). Lawrence seems to tap into the broad and universal instantiations of the vibratory among the living in general in this riff: Birth, death, celebration, pain, vulnerability all emerge from the narrator's experience of this tortoise's shout. Citing Alphonso Lingis, Grosz notes that the "first vocalizations in any articulated life are those of a cry: sobbing, gulping, breathing with a more and more intense rhythm. Pain articulates itself in many creatures, even those without vocal apparatus in roars, hisses, screams and squeals. For Lingis, expression is bound up with the rhythmic forces inhabiting and transforming bodies, the pleasures and pains the body comes to articulate" (Grosz 2008:51).

Grosz reminds us of Deleuze's conceptualization of the refrain (Deleuze and Guattari 1987:232–350) and helps to clarify its function. She writes, "The refrain is a kind of rhythmic regularity that brings a minimum of livable order to a situation in which chaos beckons" (p. 52). In music, for instance, the refrain "wards off chaos by creating a rhythm, tempo, melody that taps chaos by structuring it through the constitution of a territory" (p. 53). It is in her discussion of the Deleuzian refrain and the connection between cosmic and bodily forces that Grosz points to the very life rhythms that Lawrence catalogues in his renaming or rephrasing of the resonances of the tortoise's shout: "These rhythms of the body—the rhythms of seduction, copulation, birth, death—coupled with those of the earth—seasons, tides, temperatures—are the conditions of the refrain, which encapsulates and abstracts these rhythmic or vibratory forces into a sonorous emblem, a composed rhythm" (p. 55). The impact of this rhythm is most powerfully felt by bodies of the same species, but, as Grosz often points out, these refrains are transmuted and transferred from cosmos to earth, from animal to animal, from animal to human and back, etc. In the case of the poem, then, the refrain of the tortoise resonates through the body of the human narrator, connecting both of them to the "primeval rudiments of life, and the secret" (line 61). Ellmann's description of the "electric circuit" connecting human and nonhuman in the earlier quotation again seems especially insightful.

One of Grosz's (2008) comments in her discussion of rhythms and refrains is particularly salient here. She asks, "What else is both labile enough and appealing enough to slip from its material to its most immaterial effects, from the energy of the universe to the muscular oscillations that constitute pleasure and pain in living things? What else enables the body itself, the internal arrangement of its organs and their hollows, to resonate and to become instruments of sonorous expression?" (p. 55). Thus, when Lawrence talks of the "male soul's membrane / Torn with a shriek half music, half horror" (lines 31–32), we recognize the implicit reference to a Deleuzian refrain.

Moreover, in the poetic genre, rhythm itself is embedded in the communication of content. Lawrence repeats the tups, the jerks, the screams of the tortoise in coition, just as he includes variations of the refrain of the reptile's shout as they appear in other living organisms. Here is Lawrence's renaming and rephrasing that connects the poem most specifically to the Deleuzian refrain. A very long stanza begins with the opening line "I remember, when I was a boy, / I heard the scream of a frog, which was caught with his foot in / the mouth of an up-starting snake" (lines 69–72). The narrator goes on to catalogue "hearing a wild goose out of the throat / of night," "the scream of a rabbit," "the heifer in her heat, blorting and / blorting," and "a woman in / labour, something like an owl whooing, / . . . The first wail of an infant, / And my mother singing to herself / And the first tenor singing of the passionate / throat of a young collier" (lines 75–97 passim) among several other vibrations. The refrain of the tortoise's shout is connected to vocalizations that cross species lines and represent the universal rhythms of life—and especially the bringing forth of life in the birthing process—that we have already mentioned as the conditions of the refrain.

This long passage, which is situated near the center of the poem, is especially crucial for our reading of the poem in its posthumanist context. The connections that Lawrence makes here are not to be overlooked. The scream of a rabbit is collocated with the cries of the woman in labor, yes, but, perhaps even more strikingly, these vibrations are coterminous not only with the maternal singing voice but even with the young tenor's voice. These final connections to human song, to human aesthetics, make Grosz's point about the becoming-artistic of the vibratory quite clear. For Lawrence, all of these practices are fundamentally about life rhythm, about the cosmic forces in which all creatures, human and nonhuman, participate.

What is more, each of these comparisons highlights living beings at extremity, which is importantly connected to Grosz's understanding of excess and intensity. After the catalogue of refrains, Lawrence returns to the tortoise: "And more than all these, / And less than all these, / This last, / Strange, faint coition yell / Of the male tortoise *at extremity*, / Tiny from under the very edge of the farthest far-off horizon of life" (lines 101–107, my emphasis). It goes without saying that the organism is intensified through sexuality. Moreover, the biological fact of the male tortoise's coition scream brilliantly underscores this intensification. As we noted at the beginning of the essay, and as Peter Young documents, tortoises are mute *except for* the male's cries during mating. We might imagine a life of continual muteness only to be broken in this most acute and physically radical circumstance.

Grosz also reminds us that "art is where intensity is most at home" (2008:76). "Perhaps," she continues, "we can understand matter in art as matter at its most dilated, matter as it most closely approximates mind, diastole, or proliferation rather than systole and compression and where becoming is most directly in force" (p. 76). We see the proliferation of matter on some level through the voice's expressing bodily intensity in the poem. And, of course, that bodily intensity is sexual here. More generally, the extreme cry tends to underscore the idea that it is the excesses of evolution, the pressing beyond oneself and beyond the mundane routines of the body that manifests aesthetic practice.

Grosz continues to link the sexual and the artistic in her final chapter where she discusses aboriginal Australian art. Within that discussion she notes that "qualities and territory coexist, and thus both are the condition for sexual selection and for art making—or perhaps for the art of sexual selection and equally the sexuality of art production. It is this excess, of both harnessable forces and of unleashed qualities, that enables both art and sex to erupt, at the same evolutionary moment, as a glorification of intensity, as the production and elaboration of intensity for its own sake" (p. 102). Lawrence's poem seems to thematize this very eruption, since he emphasizes, in the poem's final considerations, the opening that sexuality both requires and results in: "Sex, which breaks up our integrity, our single inviolability, our / deep silence / Tearing a cry from us. / Sex, which breaks us into voice, sets us calling across the / deeps, calling, calling for the complement, / Singing, and calling, and singing again, being answered, / having found" (lines 113–117).

Referencing the Platonic myth that gender breaks some originary unity into separate beings who search for their "other half," Lawrence outlines a problem of "opening" and woundedness that critics have recognized throughout his oeuvre. Ellmann called Lawrence a "healer" who used his writing to dress personal and universal wounds (1990 [1953]:186). Kyoko Kondo (2003) has also written about the central role of the metaphor of opening in *Women in Love* and notes that Gerald in that novel experiences opening as a wound. We recognize that the ancient myth essentializes gender to some degree, but it nonetheless resonates with Grosz's claims about intensity, excess, sexuality, and art. Here the refrain—the singing, calling, and answering—acts as a signal to the mate, to the other, just as Grosz indicates. Lawrence ends the poem by emphasizing sexuality as the separation and recoordination of bodies: "That which is whole, torn asunder, / That which is in part, finding its whole again through the / universe" (lines 122–124).

In these final moments of Lawrence's text the link between past, present, and future poignantly emerges. If we had to sum up, we might say that the "far-off" and evolutionarily prior cry of the tortoise vibrates through the speaker in an intensified present that enlivens the speaker's recognition of the shared contingency of the living and, at the same time, illuminates the future processes of becoming in that contingency. We should note the intensely cosmic nature of the final image: One doesn't find one's complement only in the "other" here, whether human or tortoise. Indeed, one finds one's complement in the "universe," perhaps as that universe or cosmos is manifested in the creaturely embodiment of the other. And, in this particular poem, the creaturely is manifested in the rhythms, the vibrations of life itself becoming artistic. Grosz gives us a sense of this modality when she writes, "Art is not simply the expression of an animal past, a prehistorical allegiance with the evolutionary forces that make one; it is not memorialization, the celebration of a shared past, but above all the transformation of the materials from the past into resources for the future" (2008:103).

It seems fitting to end this discussion by addressing the poem itself as an affective becoming that bridges human and animal intensity through a shared vibratory force akin to the voice. Again taking her cue from Deleuze, Grosz concludes that affects are "man's becoming-other" (p. 77). By this she means that affects allow the human to "overcome" itself or move outside of itself through "the creation of zones of proximity between the human and those animal and microscopic/cosmic becomings the human can pass through. Affects signal that border between the human and the animal from which it has come" (p. 77). The rhythm of poetics might provide the most apt literary genre for the shared vibratory experience between human and animal minds, bodies, and being in general to be expressed. In Lawrence's poem it is the refrain of the animal voice linked to or becoming the refrain of the human poetic voice that demonstrates in full force the voice *of the living.* This generalized vibratory power connects the narrator to the tortoise through a deep and almost seamless vibration by the poem's end. As is often the case for Lawrence, his work refuses to cordon off the mind from the rest of the corpus, and he delves into the bodiliness and embeddedness not of the human or animal but of the living in general. Thus "Tortoise Shout" performs, in a sense, the interpenetration of cosmic intensities in which all creatures participate.

References

Becket, F. 1997. *D. H. Lawrence: The Thinker as Poet.* New York: St. Martin's.

Deleuze G., and F. Guattari. 1987. *A Thousand Plateaus: Capitalism and Schizophrenia.* Trans. B. Massumi. Minneapolis: University of Minnesota Press.

Ellmann, R. 1990 [1953]. "Lawrence and His Demon." In A. Banerjee, ed., *D. H. Lawrence's Poetry: Demon Liberated,* pp. 186–199. London: Macmillan.

Grosz, E. 2005. "The Creative Impulse" [interview with Julie Copeland]. *Sunday Morning Radio National,* August 14. Retrieved from http://www.abc.net.au/rn/arts/sunmorn/stories/s1435592.htm.

——. 2008. *Chaos, Territory, Art: Deleuze and the Framing of the Earth.* New York: Columbia University Press.

Kondo, K. 2003. "Metaphor in *Women in Love*." In K. Cushman and E. Ingersoll, eds., *D. H. Lawrence: New Worlds,* pp. 168–182. Madison: Fairleigh Dickinson University Press.

Lawrence, D. H. 1921. "Tortoise Shout." In *Tortoises,* pp. 45–50. New York: T. Seltzer. Retrieved from www.archive.org/details/tortoisesoolawr.

Rohman, C. 2009. *Stalking the Subject: Modernism and the Animal.* New York: Columbia University Press.

Young, P. 2003. *Tortoise.* London: Reaktion.

Tortoise Shout

D. H. LAWRENCE

I thought he was dumb,
I said he was dumb,
Yet I've heard him cry.

First faint scream,
Out of life's unfathomable dawn, 5
Far off, so far, like a madness, under the horizon's
dawning rim,
Far, far off, far scream.

Tortoise *in extremis.*

Why were we crucified into sex?
Why were we not left rounded off, and finished 10
in ourselves,
As we began,
As he certainly began, so perfectly alone?

A far, was-it-audible scream, 15
Or did it sound on the plasm direct?

Worse than the cry of the new-born,
A scream,

A yell,
A shout, 20
A pæan,
A death-agony,
A birth-cry,
A submission,
All tiny, tiny, far away, reptile under the first 25
dawn.

War-cry, triumph, acute-delight, death-scream
reptilian,
Why was the veil torn?
The silken shriek of the soul's torn membrane? 30
The male soul's membrane
Torn with a shriek half music, half horror.

Crucifixion.
Male tortoise, cleaving behind the hovel-wall of
that dense female, 35
Mounted and tense, spread-eagle, out-reaching
out of the shell
In tortoise-nakedness,
Long neck, and long vulnerable limbs extruded,
spread-eagle over her house-roof, 40
And the deep, secret, all-penetrating tail curved
beneath her walls,
Reaching and gripping tense, more reaching
anguish in uttermost tension
Till suddenly, in the spasm of coition, tupping 45
like a jerking leap, and oh!
Opening its clenched face from his outstretched
neck
And giving that fragile yell, that scream,
Super-audible, 50
From his pink, cleft, old-man's mouth,
Giving up the ghost,
Or screaming in Pentecost, receiving the ghost.

His scream, and his moment's subsidence,

The moment of eternal silence, 55
Yet unreleased, and after the moment, the sudden,
startling jerk of coition, and at once
The inexpressible faint yell—
And so on, till the last plasm of my body was
melted back 60
To the primeval rudiments of life, and the secret.

So he tups, and screams
Time after time that frail, torn scream
After each jerk, the longish interval,
The tortoise eternity, 65
Agelong, reptilian persistence,
Heart-throb, slow heart-throb, persistent for the
next spasm.

I remember, when I was a boy,
I heard the scream of a frog, which was caught 70
with his foot in the mouth of an up-starting
snake;
I remember when I first heard bull-frogs break
into sound in the spring;
I remember hearing a wild goose out of the throat 75
of night
Cry loudly, beyond the lake of waters;
I remember the first time, out of a bush in the
darkness, a nightingale's piercing cries and
gurgles startled the depths of my soul; 80
I remember the scream of a rabbit as I went
through a wood at midnight;
I remember the heifer in her heat, blorting and
blorting through the hours, persistent and
irrepressible; 85
I remember my first terror hearing the howl of
weird, amorous cats;
I remember the scream of a terrified, injured
horse, the sheet-lightning
And running away from the sound of a woman in 90
labor, something like an owl whooing,

And listening inwardly to the first bleat of a
lamb,
The first wail of an infant,
And my mother singing to herself, 95
And the first tenor singing of the passionate
throat of a young collier, who has long since
drunk himself to death,
The first elements of foreign speech
On wild dark lips. 100

And more than all these,
And less than all these,
This last,
Strange, faint coition yell
Of the male tortoise at extremity, 105
Tiny from under the very edge of the farthest
far-off horizon of life.

The cross,
The wheel on which our silence first is broken,
Sex, which breaks up our integrity, our single 110
inviolability, our deep silence
Tearing a cry from us.

Sex, which breaks us into voice, sets us calling
across the deeps, calling, calling for the
complement, 115
Singing, and calling, and singing again, being
answered, having found.

Torn, to become whole again, after long seeking
for what is lost,
The same cry from the tortoise as from Christ, 120
the Osiris-cry of abandonment,
That which is whole, torn asunder,
That which is in part, finding its whole again
throughout the universe.

Typescript modeled on 1921 edition, published by T. Seltzer. Available at
www.archive.org/details/tortoisesoolawr.

12

Unique Attributes of the Elephant Mind

Perspectives on the Human Mind

BENJAMIN L. HART AND LYNETTE A. HART

Among the terrestrial mammals, elephants share the unique status, along with humans and great apes, of having large brains, being long lived, and having offspring that require long periods of dependency. Elephants have the largest brains of all terrestrial mammals, including the greatest volume of cerebral cortex. Even when body-size-related functions of the cerebral cortex are subtracted, neurobiological modeling reveals that elephants still have approximately double the volume of associative cerebral cortex available for "mental" activities as humans, and about ten times that of chimpanzees (Hart and Hart 2007; Hart, Hart, and Pinter-Wollman 2008).

Obviously, with such a brain one expects a high level of intelligence or mental processing, even approaching that of humans. Even without the current neurobiological picture, the reputation for intelligence of elephants extends back to the ancient philosophers. Speaking of animals, the Greek philosopher Pliny the Elder (1890 [AD 79]) says, "the elephant is the largest of them all, and in intelligence approaches the nearest to man" (p. 244).

Given what we know about the elephants, the question of their intelligence and cognition in comparison to humans and great apes turns out to be very complex. In some respects their cognitive performance falls quite short of large-brained primates or even some smart birds. But in other respects, such as detailed, very long-term memory, their mental performance is arguably beyond that of humans. This paradox in understanding the elephant mind can be explained by reference to some very basic differences in the cytoarchitecture of the cerebral cortex of the elephant as compared with human and other large-brained primates. In this context, we use the term *cytoarchitecture* to refer to the microscopic anatomy of neurons and the types of connections between the neurons. Given that the elephant's brain evolved through an aquatic line of ancestors, as opposed to the terrestrial line of primates (DeJong 1998; Glickman, Short, and Renfree 2005), the differences in primate and elephant brain cytoarchitecture are not surprising. Recognizing these differences adds a perspective to our understanding of the extraordinary mental attributes of the human mind, which is a secondary theme of this chapter.

In the first part of this chapter we shall discuss aspects of elephant behavior relating to mental or cognitive activity. The discussion will begin with behaviors involved in tool use and insight behavior, where the large-brained elephant does not approach even the chimpanzee that has a brain one-tenth the size. Then we will discuss long-term spatial-temporal memory where elephants may surpass humans. A third category of mental activity, that of social-empathic responses, elephants appear to share with humans and chimpanzees. In the second part we shall relate these similarities and differences between elephants and large-brained primates (including humans) to the cytoarchitecture of the brain for insight not only into what is special about elephants but also into the biological basis of the human mind.

Similarities and Differences Between Elephants and Great Apes

TOOL USE

Tool use is a commonly used measure of animals' cognitive activity or intelligence. Like other highly intelligent, large-brained species, elephants use tools and modify them for a particular use. In fact, the first recorded example of animal tool use appears to be switching off flies with branches by elephants (Harris 1967 [1838]; Darwin 1871). Fly switching with branches is the most frequently observed example of spontaneous tool use in wild Asian elephants and in captive ones with ready access to branches (Hart et al. 2001). Such fly switching, shown in a captive elephant in figure 12.1, is

FIGURE 12.1. Fly switching by captive elephants, as in wild elephants, is common when flies are around and when branches are readily available. Photo courtesy of the authors.

quite effective in repelling flies (Hart and Hart 1994). When given a branch too big to function as a switch, captive elephants readily modify it into something switch-friendly by breaking off side branches (Hart et al. 2001). Other types of tool use in elephants include throwing branches over electric fences so they can cross over the fence, throwing sticks at rodents competing for fruit on the ground, and scratching with a branch (Chevalier-Skolnikoff and Liska 1993; Hart and Hart 1994).

As impressive as elephant tool use is, it falls short when compared with other species' more fine-grained tool use. For example, chimpanzees use sticks to extract termites from mounds and marrow from bones (Matsuzawa 2003; Sanz, Call, and Morgan 2009; van Schaik, Deaner, and Merrill 1999) and stone hammers to open nuts held against an anvil stone (Humle and Matsuzawa 2001), and New Caledonian crows spontaneously bend wires to pull food out of a tube (Weir, Chappell, and Kacelnik 2002). One must acknowledge that, for elephants, cracking nuts or fishing termites has little functional value given the amount of food they must consume. Some other tests of cognitive behavior, which will be described, are more revealing.

INSIGHT BEHAVIOR

With captive elephants one can contrive experiments to test for insight behavior. One such example is a test that requires the subject to catch on to pulling a retractable cord thrown over a ledge to obtain a visible and desirable food treat. Chimpanzees, rhesus monkeys, and even birds easily catch on, but elephants again fall far short, even when the test offers very large food treats (Nissani 2004).

An illustration of the disappointment one finds in trying to document insightful tool use in elephants is an unpublished experiment we conducted several years ago on captive Asian elephants. We placed a favorite food used by elephant handlers on a platform suspended from a tree, initially allowing elephants to tip the platform with their trunks and thus be rewarded with access to the food. We then raised the platform just beyond the reach of the trunk and put a long stick in a prominent place that we assumed the elephant could see. We expected they might grasp the stick and tip the platform, thus releasing the food reward. Several attempts at offering this puzzle to five elephants yielded no success.

We conclude that, despite their impressive brains, elephants simply are no match for some monkeys, chimpanzees, and humans in solving some insight problems or in tool use requiring fine-grained muscle coordination. Next we turn to long-term multimodal memory where elephants appear to excel beyond great apes, approaching or even exceeding the capacities of humans by some measures.

EXTENSIVE LONG-TERM MEMORY

Elephants are legendary for their memory. Older elephants, particularly long-lived matriarchs, accumulate and retain memories from a lifetime of varied experiences that have adaptive consequences for their families. When investigators played recordings of lion roars of one to three lions to thirty-nine elephant families in Kenya's Amboseli National Park, matriarchs over sixty years of age were much more likely than younger matriarchs to accurately access the potential danger revealed in the three-lion roars and lead their families to defensive behaviors (McComb et al. 2011).

Probably most noteworthy of elephant memory is their long-term spatial-temporal memory, which is important because they can range extensively. One study of African elephants documented movement of 625 km over a period of five months (Leggett 2006). The most advanced type of spatial memory is that which is location independent, where the mind works like a GPS instrument with a memory of how to find location X while starting from location A, B, or C. In adding the temporal element, the individual remembers how characteristics of the distance locations change on a seasonal basis. For example, during a severe drought that threatened the welfare of all elephants in the Tarangire National Park in Tanzania, elephants were forced to find forage and water outside the park in unfamiliar territory. Members of two clans led by old matriarchs followed the matriarchs to areas outside the park where the matriarchs had experienced a similar drought thirty-five years previously and retained the memory of where forage and water could be found (Foley, Pettorelli, and Foley 2008). The survival of elephant calves in these clans depended on the matriarchs' memories of places to find water and food outside the park. Another study on detailed, spatial-temporal memory plotted elephant movements from a foraging environment that was drying up, to an area 200 km away where there had been rainfall; the elephants arrived within three days after the start of rains (Viljoen 1989).

Detailed social-acoustic memory is another area in which elephants arguably outperform humans. Playback experiments of recordings from family members at Amboseli revealed that elephants could recognize individual calls of one hundred elephants from 1.5 km away. What is especially noteworthy is that these calls were recognized even with degradation of the individual acoustic signatures to a fraction of that present at close range (McComb et al. 2000). For reference, try recognizing someone's voice with 90 percent of the sound frequencies deleted.

Elephants also have impressive social-chemical memory with regard to individual urine signatures. Now here is an area where even your pet dog or cat outperforms humans. In a remarkable display of memory, male elephants recognized their mother's urine decades after they had been separated (Rasmussen 1995; Rasmussen and Krishnamurthy 2000)—a memory feat most likely beyond that of the best sniffer dogs.

SOCIAL-EMPATHIC BEHAVIORS

In this third category of mental activity, elephants share with humans and chimpanzees rather unique social-empathic or compassionate responses. These responses, opportunistically observed in wild elephants, usually in Africa, are documented by experts who happened to be present to record the events. Numerous examples of behaviors that one would associate with empathy or compassion in African elephants occurred over a thirty-five-year period of study at the Amboseli study site. The array of types of behaviors included defending vulnerable family members from a dangerous situation, comforting others, babysitting, removing foreign objects from another elephant, and assisting another elephant whose mobility had been impaired (Bates et al. 2008). A dramatic example of elephants helping a disabled family member, occasioned by immobilization of an African elephant for translocation, followed the darting of a young elephant: "There was an indescribable melee of screaming, trumpeting beasts. The young immobilized animal was lifted repeatedly on the tusks of the big older cows until, after two hours, it began to stand and eventually was marched off into the forest" (Harthoorn 1970:205; for other examples of assistance, see Hart, Hart, and Pinter-Wollman 2008).

The other category of frequently observed social-empathic behaviors is that of grieving and standing vigilant over the body of a recently deceased elephant (see Hart, Hart, and Pinter-Wollman 2008 for examples). The most persuasive example is described by Douglas-Hamilton and colleagues (2006). They feature several poignant photographs, one of which we include as figure 12.2. The matriarch of a family, Eleanor, had died following an injury. On the day of her death, a family member spent seven hours in the vicinity of the corpse. During that time a female relatively unfamiliar to Eleanor from another family hesitantly approached Eleanor's body, extended her trunk, sniffed the body, and then touched it, as if to confirm that she was dead. She hovered her right foot over the body and nudged it; after doing this, she rocked the body to and fro with her left foot and trunk. Eleanor's family members visited the body on the second and fourth days after her death. During the five days after her death, unrelated elephants from three other families visited the body.

With such vivid examples of social-empathic behaviors, it is not surprising that in certain circumstances behavior strikingly similar to the human posttraumatic stress disorder (PTSD) has been noted in elephants. An all-too-common precipitating factor is an infant elephant watching as his or her mother and close-knit family are gunned down by poachers and hacked up for removal of the tusks. The infants survive sometimes but, as researchers describe their case, they face a lifelong struggle with "abnormal startle responses, depression, unpredictable social behavior and hyper-aggression" (Bradshaw et al. 2005:807). These are, of course, some of the defining elements of human PTSD.

FIGURE 12.2. Visit to the deceased matriarch, Eleanor, by a family member days after her death. Photo courtesy of Shivani Bhalla—Save the Elephants Foundation.

Just as with tool use behavior, social-empathic behaviors of elephants were reported long before comparable behaviors in other nonhuman species. It is now established that chimpanzees display empathic behaviors that bear some resemblance to what has been discussed in elephants. Most noteworthy is a detailed account of the response of chimpanzees in a naturalistic environment to a dying mother chimp. As the mother was dying, the son and daughter as well as another adult gathered around her. After she died they inspected her mouth and moved her limbs, as if to confirm that she was dead. At first they all moved away from the body, but the daughter came back and lay next to it, maintaining a night-time vigil. For weeks after her death, all the survivors remained lethargic and quiet and ate less than they did normally, clearly grieving and mourning the passing of the elderly female (Anderson, Gillies, and Lock 2010).

The Cerebral Cortical Cytoarchitecture of Elephants, Humans, and Great Apes

In this section we explore how differences and similarities in mental activities in elephants and large-brained comparison species reflect the concept that natural selection has influenced brain organization to support behaviors with the greatest fitness consequences to the species in question (Balda and Kamil 2002; Hofman 2003). The elephant brain elegantly allows for extensive spatial-temporal memory, which is important for a species that spends up to eighteen hours a day foraging and must remember where forage and water resources are located during different seasons. For a highly social species wandering far and wide, long-term extensive acoustic and chemical memory also pays off in fitness consequences. In contrast, in great apes and humans detail-oriented tool use and the ability to solve insight problems relatively quickly play a more vital role in overall fitness than extensive long-term spatial-temporal memory of the type displayed by elephants.

Following the paradigm that natural selection has influenced brain organization to support behaviors with the greatest fitness consequences, one expects that the organization of the associative or nonsomatic cerebral cortex of elephants is different than that of humans and great apes. However, the occurrence of social-empathic behaviors such as grieving or mourning that elephants share with chimpanzees and humans suggests some important shared neurological attributes. These issues will be explored in our attempt to link behavior directly to brain characteristics.

Among the large-brained primates, the emphasis is on connectivity among nearby neurons, with a minimal nerve impulse conduction time and very efficient information processing. The cerebral cortex has evolved with an emphasis on compartmentalization into modules (Changizi and Shimojo 2005; Kaas 2000, 2007), and as the brain becomes larger there is an increase in the number of modules (Krubitzer 1995; Krubitzer and Hoffman 2000; Krubitzer and Hunt 2007). At the same time, the proportion of cortical neurons that are local circuit neurons within the modules increases. In a comparison of chimpanzees to humans, for example, the proportion of local circuit neurons increases from 93 percent to 98 percent (Hofman 1985). In short, the primate brain is characterized by relatively small neurons that can be packed together with relatively few of these neurons engaged in sending axons across the cortex for time-consuming global connections. The neural circuitry for fast, complex neural processing serves the large-brained primates well for their lifestyle, especially the circuitry for humans, which has no match for sheer number of cortical neurons—some twenty billion—almost all of which are fast, information-processing, local circuit neurons.

A much different picture emerges from what is known about the elephant brain. In elephants the neurons of the cerebral cortex are large—larger than any other species on the planet except the sperm whale—and primate-like modularity is not as emphasized. Elephant cortical neurons are spread much farther apart than is typical of human and chimpanzee cortical neurons (Haug 1987), and a high proportion of these neurons send large axons to distant cortical areas, providing a neuronal network rich in global connections (see figure 12.3). With the emphasis on global connections, one finds that only 91 percent of cortical neurons reside within modules, compared with the 98 percent in humans (Hofman 1985).

The poor showing that elephants make in the areas of primate-relevant tests of insight and tool use can be understood in the context of a bias toward transcortical, global connections rather than local circuit connections (Hart and Hart 2007; Hart, Hart, and Pinter-Wollman 2008). Information processing time reflects the distance between neurons because it is the impulse conduction time across axons that accounts for most of the time required for information processing. *Elephants are slow at thinking!* The widespread global connections also restrict the amount of connectivity between neurons because the long axons take up lots of space, meaning each neuron has fewer connections with other neurons than if axons did not so extensively traverse the cerebral cortex. *Elephants do not think in detail; they think globally*!

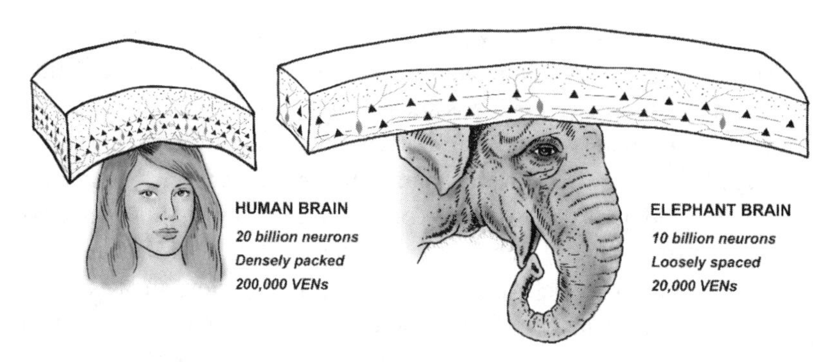

FIGURE. 12.3. Comparative neural cytoarchitectural aspects of the human and elephant nonsomatic cerebral cortex available for mental functions. The nature of information processing is a reflection of the type, size, and number of neurons, type of interconnections, and average distance between interacting neurons. The information-processing pyramidal neurons, profiled here as black triangles in two layers 3 and 5 (there is no layer 4 in these brain regions), are much less densely packed, but larger and more globally connecting, in the elephant cerebral cortex than in humans. Thus information processing in the elephant brain is much slower, with less local interconnectivity than in humans who have primarily compartmentalized, local-circuit interactions. Both humans and elephants have the unique von Economo neurons (VENs), illustrated as light gray, spindle-shaped cells that are believed to mediate social-empathic behaviors. The broadly reaching VENs each have a dendrite extending into upper layers and a descending axon branching into widely dispersed neurons. The basis for the estimate of neuron density and interconnectivity can be found in Hart et al. (2008). See text for more details.

With regard to the mind-boggling memory feats of elephants, keep in mind that for animals living in nature, the brain is where all information is stored, processed, and accessed—no library of books or desktop computer to consult! More than that of any other terrestrial species, the elephant's brain appears to provide the elephant's mind with an exceptional ability to remember and integrate information from a wide variety of spatial, temporal, acoustic, chemical, and social domains. And the long-lived matriarch, with a lifetime of cumulative experiences, is the repository of this living library of lifesaving information. Contrast elephant memory with human memory, which decreases with age (Andrews-Hanna et al. 2007).

The frequently observed, complex social-empathic behaviors that elephants share with humans and chimpanzees—empathy, compassion, grieving, mourning—are difficult to relate to the neural cytoarchitectural features discussed earlier. However, the recent discovery in elephants of special cortical neurons, von Economo neurons (VENs), which are virtually identical to the VENs of humans and chimpanzees (Hakeem et al. 2009), suggests that elephants share an important mediating substrate for social-empathic behaviors. These neurons have broad connections to neuronal targets across long distances of the cortex, presumably integrating stored information from diverse domains. It is believed that VENs subserve the mental attributes of empathy, compassion, social cognition, and awareness (Butti, Sherwood, Hakeem, Allman, and Hof 2009; Hof and Van der Gucht 2007; Nimchinsky et al. 1999; Nimchinsky et al. 1995). With the exception of cetaceans (whales, dolphins, etc.), VENs are found in significant numbers in just humans, elephants, and great apes. In terms of proportion of total cortical neurons, humans have by far the greatest number of VENs: about two hundred thousand. Elephants have proportionately fewer VENs—twenty thousand—while chimpanzees have about seven thousand, proportionately somewhat fewer than elephants. Figure 12.3 portrays the relative number and position of VENs in humans and elephants.

Elephants already have a neural bias toward global connections that could synergize with the globally connecting VENs in the motivation toward social-empathic responses. We hypothesize that the prominence of social-empathic responses in elephants, especially the grieving and standing vigilant over dead conspecifics—perhaps beyond that typical of humans and chimpanzees—reflects this synergism between VENs and the already globally biased cytoarchitecture of the elephant brain.

What the Elephant's Brain and Mind Tell Us About the Human Mind

Over evolutionary time, large-brained great apes and humans, with an increasing brain size, incorporated a bias toward fast-action, detailed information processing valuable for securing high-energy and protein-rich food in an environment with other species competing for the same nutrient-rich resources. The expansion of the brain with the modular, local-circuit template evidently reached a maximum in humans to the extent physiologically possible (Hofman 2003). The myriad of human creative accomplishments stemming from writing, painting, music composition, and scientific work would not have been possible with a more globalized neural cytoarchitecture. Comparison of the human brain with the elephant brain reminds us that the human mind's intellectual prowess is not based on the size of the brain per se but on a particular neural cytoarchitecture infrastructure that expanded to the extent physiologically possible. For elephants and humans, each representing a different type of cerebral cortical cytoarchitecture, the superimposition of von Economo neurons lends an empathic dimension to the minds of both.

Note

Preparation of this chapter was supported in part by allocation #03–65-F from the Center for Companion Animal Health, University of California, Davis. We would like to thank Patrick Hof, the late Donald Owings, and the editors of this volume, Robert Mitchell and Julie Smith, for their valuable suggestions.

References

Anderson, J. R., A. Gillies, and L. C. Lock. 2010. "*Pan* Thanatology: Chimp Grieving." *Current Biology* 20:R349-R351.

Andrews-Hanna, J. R., A. Snyder, J. Vincent, C. Lustig, D. Head, M. Fox, M. Raichle, and R. Buckner. 2007. "Evidence for Large-Scale Network Disruption in Advanced Aging." *Neuron* 56:924–35.

Balda, R. P., and A. C. Kamil. 2002. "Spatial and Social Cognition in Corvids: An Evolutionary Approach." In M. Bekoff, C. Allen, and G. M. Burghardt, eds., *The Cognitive Animal*, pp. 129–134. Cambridge: MIT Press.

Bates, L. A., P. C. Lee, N. Njiraini, J. H. Poole, S. Sayialel, C. J. Moss, and R. W. Byrne. 2008. "Do Elephants Show Empathy?" *Journal of Consciousness Studies* 15:204–225.

Bradshaw, G. A., A. N. Schore, J. L. Brown, J. H. Poole, and C. J. Moss 2005. "Elephant Breakdown: Social Trauma—Early Disruption of Attachment Can Affect the Physiology, Behaviour, and Culture of Animals and Humans Over Generations." *Nature* 433 (February 24): 807.

Butti, C., C. C. Sherwood, A. Y. Hakeem, J. M. Allman, and P. R. Hof. 2009. "Total Number of and Volume of von Economo Neurons in the Cerebral Cortex of Cetaceans." *Journal of Comparative Neurology* 515:243–259.

Changizi, M. A., and S. Shimojo. 2005. "Principles of Connectivity and Parcellation in Neocortex." *Brain, Behavior and Evolution* 66:88–98.

Chevalier-Skolnikoff, S., and J. Liska. 1993. "Tool Use by Wild and Captive Elephants." *Animal Behaviour* 46:209–219.

Darwin, C. 1871. *The Descent of Man and Selection in Relation to Sex.* Vol. 1. London: Murray.

DeJong, W. W. 1998. "Molecules Remodel the Mammalian Tree." *Trends in Evolution and Ecology* 13:270–275.

Douglas-Hamilton, I., S. Bhalla, G. Wittemyer, and F. Vollrath. 2006. "Behavioural Reactions of Elephants Towards a Dying and Deceased Matriarch." *Applied Animal Behaviour Science* 100:87–102.

Foley, C., N. Pettorelli, and L. Foley. 2008. "Severe Drought and Calf Survival in Elephants." *Biology Letters* 4, no. 5 (October): 541–544.

Glickman, S. E., R. V. Short, and M. B. Renfree. 2005. "Sexual Differentiation in Three Unconventional Mammals: Spotted Hyenas, Elephants, and Tammar Wallabies." *Hormones and Behavior* 48:403–417.

Hakeem, A. Y., C. C. Sherwood, C. J. Bonar, C. Butti, P. R. Hof, and J. M. Allman. 2009. "Von Economo Neurons in the Elephant Brain." *Anatomical Record* 292:242–248.

Harris, W. C. 1967 [1838]. *Narrative of an Expedition Into Southern Africa.* New York: Arno.

Hart, B. L., and L. A. Hart. 1994. "Fly Switching by Asian Elephants: Tool Use to Control Parasites." *Animal Behaviour* 48:35–45.

——. 2007. "Evolution of the Elephant Brain: A Paradox Between Brain Size and Cognitive Behavior." In J. H. Kaas and L. A. Krubitzer, eds., *Evolution of Nervous Systems, a Comprehensive Review: Mammals,* 3:491–497. Amsterdam: Elsevier.

Hart, B. L., L. A. Hart, M. McCoy, and C. R. Sarath. 2001. "Cognitive Behaviour in Asian Elephants: Use and Modification of Branches for Fly Switching." *Animal Behaviour* 62:839–847.

Hart, B. L., L. A. Hart, and N. Pinter-Wollman. 2008. "Large Brains and Cognition: Where Do Elephants Fit In?" *Neuroscience and Biobehavioral Reviews* 32:86–98.

Harthoorn, A. M. 1970. *The Flying Syringe: Ten Years of Immobilizing Wild Animals in Africa.* London: Geoffrey.

Haug, H. 1987. "Brain Sizes, Surfaces, and Neuronal Sizes of the Cortex Cerebri: A Stereological Investigation of Man and His Variability and a Comparison with Some Mammals (Primates, Whales, Marsupials, Insectivores, and One Elephant)." *American Journal of Anatomy* 180:126–142.

Hof, P. R., and E. Van der Gucht. 2007. "Structure of the Cerebral Cortex of the Humpback Whale, *Megaptera novaeangliae* (Cetacea, Mysticeti, Balaenopteridae)." *Anatomical Record* 290:1–31.

Hofman, M. A. 1985. "Neuronal Correlates of Corticalization in Mammals: A Theory." *Journal of Theoretical Biology* 112:77–95.

——. 2003. "Of Brains and Minds: A Neurobiological Treatise on the Nature of Intelligence." *Evolution and Cognition* 9:178–188.

Humle, T., and T. Matsuzawa. 2001. "Behavioral Diversity Among Wild Chimpanzee Populations of Bossou and Neighboring Areas, Guinea and Cote d'Ivoire, West Africa." *Folia Primatologica* 72:57–68.

Kaas, J. H. 2000. "Why Is Brain Size So Important: Design Problems and Solutions as Neocortex Gets Bigger or Smaller." *Brain and Mind* 1:7–23.

——. 2007. "Reconstructing the Organization of Neocortex of the First Mammals and Subsequent Modifications." In J. H. Kaas and L. A. Krubitzer, eds., *Evolution of Nervous Systems, a Comprehensive Review: Mammals,* 3:27–48. Amsterdam: Elsevier.

Krubitzer, L. A. 1995. "The Organization of Neocortex in Mammals: Are Species Differences Really So Different?" *Trends in Neuroscience* 18:408–417.

Krubitzer, L. A., and K. J. Hoffman. 2000. "Arealization of the Neocortex in Mammals: Genetic and Epigenetic Contributions to the Phenotype." *Brain, Behavior, and Evolution* 55:322–355.

Krubitzer, L. A., and D. L. Hunt. 2007. "Captured in the Net of Space and Time: Understanding Cortical Field Evolution." In J. H. Kaas and L. A. Krubitzer, eds., *Evolution of Nervous Systems, a Comprehensive Review: Mammals,* 3:49–72. Amsterdam: Elsevier.

Leggett, K. E. A. 2006. "Home Range and Seasonal Movement of Elephants in the Kunene Region, Northwestern Namibia." *African Zoology* 41:17–36.

McComb, K., C. Moss, S. Sayialel, and L. Baker. 2000. "Unusually Extensive Networks of Vocal Recognition in African Elephants." *Animal Behaviour* 59:1103–1109.

McComb, K., G. Shannon, S. M. Durant, K. Sayialel, R. Slotow, J. Poole, and C. Moss. 2011. "Leadership in Elephants: The Adaptive Value of Age." *Proceedings of the Royal Society B: Biological Sciences, 278,* no. 1722 (November 7): 3270-3276.

Matsuzawa, T. 2003. "The Ai Project: Historical and Ecological Contexts." *Animal Cognition* 6:199–211.

Nimchinsky, E. A., B. A. Vogt, J. H. Morrison, and P. R. Hof. 1995. "Spindle Neurons of the Human Anterior Cingulate Cortex." *Journal of Comparative Neurology* 355:27–37.

Nimchinsky, E. A., E. Gilissen, J. M. Allman, D. P. Perl, J. M. Erwin, and P. R. Hof. 1999. "A Neuronal Morphologic Type Unique to Humans and Great Apes." *Proceedings of the National Academy of Science USA* 96:5268–5273.

Nissani, M. 2004. "Theory of Mind and Insight in Chimpanzees, Elephants, and Other Animals." In R. J. Lesley and G. Kaplan, eds., *Developments in Primatology—Progress and Prospects: Comparative Vertebrate Cognition,* 4:227–261. New York: Academic.

Pliny the Elder. 1890 [AD 79]. *The Natural History of Pliny.* Vol. 2. Trans. J. Bostock and H. T. Riley. London: George Bell.

Rasmussen, L. E. L. 1995. "Evidence for Long-Term Chemical Memory in Elephants." *Chemical Senses* 20:762.

Rasmussen, L. E. L., and V. Krishnamurthy. 2000. "How Chemical Signals Integrate Asian Elephant Society: The Known and the Unknown." *Zoo Biology* 19:405–423.

Sanz, C., J. Call, and D. Morgan. 2009. "Design Complexity in Termite-Fishing Tools of Chimpanzees (*Pan troglodytes*)." *Biology Letters* 5, no. 3 (June): 293–296.

van Schaik, C. P., R. O. Deaner, and M. Y. Merrill. 1999. "The Conditions for Tool Use in Primates: Implications for the Evolution of Material Culture." *Journal of Human Evolution* 36:719–741.

Viljoen, P. J. 1989. "Spatial Distribution and Movements of Elephants (*Loxodonta africana*) in the Northern Namib Desert Region of the Kaokoveld, South West Africa/ Namibia." *Journal of Zoology (London)* 219:1–19.

Weir, A. A. S., J. Chappell, and A. Kacelnik. 2002. "Shaping of Hooks in New Caledonian Crows." *Science* 297, no. 5583 (August 9): 981.

13

Brains, Bodies, and Minds

Against a Hierarchy of Animal Faculties

DAVID DILLARD-WRIGHT

What good is a brain? By itself a brain doesn't do anything at all. In a jar with formaldehyde, a brain makes a good paper weight or teaching tool, but it can't think. Magnetic Resonance Imaging (MRI) technology, used to map processing regions and draw the tenuous line between life and death, can only detect mental activity in living brains, which means that the diagnostic tool also measures the contributions of a functioning body. Brains, human or nonhuman, mammal or reptilian, begin to die within a few minutes of oxygen deprivation. The brain only accomplishes its work when intricately linked to a living creature, itself tied by innumerable fluid connections to its environment. Having a mind requires having a brain, to be sure, but it requires much more than that. It requires all of those things a brain requires: circulation, respiration, and sensation, among other factors, like a suitable and stimulating environment. Understanding minds requires including all the processes that make brains possible. If the body is instrumental to the brain, it is equally instrumental to the mind.

Traditionally, we think of mind as consciousness, as self-awareness and language, but these functions of mind, the *positing* functions so prized by human beings, are a by-product of mind's features, chief of which are to keep the body alive and moving around in the environment, alert to the exigencies of situations. Language and self-consciousness, traditional markers of the human, should not be termed "higher" functions of mind, because they are not more complex or more necessary than what are labeled biological or vital processes. Human beings can survive perfectly well without words, but they can't survive at all without a steady supply of oxygen flowing toward their cells, without nourishment and water. The process of language acquisition, though certainly a remarkable cohesion of social, neurological, and motor faculties, is no more remarkable than the transcription of DNA or the chemical steps of glycolysis. Self-awareness, language, and other "macro" processes are just the flip side of the coin to the "micro" processes of cell biology and neurology. Neither end of the spectrum should claim primacy as preeminently real, superior, complex, abstract, or another modifier that adds "specialness." Rational thought may give intricacy to human meaning, but it could not arise at all without the body's engagement with its environment.

Some philosophers, especially in transplantation ethics, believe that the essential part of a human being is the mind, incautiously equated with the brain, and that this is where the person is really located (e.g., Baron 2006; Cherry 2005; Taylor 2005; Wilkinson 2003). This cluster of recent thinking on transplantation views the rest of the body as more or less interchangeable, and new medical technologies seem to confirm this view. After all, surgeons, aided by a bevy of pharmaceuticals, can now transplant the liver, the kidneys, the heart, and the lungs. Because transplantation of the brain still exceeds present capabilities, it seems natural to conclude that the brain is unmistakably the seat of unique personhood. Personhood—an idea often directly or indirectly wed to anthropocentricism, even when applied to cetaceans and primates ("higher" mammals)—does not, in fact, reside in the brain but rather encompasses a wide range of factors.[1]

The notion of personhood residing in the brain should be complicated and explored because it has a great impact on the way that we think about mind in humans and other animals. Mark Cherry, one of many theorists to take up the question of personhood and brain, writes, in a way typical of recent bioethics: "[The] higher brain [is the] . . . seat of human consciousness, memory, and the cognitive capacities that sustain personhood. . . . As [philosopher of science Roland] Puccetti put it, where the brain goes, there too goes the person. Persons are distinguishable from all their body parts; however, they are only separable from most. This implies that, in strictly secular terms, a person's body can be regarded as a collection of things with which the person is more or less intimately associated" (2005:25).

Here Cherry argues that only the brain "counts" as the seat of personhood, because the memory and personality reside in the brain. Cherry attributes a sentimental attachment to the body and makes it into a possession, a special type of property: bodies and body parts may be treasured possessions, but they are possessions nonetheless. The insistence on "human" consciousness in the previous quote itself rests on the Judeo-Christian ideal of the image of God. Only the "higher" brain (implicitly serving as the site of the *imago Dei*) really counts in Cherry's vision: All else is exchangeable, and, by this logic, animals without "higher" minds (however this questionable line might be drawn) also do not count and must be entirely valued at exchange rates.

I disagree with Cherry. I believe that personhood, however fuzzy the term, must be continually renewed by myriad processes, which take place in interaction with information and not merely in its storage. Memory itself is a complex philosophical topic and should not be confused with mere information storage; it too interacts with the surround. It can be externalized through books and computers, shared in storytelling and writing, linked to people and places, carried in posture and gesture. Moreover, the body is not a "collection of things" only loosely associated with the person: To have a different body is to experience the world differently, to be a different person. Yes, the body is plastic and its parts are "interchangeable" to a certain extent, but these interchanges matter and should not be dismissed as incidental, as though some hidden core or essence would remain unaffected beneath these alterations (e.g., transplantation, implantation, plastic surgery, and the like).

To Cherry's credit, he says that the processes of the "higher brain . . . *sustain* personhood," meaning that they should not be *equated* with personhood. On one level, the body might properly be regarded as an object, as property, so long as it is equally conceded that a second perspective goes along with this one, a perspective in which the body is not merely *mine* but is also *me*. The body is an object and yet not an object: It is "that by which there *are* objects" (Merleau-Ponty 2004 [1945]:105, emphasis added).

This leads to considerations of whether a brain in a vat would qualify as a person. Even if scientists could keep a brain alive in a vat, it would hardly qualify as a mind, unless the vat were to become a prosthetic body. Mind requires body, requires a network of relationships with the outside world. *Mind is inherently social,* and I mean this in the sense of a "general sociality" that includes any sort of vital or semiotic exchange with the surround (Dillard-Wright 2009a, 2009b). The simplest one-celled organisms have a form of mind, here taken as a kind of structuring or organization, in the form of nutrition and reproduction, absent any kind of neural network.[2] Their "bodies" network with the surround through these processes, and we can follow this same embodied logic all the way to the most complex mammals. Even computers have a sort of body—a processor, a hard drive, a monitor, etc.—and network socially with other computers and human users. Even sentient robots would have to have a body, and abstracting mind from the processes needed to produce that mind is an artificial exercise. This would still hold true in the science fiction scenario of completely self-aware robots. Indeed, the processes themselves, whether living or nonliving, "wet" or "dry," already contain mindedness, inasmuch as they organize inputs into a new whole. Even if we concede the interchangeability of "parts," *different bodies would lead to different minds*, different styles of thinking and problem solving. Embodiment constitutes mind and its so-called higher functions, and the body is not merely appended to the mind. Cherry says that the "higher" functions of mind sustain personhood, but he neglects to say that cognition alone does not, in fact, sustain personhood.

Once again, if it is possible to view the body as instrumental to the brain, it is equally possible to view the brain as instrumental to the body. The brain, of course, should not be situated as the polar opposite of body, since it belongs to body. The feedback loop between body and brain, or between the different sectors of an overall sensitivity, makes for the mind, a point some theorists miss by insisting on the "higher" mind as the "seat" of consciousness. Consciousness or mind, as a *process* involving many factors, doesn't *sit* anywhere but rollicks along with the organism into its environment, which also constitutively creates "mind." The maze is part of the lab rat's mind; the thoughts of a driver on the highway include the lanes of traffic. Minds propel themselves forward through the world and take their cues from that world. Making consciousness a static entity or essence in-residence decapitates it, obscures the temporal, perceptual, osmotic flows that make minds. The functions of body that produce mindedness should include not only perception and motility, as Merleau-Ponty (1963 [1942]; 2004 [1945]) demonstrated, but also the less intentional processes of circulation and respiration, together with the conditions that make them possible. The yardstick of self-awareness, which seemed the only proof of "higher" functions, has been overdetermined as a measure of cognitive, and hence moral, worth. If we accept the Darwinian paradigm of natural selection and variation over time, it becomes impossible to generate such ontological dividing lines that would separate functions into higher and lower groupings. In other words, a neuron is a neuron, in an earthworm, a chicken, or a human: One experiment even uses leech neurons for biological computing (Chase 1999). By privileging the human platform of intelligence, specifically the arrangement of the human brain, investigators historically missed the intelligence of other species. Now important advances are being made that recognize the intelligence of birds, octopi, and reptiles, nonmammal species with widely differing anatomies (Gaidos 2009). Respiration and circulation link together the various creaturely and cybernetic architectures: Whether mounted on a motherboard or within a fleshly matrix, neurons, as living cells, must breathe and consume.

The privileged role afforded contemplation in the classical and Christian traditions carries over into scientific pedagogy on cognition: We have learned to speak of "higher" and "lower" functions for the brain. The brain stem is a more "primitive" region, controlling the baser functions, while the rest of the brain takes the more weighty work of reason itself. Dividing the body into systems (circulatory, respiratory, nervous, etc.) makes possible the abstraction of parts or sets of parts from each other. While this subdivision serves as an important heuristic device, these cleavages should not be taken as literal descriptions if presumed to mean that "parts" and "systems" function independently. The pervasiveness and persistence of the mechanical metaphor of parts and systems can be explained by its political usefulness. By dividing the body into systems—"Divide and conquer" goes the Roman maxim—the animal (human or nonhuman) is subjected to various forms of control and loses a sense of the body as lived from within. As opposed to the phenomenological perspective, or what Deleuze and Guattari (1987:159) call the "Body without Organs" (BwO), the perspective of science turns the individual into a medical or research subject that becomes a conglomeration of parts or bits of information, a sort of dissolving deployed for economic/political purposes (see also Weisz 2006). Known from the inside, experientially, the body has no organs, or the organs are only dimly, obscurely felt, as in a stomachache, asthma, or appendicitis. Even then, the boundary between say, the feeling of the stomach and the feeling of the small intestine cannot be securely located, just as, in the children's game, one cannot say, with eyes closed, when another person's finger has reached the inside of the elbow on the surface of the skin. The view from the outside, the third-person point of view, the God's-eye, theological point of view, which is also the scientific, medical way of seeing, makes dim awareness of the organs explicit. This specialized knowledge then extracts value from the patient in the form of treatments and procedures: "The organism is already . . . the judgment of God, from which medical doctors benefit and on which they base their power. The organism is not at all the body, the BwO; rather, it is a stratum on the BwO . . . that, in order to extract useful labor from the BwO, imposes upon it forms, functions, bonds, dominant and hierarchized organizations, organized transcendences" (Deleuze and Guattari 1987:159). The biomedical researcher takes over the position of God in having the privileged point of view on the living body and, by deploying this privileged or "objective" point of view, can then claim access to power over the living subject (the BwO). Biomedicine has inherited a theological tradition it does not acknowledge, a tradition that subsumes all ways of knowing under the one correct or orthodox positivism. "Having the organism be first" means claiming the body as composed of systems, as mechanical, to be the primary and most correct way of viewing the body. Other ways of knowing the body are not necessarily decried as false, but they must take their place in line, so to speak, behind this most forceful and politically effective hermeneutic.

By contrast, the lived body knows no distinctions between respiration and circulation, thinking and movement, and immerses itself in its tasks. In fact, these processes absolutely depend upon one another in the formation of what can only be called "thinking" after the fact.[3] The body is the unification of diverse sensations, and, as a "nexus of living meanings," leads to positing consciousness, which then returns to the world through intentional activity (Merleau-Ponty 2004 [1945]:171–175). No special center coordinates the body's processes: A more chaotic, ad hoc organization leads to what we call mind. Looking for mind, then, requires tracing the patterns of organization that give rise to it. The ecological niche calls forth the appropriate manner of mindedness for that particular evolutionary situation, and an advantage is only an advantage so long as that particular niche remains relatively stable. Popular markers of "thinking" as construed from an anthropocentric perspective—language, self-consciousness, etc.—are not absolute but contingent definitions of intelligence that have little bearing on mindedness as such. A complete theory of mind would need to take into consideration differing types of mindedness that account for the wide array of intelligences on the planet. While it may flatter human egos to think of ourselves as the only intelligent life in the universe, such viewpoints ignore the many brilliant faculties of our nonhuman (and even inanimate, e.g., plants, computers) kin.

This cosmological view of thought poses a threat to the human sense of uniqueness, but it better matches the Darwinian continuity of species and gives human thought not just precursors but parallels in nonhuman modes of knowing. To acknowledge continuities between human thought and the thinking of other animals does not amount to positing sameness—rather, such a viewpoint contextualizes mindedness within the somatic and environmental niches that support it. There are as many possible modes of mind as there are species, and individual variations and histories also come into play. Nevertheless, processes like circulation and respiration, motility and communication overlap species boundaries and inform transspecific theory and practice. Going into this territory of relatedness requires addressing the underlying fears that would prevent such an exploration from happening.

Much horror and supernatural writing plays on the fear of the animate body, which comes from a discomfort with the animal nature of the human life. In Washington Irving's classic tale, "The Legend of Sleepy Hollow," a headless horseman torments poor itinerant schoolmaster Ichabod Crane, chasing the young teacher through the small New York town and finally dashing Crane to the ground by using his pumpkin head as a projectile. Crane himself, like the famous apparition, is all body and no head. "The cognomen of Crane was not inapplicable to his person. He was tall, but exceedingly lank, with narrow shoulders, long arms and legs. . . . His head was small, and flat at top, with huge ears, large green glassy eyes, and a long snipe nose, so that it looked like a weather-cock perched upon his spindle neck to tell which way the wind blew" (Irving 1820). A gullible fellow with a penchant for the fantastic, Crane's mental life consists mostly of ghost stories garnered from Mather's *History of New England Witchcraft*.[4] Crane also spends his free time fantasizing about food, plotting ways to fill his stomach from the larders of the parents of his pupils. Crane, the scarecrow with a small head, can never seem to fill his straw stomach. At last he falls victim to a body without a head, as though Crane's fantasies turned against him. The story tacitly identifies witchcraft and superstition with the body and the stomach, the flip side of the head's rationality. Ichabod and his plight symbolize discomfort with the body and its materiality (somatophobia), as the feminine-gendered processes of the body threaten the masculine rationality of the head (Acampora 2006; Schusterman 2008).[5]

Washington Irving's headless horseman, like the zombie hordes popular in fiction and movies today, has a body but no brain. Rather than being controlled by the mind, in these forms of popular entertainment the body has broken away from the mind and gone along without it. The headless horseman actually wields his own head as a weapon, reversing the usual direction of philosophical and popular thinking: We expect the head to control the body and not the other way around. A corpse strikes fear for the same reason: although vacated of mind, it carries the vestiges of personhood in face and limbs. Even though it has lost its animation, a corpse still wears an expression, still has the outward form associated with living. This specter scares because of its loss of unity and the lingering suspicion that perhaps rationality represents a false sense of security and control. Even without Buddhist philosophy or Husserlian phenomenology, we all suspect that maybe we aren't the contained selves our personalities and bodies advertise: Internal dialogue becomes a chorus of competing voices, and the body a medium for conveying competing identities.[6] In contrast to the classical and Christian traditions, in which sensation and passion had to be subordinated to rational thought, the body too has its intelligence, as anyone who has ever stubbed a toe or suffered from food poisoning knows. The members, the organs assert themselves more forcefully than a syllogism. Thought can work its way "up" as well as "down": we don't treat our bodies as mere stumps or pedestals for the head and the brain. Even our somewhat sedentary, office-bound bodies know the feeling of a sore butt, a crick in the neck, a foot gone to sleep. This knowledge below rational thought and the primary senses contributes as much or more to our sense of "being-in-the-world," to use the Heideggerian description, as any propositional thinking ever could.[7] Merleau-Ponty has called attention to the domain of embodiment neglected by Heidegger and insisted that we think through the dialogical relations, the constitutive interplay, of body and world.[8] Minds do not govern bodies from above, as a puppetmaster controls a puppet, he insists. Rather, mindedness unfolds as the body navigates around in the world and propels its intentions outward into concrete actions.

Merleau-Ponty's embodied phenomenology has left its mark on a number of disciplines. Over the last few decades, it has become de rigueur in academic circles, especially feminist, environmental, and animal studies, to write about the lived body and the ways our perceptual and sensual lives contribute to the formation of knowledge. Even engineers seem to have gotten the message, perhaps by a different route, as the new disciplines of social robotics and embodied AI demonstrate (Mazis 2008). Allan Hobson (2002:466), summarizing the ideas of neurologist Rodolfo Llinás (2001), might have been channeling Merleau-Ponty when he wrote, rephrasing

Descartes's famous *cogito*, "I move, therefore I am a self, (I think)." In other words, motility, for Llinás specifically the nerve cells, underlies the "higher" processes of cognition. Respiration and circulation, among other processes, make motility possible and cannot be viewed as "beside the point," or outside the scope of mind. If we confine what "counts" as mindedness to the so-called higher processes of conceptual and linguistic thinking, we run the risk of trying to make consciousness self-supporting, like a house of cards suspended in midair. A better theory of consciousness or mind will be dispersed into a number of different processes and will not posit a break or a hierarchy between the "higher" and "lower" functions of mind. We might rephrase the *cogito* even farther, to say, "I breathe (or I digest), therefore I move, therefore I think, therefore I think I am," but this isn't a very snappy phrase. At this moment we realize the appeal of René Descartes's (1998 [1637]) original formulation and at the same time realize its limitations: The price it pays for its simplicity is the loss of accuracy and complexity. In order to be more accurate and complete, we will have to think all the way "down," to all of the processes that make minds possible. Thinking through all the body's processes will help reattach mind and body, preventing philosophy from suffering the same fate as Ichabod Crane, who was haunted by the headless body.

Insights from voice theory help to clarify the relationship between the vital body and the body as representation and will provide a bridge between human mindedness and the minds of other creatures. Voice theory is currently undergoing a shift in emphasis under the aegis of phenomenology and Eastern philosophies. Phenomenologist and singer Päivi Järviö speaks of two paradigms of voice pedagogy, one that views the "singer's body [as] an instrument that is used as a tool to produce a sound" and a second paradigm in which "the singer with his voice is a living human being" (Järviö 2006). The first perspective uses anatomy and physiology, along with "drawings, plastic models, and mirrors" to make the singer or actor more aware of the things she does with her larynx, diaphragm, and vocal chords, becoming aware of the good and bad habits that she has developed over time (pp. 68–69; see also McAllister-Viel 2009). However, in order for the process to continue, this objective, third-person perspective must merge into a greater synthesis, in which breathing and thinking become one, or, stronger still, a perspective in which "mind can be understood as a manifestation of breath" (Järviö 2006:169; see also McAllister-Viel 2009:172). Järviö gives a first-person account of what this synthesis feels like from the inside: "My body reads the score, singing silently. My body is this ability to read music and to sing. . . . Hitting that weightless moment, jumping into an empty space, is a frightful moment of almost limitless freedom, producing a voice that is one with the body and not in a continuous state of imbalance" (p. 66). Järviö here identifies freedom with silently plunging into the lived body, not with consciously carrying out a plan of action. In the immersed perspective that thinking through breathing affords, all the entanglements of positing consciousness fall away in that "frightful moment" of union with the bodymind.

So used to thinking of our heads, à la Sleepy Hollow, as detached from the body, the mindedness of the chest region can indeed be scary. Since we have accustomed ourselves so much to the plotting, planning, and abstraction of propositional understanding, going along with the flow of the lived body can feel like rushing downstream in a whitewater raft or feeling weightlessness on a roller-coaster ride. But this current beneath the currents of thought, once recognized and appreciated, becomes nothing less than a new way of approaching the world.[9] The surface interaction of the senses and the conceptual wrangling of thought processes give way to an enveloping unity, which includes these other perspectives and yet transcends them. So it is that singers and actors often describe themselves feeling a unity with the audience, a kind of "absorption" or "communion" in which interpersonal barriers dissolve (McAllister-Viel 2009:173).[10] Järviö continues, "Singing is not about the material factors of music: sounds, timbres, pitches, durations, scores, notations. . . . Rather, it is about a living singer opening her being for others in the moment of letting go, about allowing the listeners to feel the invisible infinity of shared Life, the silent experience of being, opening in music" (2006:74). By describing singing as a kind of silence, Järviö intimates that the idea of singing or performance as a conveyance of content from one party to another gives way, in the better moments, to a whole in which singer and audience merge into the song. The song reverberates through all of them, and a life beneath or before conscious thought makes the experience worthwhile. The breath that courses through the singer's body becomes her life, is literally life-giving and yet also eclipses her other concerns, and it is the life and sole focus of the listeners as well.

This enraptured description surely does not happen in all musical experience (otherwise we would all rush immediately to the opera house), but it does disclose something of the daily life of our bodies in their congress with the environment and with other subjects. Since it does not rely solely on discursive reasoning, I believe this way of seeing and understanding the body can provide glimpses of what the minds of other animals must be like. The language and concepts that we prize so much as human beings may just be a complicated set of blinders that prevent us from seeing our emplacement in a completely different domain of life, equally available to us but stunted by the frenetic pace of infotainment culture. Concentrating on breath and body, whether through meditative practices, physical exercise, or artistic expression, becomes a way to bring awareness down the spinal axis to the chest. This recentering places the head along the periphery of the body as just another extremity. We may recoil at such a realignment because it challenges so many of our assumptions about what it means to be human, but the results can be quite welcome. This displacement creates an opening to the world previously unnoticed and makes known a fundamental kinship with other beings not bound by genera and species. We need not posit a World Soul or a group mind in order to make such co-constitution possible, because similar processes make all of the various animal minds function.

Co-constitution through similar processes allows at least a partial glimpse into the minds of nonhuman animals. To turn from Ichabod Crane to whooping cranes, the endangered species that helped to spur the American conservation movement, we can see large similarities and differences between species that mark the possible iterations of mind.[11] Cranes can stand stock still indefinitely, as anyone who has seen them or other shoreline birds can attest, and yet they also have famously elaborate mating dances. They can be quieter than thought and they can call loud enough to be heard for several miles. Eurasian cranes fly at thirty thousand feet over the Himalayas, and their close relatives can be found near sea level in the marshes of Texas (Beletsky 2006). Cranes breathe like we breathe, albeit with the specialized lungs of birds, and yet they can sleep while standing on one leg. To the untrained eye they all look the same, yet they can be recognized individually using Wessling "voice prints" (Hutchins 2003:23–39). What do we make of this wondrous difference of capacities, both within the family of cranes and between cranes and human beings?

Spoken and written language becomes a specialized type of human vocalization, akin to the mournful cries—indeed the wide variety of purposive vocalizations—of the crane. The shared phenomena of vocalization, respiration, circulation, and cultural features like mating rituals allow us to affirm kinship with the crane and yet to recognize that significant differences exist that we may never be able to precisely access from a subjective viewpoint. Style of embodiment matters—what counts about the crane is its unique mindedness—not the crane's ability to measure up to an invented and artificial anthropocentric yardstick of intelligence. A broader theory of mind will value the crane's intelligence per se and not only by comparison to human capabilities. Differing species—different individuals as well—live their lives with different mind-body architectures styled for differing modes of problem solving. Human beings certainly have many wonderful, unique capabilities, but these capabilities should not be regarded as the endgame, the sine qua non of mindedness. The earth exhibits a queer proliferation of forms, each productive in its own right, each capable of ambling, playing, mating, thinking in its own way.

The different layers of organization—motility, language use, breathing and respiration—occur side by side in a lateral relationship. Sometimes they mutually affect one another, as when a morning run triggers a higher pulse or a heated conversation leads to actual body heat. At other times they may function in a more parallel fashion, with little obvious interaction or mutual influence. All these layers of organization occur with more or less specificity in humans and in other animal species, and preferring one of these layers over another is a value judgment and not a neutral statement of fact. We may say that the capacity to use language or the capacity for self-awareness is a higher or better kind of consciousness, but such a judgment rests on no ultimate criteria. A computer cannot continue to function without the fan that cools its motherboard or a source of electricity, and an animal body cannot function without breath and blood. I suggest, then, that the value judgment of higher and lower is not justified and may actually obscure the way that minds arise from embodiment. Turning awareness to processes of embodiment that may usually go unnoticed makes possible a broader and deeper theory of mind that takes better account of how humans and other animals interact with their surroundings.

The very idea of mind can become an "organized transcendence" of the sort Deleuze and Guattari discuss (e.g., in the earlier quote about BwO). Isolating a single trait or a single way of being from the diversity of species on the earth risks a totalizing kind of discourse that would privilege some forms of embodiment over others.[12] A pro-animal perspective—one that combines the best impulses of ethology and animal studies on the one hand and animal welfare and animal rights on the other—must be suspicious of the category of "mind" in general, because it usually implies human minds as a standard of achievement or proficiency. Mindedness is a moving yardstick, one that animals are often found to surpass only to have the achievement removed retroactively, depending on the mood or agenda of the investigator. If discourse on minds is to prove useful in the future, it must take into account a broad range of capacities and acknowledge differing types of minds for differing ecological niches. Otherwise, mindedness becomes just another wedge to justify continued human exploitation of animals and the earth. Further exploring the lived body—even its unconscious processes—brings to awareness a level of embodiment that functions alongside motility and rationality as co-constitutive of "mind." In speaking of mind, whether in the humanities or the sciences, investigators must beware of what "counts" as mind, because so much cultural baggage hinges on that question. Including processes like circulation and respiration in the discussion of mind does not just satisfy the ethical requirement of including other species, it also promises to more accurately portray the ways in which minds arise from embodiment.

Notes

Part of the research for this project was conducted through a summer residential fellowship at Michigan State University in 2008, funded by the Animals and Society Institute.

1. A full examination of "personhood" and its limits and implications is beyond the scope of the present essay. As a provisional definition, personhood is a state belonging to any entity capable of self-regulating, independent action in concert with the surround. This would include human and nonhuman animals as well as many plants and machines.
2. Here I follow Charles Sanders Peirce's definition of mind as any kind of structure or organization in nature, parallel to what he calls "thirdness" (see Corrington 1993; Weiner 1958).
3. I align myself here with the bundle theory of the person, developed by Derek Parfit (1989).
4. I was unable to find the original document. The title could be a creation of Irving's, perhaps a conflation derived from Cotton Mather's *Magnalia Christi Americana: The Ecclesiastical History of New England* (1702) and his *Wonders of the Invisible World* (1970 [1693]).
5. Richard Shusterman's ongoing project of "somaesthetics" more fully explores this phenomenon, and Ralph Acampora articulates a somaesthetics of human-animal relationships.
6. The Buddhist rejection of a stable core of the self has many commonalities with postmodern relational and narrative ideas of selfhood (see Klein 1994; Lyotard

1984; McGhee 1995). Husserl's (1977 [1931]) study of phenomena or appearances paradoxically became a study of interiority or consciousness.

7. "The compound expression 'Being-in-the-world' indicates in the very way we have coined it, that it stands for a unitary phenomenon. This primary datum must be seen as a whole. But while Being-in-the-world cannot be broken up into contents which may be pieced together, this does not prevent it from having several constitutive items in its structure" (Heidegger 1962:78, H. 53).
8. Merleau-Ponty (2004 [1945]) critiques Heidegger's *Dasein* and also continues Heidegger's project by saying, "The body is our general medium for having a world" (p. 169).
9. See also Carrie Rohman's Deleuzian discussion of vibration and voice in this volume. The vibrational has an intensity in its own right, before or otherwise than the currents of thought, an aesthetic bridge between the orders of life.
10. Merleau-Ponty (2004 [1945]:373) uses the metaphor of "coition" to describe perception, a theme discussed in Rohman's essay in this volume.
11. Only a few hundred whooping cranes remain in the wild (Sibley 2000:156).
12. Totalizing in the Levinasian sense of an imperialist homogenization that ignores infinite variation beneath the surface (Levinas 1969). Making the infinitizing movement, when it comes to animals, means revealing the varied and complex worlds of other species, which are reduced to bland sameness in anthropocentric thinking.

References

Acampora, R. 2006. *Corporal Compassion: Animal Ethics and Philosophy of Body.* Pittsburgh: University of Pittsburgh Press.

Baron, J. 2006. *Against Bioethics.* Cambridge: MIT Press.

Beletsky, L. 2006. *Birds of the World.* Baltimore: Johns Hopkins University Press.

Chase, V. D. 1999. "Team Develops Biological Computer." *Nature Medicine* 5, no.7: 722.

Cherry, M. J. 2005. *Kidney for Sale by Owner: Human Organs, Transplantation, and the Market.* Washington, DC: Georgetown University Press.

Corrington, R. S. 1993. *An Introduction to C. S. Peirce.* Lanham, MD: Rowman and Littlefield.

Deleuze, G., and F. Guattari. 1987. *A Thousand Plateaus: Capitalism and Schizophrenia.* Trans. B. Massumi. Minnesota: University of Minnesota Press.

Descartes, R. 1998 [1637]. *Discourse on Method.* 3d ed. Trans. D. A. Cress. Indianapolis: Hackett.

Dillard-Wright, D. 2009a. "Thinking Across Species Boundaries: General Sociality and Embodied Meaning." *Society and Animals* 17, no. 1: 53–71.

——. 2009b. *Ark of the Possible: The Animal World in Merleau-Ponty*. Lanham, MD: Lexington.

Gaidos, S. 2009. "Humans Wonder, Anybody Home? Brain Structure and Circuitry Offer Clues to Consciousness in Nonmammals." *Science News* 176, no. 13: 22.

Heidegger, M. 1962. *Being and Time.* Trans. J. Macquarrie and E. Robinson. New York: Harper.

Hobson, A. 2002. "Book Review [*I of the Vortex: From Neurons to Self* by R. Llinás]." *Perspectives in Biology and Medicine* 45, no. 3: 466–468.

Husserl, E. 1977 [1931]. *Cartesian Meditations: An Introduction to Phenomenology.* Trans. D. Cairns. Leiden: Nijhoff.

Hutchins, M. 2003. *Grzimek's Animal Life Encyclopedia.* Vol. 9. 2d ed. Detroit: Thomson.

Irving, W. 1820. "The Legend of Sleepy Hollow." Retrieved from http://www.gutenberg.org/files/41/41-h/41-h.htm.

Järviö, P. 2006. "The Life and World of a Singer: Finding My Way." *Philosophy of Music Education Review* 14, no. 1: 65–77.

Klein, A. C. 1994. "Presence with a Difference: Buddhists and Feminists on Subjectivity." *Hypatia* 9, no. 4: 112–130.

Levinas, I. 1969. *Totality and Infinity: An Essay on Exteriority.* Trans. A. Lingis. Pittsburgh: Duquesne University Press.

Llinás, R. 2001. *I of the Vortex: From Neurons to Self.* Cambridge: MIT Press.

Lyotard, J. 1984. *The Postmodern Condition.* Trans. G. Bennington and B. Massumi. Minneapolis: University of Minnesota Press.

McAllister-Viel, T. 2009. "(Re)considering the Role of Breath in Training Actors' Voices: Insights from Dahnjeon Breathing and the Phenomena of Breath." *Theater Topics* 19, no. 2: 165–180.

McGhee, M. 1995. "The Turn Towards Buddhism." *Religious Studies* 31, no. 1: 69–87.

Mather, C. 1702. *Magnalia Christi Americana: The Ecclesiastical History of New England.* London: Parkhurst.

——. 1970 [1693]. *Wonders of the Invisible World.* New York: Franklin.

Mazis, G. A. 2008. *Humans, Animals, Machines: Blurring Boundaries.* Albany: SUNY Press.

Merleau-Ponty, M. 1963 [1942]. *The Structure of Behavior.* Trans. A. L. Fisher. Boston: Beacon.

——. 2004 [1945]. *Phenomenology of Perception.* Trans. C. Smith. London: Routledge.

Parfit, D. 1989. "Divided Minds and the Nature of Persons." In C. Blakemore and S. Greenfield, eds., *Mindwaves: Thoughts on Intelligence, Identity, and Consciousness,* pp. 19–28. London: Blackwell.

Shusterman, R. 2008. *Body Consciousness: A Philosophy of Mindfulness and Somaesthetics.* Cambridge: Cambridge University Press.

Sibley, D. A. 2000. *Sibley Guide to Birds.* New York: Knopf.

Taylor, J. S. 2005. *Stakes and Kidneys: Why Markets in Human Body Parts Are Morally Imperative.* Aldershot: Ashgate.

Weiner, P. P., ed. 1958. *Values in a Universe of Chance: Selected Writings of Charles S. Peirce.* Stanford: Stanford University Press.

Weisz, G. 2006. *Divide and Conquer: A Comparative History of Medical Specialization*. Oxford: Oxford University Press.

Wilkinson, S. 2003. "Bodies for Sale: Ethics and Exploitation in the Human Body Trade." London: Routledge.

PART IV

Animal Versus Human Consciousness

14

Rethinking the Cognitive Abilities of Animals

GARY STEINER

For a long time in the history of Western philosophy it was assumed that nonhuman animals are incapable of rationality. Aristotle observes in his zoological writings that many animals appear to exhibit rational capacities, yet in his psychological and ethical writings he excludes nonhuman animals from community with humans on the grounds that animals are incapable of *phronesis*, the notion of practical wisdom that Aristotle considers crucial for active citizenship. The Stoic philosophers formalize Aristotle's position in the psychological and political writings into a cosmic principle according to which animals were created expressly to satisfy human needs and lack the cognitive capacities necessary to occupy the cosmopolitan standpoint from which rational beings contemplate the divine logos alongside the gods.[1] Saint Augustine follows this Stoic line of thinking and maintains that even though animals have sufficient subjective states of awareness to feel pain, this pain has no moral significance inasmuch as animals do not participate in *koinoia* or community with human beings. René Descartes refuses to acknowledge that animals are sentient and argues for the legitimacy of experimenting on animals, as well as killing and eating them, on the grounds that animals are mere machines with no inner states of awareness whatsoever. Kant returns to the Aristotelian-Stoic-Augustinian view according to which animals do have subjective states of awareness, but he argues that, because animals lack the capacity for rational choice, they are neither subjects nor objects of direct moral concern.

Already in antiquity there were proponents of the view that animals are rational or at least that animals have complex inner states of awareness in virtue of which they should be considered direct beneficiaries of moral concern. The most conspicuous animal advocates in ancient philosophy are Plutarch and Porphyry.[2] Plutarch wrote three texts on animals in which he criticizes the Stoic position by offering a wide array of examples of animal thought and ingenuity and in which he deplores the human practice of meat eating. Porphyry argues in *On Abstinence from Killing Animals* that there is no need to kill and eat animals (indeed, that the practice interferes with the process of soul purification), that the practice of animal sacrifice is an offense against the gods (for whom animals are beloved), that we owe duties of justice to animals, and that animals exhibit at least imperfect rationality. But many of the anecdotes offered by these ancient thinkers bear the traces of an unmistakable anthropomorphism that ultimately detracts from the credibility of their arguments. One such anecdote is offered by Plutarch (1995 [AD 70]):

> Cleanthes [a Stoic philosopher], even though he declared that animals are not endowed with reason, says that he witnessed the following spectacle: some ants came to a strange anthill carrying a dead ant. Other ants then emerged from the hill and seemed, as it were, to hold converse with the first party and then went back again. This happened two or three times until at last they brought up a grub to serve as the dead ant's ransom, whereupon the first party picked up the grub, handed over the corpse, and departed.
>
> (p. 369)

Another such anecdote concerns Chrysippus's dog. Sextus Empiricus (1996 [ca. AD 190]), who, like Plutarch, is at pains to criticize the Stoic position concerning animals, offers the following account and interpretation of this anecdote:

> According to Chrysippus, who was certainly no friend of non-rational animals, the dog even shares in the celebrated Dialectic. In fact, this author says that the dog uses repeated applications of the fifth-undemonstrated argument-schema, when, arriving at a juncture of three paths, after sniffing at the two down which the quarry did not go, he rushes off on the third one

> without stopping to sniff. For, says this ancient authority, the dog in effect reasons as follows: the animal either went this way or that way or the other; he did not go this way and he did not go that; therefore, he went the other. (p. 98)

Lest we suppose that this tendency to anthropomorphize is a relic of antiquated prejudices that we wise postmoderns have long since outgrown, it is worth bearing in mind that this tendency manifests itself even in contemporary ethology. One illustrative example is the work of Donald Griffin, a prominent cognitive ethologist and the codiscoverer of echolocation in bats. In an effort to counter the influence of behavioral ethology, according to which animal behavior can be adequately understood without any reference to subjective inner states of awareness, Griffin maintains that a wide variety of animals engage in conceptual abstraction. He suggests, for example, that "it seems reasonable to suppose that when pigeons are working hard in Skinner boxes to solve these challenging problems [of making fine discriminations between different types of objects], they are thinking something like: 'Pecking that thing gets me food'" (Griffin 1992:131). On Griffin's view, even invertebrates such as honeybees "may think in terms of concepts" (p. 139).

Griffin's motivation is to arrive at more edified conceptions of the subjective lives of animals than either the Western philosophical tradition or behavioral ethology allow. Griffin would have us believe that animals, all the way down to at least some invertebrates, engage in conscious thought that is conceptual and predicatively structured. To say that nonhuman animals are capable of conceptual abstraction can mean many things; indeed, a survey of the literature on concepts shows that there is nothing even remotely like a consensus on the question what a concept is. In my own work I follow the definition offered by Colin Allen and Marc Hauser (1991): "To have a concept of *X* where the specification of *X* is not exhausted by a perceptual characterization, it is not enough just to have the ability to discriminate *X*'s from non-*X*'s. One must have a representation of *X* that abstracts away from the perceptual features that enable one to identify *X*'s" (p. 227). On this view, conceptual ability is more than and something essentially different than discriminatory ability; it is the ability to form representational content that is separate from any particular perceptual experience to which that representational content pertains. The ability to distinguish, say, black from white does not require any conceptual ability. To recognize *that black and white are both colors* does require conceptual ability, and, in attributing conceptual ability to animals, Griffin is attributing to them precisely this kind of sophisticated mental operation. Moreover, to the extent that intentional states such as beliefs and desires are predicatively structured and employ concepts, as when I state or recognize *that* something is the case, Griffin, in attributing conceptual ability and thought to animals, is implicitly attributing intentional capacity to them. Thus Griffin would presumably say that the piping plover, in performing its broken wing display to lure predators away from its young, has a more or less specific set of beliefs: that the predator is endeavoring to catch the plover's young, that the predator will be fooled by the broken wing display into thinking that the plover will be easier to catch than the plover's young, etc. Intentionality in this sense involves a variety of sophisticated mental capacities, including the ability to recognize that one's own mind is one mind among many minds, that other minds operate in essentially the same ways as my mind, that these operations involve phenomena such as beliefs and desires, etc. And to the extent that beliefs and desires are by their very nature predicatively structured—a belief is a belief that something is the case, and a desire is a desire that some state of affairs obtain—the attribution of intentionality to nonhuman animals is the attribution of all these abilities to them.[3]

To adopt such a view of many if not all nonhuman animals is to seek as far as possible to erode what was traditionally taken to be a difference in kind between humans and nonhuman animals and to represent that difference as nothing more than one in degree. Griffin and others who take this tack on behalf of animals appeal to physiological similarity and evolutionary continuity between humans and animals as a basis for concluding that animals, like humans, possess conceptual and in some cases predicative abilities. But even Griffin acknowledges the claim of the tradition, that the human possession of language constitutes an important difference between humans and animals (Griffin 1992:4). Thus he finds himself caught in a peculiar contradiction: He wants to attribute to many animals the capacity for conceptual thought and acts of predication and he wants to acknowledge that there is something distinctive about human language. But if the distinctiveness of human language does not consist in, or at least include, conceptual and predicative abilities, in what *does* it consist? It seems to me that thinkers such as Griffin, in their zeal to assert that animals have rich inner lives, go too far from one extreme to the other in attributing to animals capacities that animals simply do not seem to possess. Specifically, it seems implausible to attribute predicative intentionality and conceptual ability to any but perhaps the most sophisticated nonhuman animals, such as higher primates, dolphins, and some birds.

I say "perhaps" because the results of much of the observation that has been performed on animals that lie closest to the human-animal cognitive dividing line are highly ambiguous. Researchers such as Griffin are forced by the very nature of their vocation to reject Thomas Nagel's (1974) challenge to cognitive ethology, a challenge according to which we can never know what it is like to be an animal of another species inasmuch as differences in perceptual apparatus make for fundamentally different sorts of subjective encounters with the world. Griffin (1992:237–238) argues that while we may not be able to arrive at absolutely precise accounts of the inner lives of other animal species, we can nonetheless learn a great deal about the mental lives of animals by analogy to our own experience. It is widely accepted in the community of cognitive ethologists that analogy is an indispensable tool in the endeavor to understand the subjective experience of animals. My colleague at Bucknell, the ethologist Douglas Candland (1993), has noted our strong tendency to conceive of the mental lives of animals in terms of our own intentionality, our capacity to form beliefs and desires: "When my dog barks at me, do I not attribute to the dog some purpose, a purpose reflected in my reflecting about the contents of the dog's mind? In this way, the mind of my dog is inextricably bound to my mind, for I have no other way to create its mind other than by applying the categories and concepts of my own" (p. 369). But this by itself is no guarantee that what we infer by means of analogy is an accurate representation of the inner states of other animal species; and it certainly constitutes no proof that animals employ the same cognitive apparatus as we do. Even Candland urges us to "deny the arrogance of thinking that we are objective and [to] devote our attention to examining our own categories, and thereby the power and weakness of our own natures" (p. 369).

I take one potentially crucial weakness in this connection to be the tendency to assume that, because of our evolutionary continuity with other species, nonhuman animals must possess the same sophisticated capacities for linguistic intentionality and concept formation that we possess. I see nothing wrong with granting that animal species that lie near the human-nonhuman divide are capable of some conceptual abstraction and at least the rudiments of linguistic communication; one positive lesson of evolutionary theory is that humans do lie on some sort of continuum with other animal species and that the historical supposition that we, unlike all other animals, are created in God's image is the expression of a wish to be unique and to be superior to all other earthly beings. But some differences in degree are so significant as to constitute differences in kind; and I believe that the possession of conceptual ability and predicative intentionality, both of which are needed to transform communication into language, distinguish humans from all but perhaps the most sophisticated nonhuman animals. Even in the face of some extraordinary ethological discoveries, it should still be apparent to us that there are some differences in kind between humans and nonhuman animals. In particular—and I accept Nagel's proposition that we cannot ultimately be certain about this sort of thing—nonhuman animals do not appear to be capable of contemplating the distant future or the remote past. To do things that are somehow *related* to the distant future or remote past is not to be able to *contemplate* past or future moments; I believe that animals that do exhibit some sort of relation to the remote past or future may well do so in a fundamentally different way than humans do. Specifically, I believe that all but perhaps the most sophisticated animals relate to past and future nonconceptually, and this mode of relating to past and future significantly restricts the capacity of animals to engage in practices such as long-range planning. Bear in mind here that practices such as storing nuts for the winter can be explained without any reference to conceptual abstraction and thinking about the future as such.[4] I also believe it is in virtue of the lack of conceptual ability in most animals that it does not make sense to attribute anything like moral agency to animals. For activities such as long-range planning and moral choice involve the capacity to represent oneself to oneself as a self among other selves, to take an explicit inventory of one's needs and desires as well as the needs and desires of other agents, and to represent and relate to one another specific parts of space and moments of time that extend far beyond what is immediately perceivable. Such capacities involve formal, predicatively structured intentionality, which is the ability to relate logical subjects and predicates to one another in ways that represent the world; this activity, in turn, requires conceptual ability, inasmuch as the logical subjects and predicates, as well as the possible relations between them, are themselves conceptual units. The manipulation of these conceptual units enables us to grasp reality as such, which is to say that we represent states of affairs and contingencies explicitly and in relation to one another. My best sense is that any being capable of operations such as the consideration of counterfactuals must be capable of forming concepts and intentional states, and I find it hard to believe that many if any nonhuman animals are capable of contemplating counterfactual conditions such as "what if it were to rain next Tuesday?" I further consider it more than coincidental that language is predicatively structured, i.e., I proceed on the assumption that all and only those beings capable of intentionality are capable of language, and vice versa: All and only those beings capable of language are capable of intentionality. Why? Because, as the Stoic philosophers observed, human beings but not animals appear to be able to contemplate isolated moments and events, even those far removed from the present, *as such*, which is to say that humans appear to be unique among living beings in possessing the capacity to consider a moment or an object in the abstract.[5]

Of course, there is no knowing this with any degree of certainty. The importance of Nagel's challenge to the science of ethology consists in its recognition that our judgments and inferences about the mental states of nonhuman animals must ultimately remain hypothetical, which is to say that science cannot have the last word about the nature of animal cognition. Griffin himself proceeds on the basis of an open acknowledgment that he is employing a colossal and ultimately unprovable hypothesis when he attributes conceptual ability to animals. Thus pragmatic criteria become indispensable, if not ultimately dispositive, in addressing questions pertaining to animal cognition. In particular we have to ask ourselves whether, on the basis of what we experience in ourselves and what we experience in animals, it seems prudent to attribute to animals the sophisticated abilities associated with intentionality, in particular the capacity to contemplate individual units of meaning "as such." The capacity to contemplate things as such is perhaps the most sophisticated form of abstract thinking, and on my view it is in all likelihood possessed exclusively by linguistic, which is to say human, beings. An excellent example of the as such, and one of great relevance in reflecting on the moral status of animals, is the capacity to contemplate rights and duties. Human beings are capable of taking on duties in virtue of their ability to contemplate what a right and a duty are in the abstract and in turn to reflect on specific rights and duties. Thus it makes sense to say that human beings, at least those of a certain degree of cognitive and moral maturity, possess rights and duties. And while I believe it makes perfect sense to say that sentient nonhuman animals have rights (such as the right not to be killed and eaten by human beings), I consider it neither sensible nor prudent to say that animals have duties toward other sentient beings. To be able to take on a duty is to be able to contemplate that duty, which presupposes, among other things, the ability to contemplate what a duty in general is. A very apparent if ultimately unprovable difference between humans and nonhuman animals is that humans appear to possess the kind of cognitive abilities in virtue of which it makes sense to hold humans morally and legally responsible, whereas animals precisely do not. This is why, on the view that I have developed in my work on the moral status of animals, I argue that it would make no sense whatsoever, for example, to say that lions have duties not to harm gazelles, even though it makes perfect sense to say that human beings have duties not to harm nonhuman animals. It *makes sense*, even if you do not ultimately agree with it. But it would not comparably make sense to attribute moral responsibility to nonhuman animals, because they give no indication of being able to step back from their bodily desires and evaluate them in the ways in which human beings seem quite clearly to be capable of doing.

A useful occasion for reflection on the question of sophisticated conceptual abilities in animals is provided by Irene Pepperberg's work with Alex the parrot. A close reading of Pepperberg's work shows an acknowledgment on her part that we cannot ultimately know the cognitive mechanisms that were at work when Alex the parrot performed his remarkable cognitive tasks. For example, Pepperberg documented Alex's ability to answer questions about the number and color of objects present on a table; he was able to do this with up to six objects. Pepperberg (2005:197) notes that Alex's accuracy on this task was "comparable to that of chimpanzees and very young children." Given that "'number sense' requires handling abstract concepts—representations and relations," Pepperberg concluded that "Alex [used] and comprehend[ed], in appropriate situations, abstract utterances at a representational level" and that he may "have shown a numerical competence not unlike that of humans" (p. 204). But, at the same time, Pepperberg acknowledges that Alex might simply have been making some kind of elaborate associations between present sense impressions and past experience, or he might have been "subitizing," which means that he may simply have been instantaneously recognizing a pattern on the table the way one might recognize a particular die or domino without actually counting the dots.

I do not doubt Pepperberg's conclusion that Alex did as well on this task as chimpanzees or young children. What I do doubt is that an experimental setup of this kind can offer conclusive evidence that, as Pepperberg urges, Alex comprehended "abstract utterances at a representational level." At the same time, Pepperberg does offer one observation that might lend credibility to the claim that Alex was capable of abstract thought. In the experiments with colored objects, Alex spontaneously and appropriately employed the word *none*. In a previous study, he "had been taught to use 'none' to indicate absence of information in one situation and, without training, transferred its use to another when specifically queried" (p. 202). In the study in question, Alex spontaneously used the term *none* appropriately, without Pepperberg having solicited it as a possible answer. Pepperberg concluded from this that Alex employed a notion that is "abstract and relies on violation of an expectation of presence" (p. 203). Pepperberg notes that "even young children and some apes have some difficulty with ordinal use of zero" (p. 203), and hence that it would be hasty to conclude that Alex possessed an ordinal comprehension of zero. Nonetheless, Pepperberg urges the conclusion that Alex possessed a sophisticated capacity for conceptual abstraction; she acknowledges but resists the possibility that Alex was engaging in some form of association that is in no way dependent upon the capacity for abstraction.

In the case of cognitively highly sophisticated animals such as Alex, I don't see why we need to exclude categorically the possibility that they employ conceptual abstraction and intentionality. But the attribution of such capacities to other animals strikes me as highly implausible for reasons that I have touched upon already in connection with the notions of right and duty. These reasons are elaborated by Donald Davidson (1985) in his reflections on the dynamics of belief formation. Davidson presents a holistic conception of belief according to which any being capable of having one belief must be capable of having a whole array of beliefs and a grasp of the difference between truth and falsity; to have these capacities, a being must be capable of linguistic intentionality. "One belief demands many beliefs, and beliefs demand other basic attitudes such as intentions, desires and, if I am right, the gift of tongues. This does not mean that there are not borderline cases. Nevertheless, the intrinsically holistic character of the propositional attitudes makes the distinction between having any and having none dramatic" (p. 473).[6] Propositional attitudes or intentional states, like language, are predicatively structured and involve the use of concepts. Any being that possesses any of these abilities possesses them all. To suppose that a being could be capable of intentionality and concepts without being capable of language would be to suppose that language consists simply in the attachment of words to fully formed, predicatively structured thoughts—it would be to misunderstand completely the nature of language. Beings that are capable of concepts and intentionality, which is to say linguistic beings, are not simply able to relate to a complex array of objects and conditions; they are able to represent these objects and conditions as such, which is to say that they are able to relate to them as explicit objects of contemplation. Wittgenstein makes a good point when he observes that the dog cannot believe that his companion human will return the day after tomorrow. But he makes a mistake when he supposes that the dog *is* capable of believing "that his master is at the door" right now (Wittgenstein 1953:174). The dog may well have a mental state that is *equivalent* to the state of belief in a human being, but the dog's mental state, I believe, is not an intentional state with a propositional attitude and conceptual content.

The Soviet psychologist Lev Vygotsky proposed that human children are incapable of conceptual abstraction, that they relate to objects and situations by means of what Vygotsky called pseudoconcepts. These are mental tools that lack the logical ordering and unity of genuine concepts but nonetheless allow the being employing them to make complex associations between present stimuli and past experiences or future possibilities. The pseudoconcept is "only an associative complex limited to a certain kind of perceptual bond" based on a "concrete, visible likeness" (Vygotsky 1986 [1934]:119).[7] For example, when an individual is presented with a triangle and asked to pick out all the triangles in an array of objects, the selection process can be based on a concept or on an association made between the triangles in the array and the image of the triangle with which the individual was first presented. On Vygotsky's view, a human child who has not yet developed the capacity to grasp the logical relations involved in a conceptual grasp of triangularity must employ a pseudoconcept, the functional equivalent of a genuine concept, in selecting the triangles. This fact is obscured by the ability of adults and children to communicate with one another and by the fact that the adult and the child may make the same selection. Vygotsky argues that "the functional equivalence between complex and concept" has "led to the false assumption that all forms of adult intellectual activity are already present in embryo in the child's thinking and that no drastic change occurs at the age of puberty" (p. 121). It is at the age of puberty, on Vygotsky's view, that humans first become capable of genuine conceptual abstraction (p. 98).

Even if Vygotsky's claim that human children are incapable of conceptual abstraction has become implausible, his notion of the pseudoconcept and his claim that genuine abstraction is inseparable from linguistic ability have important implications for the understanding of *animal* behavior. Davidson is right to suggest that there is a dramatic difference between linguistic and nonlinguistic beings, and I see no conflict between this suggestion and the Darwinian view that most if not all differences between human and nonhuman animals are differences in degree rather than differences in kind. I think what makes us hesitant to acknowledge basic differences between human and nonhuman animal cognition is the fear that we will fall into the old trap of supposing that we are somehow not animals and that by acknowledging basic differences between human and animal cognition we will condemn animals to the kind of inferior moral status that since antiquity has made us feel entitled to exploit animals to gratify our own desires. But, once we recognize that differences in cognitive ability have no moral significance whatsoever, we should be able to overcome this fear and begin to acknowledge that animals don't think the way we do and that their ways of relating to the world, while rich and complex, are not conceptual or predicatively structured. In doing so, we will stop trying to recreate animals in our own image and begin to let animal beings be the beings they truly are.

Notes

1. I provide a more detailed discussion of Aristotle and the Stoics in Steiner 2005, chapter 3; I discuss Saint Augustine in chapter 5, Descartes in chapter 6, and Kant in chapter 7.
2. For an extended discussion of Plutarch and Porphyry, see Steiner 2005, chapter 4.
3. I discuss conceptual ability and intentionality in greater depth in Steiner 2008, chapters 1 and 2. There I explain in greater detail why I believe the structure of intentionality must be understood to be specifically predicative.
4. In chapter 3 of Steiner 2008, I present an account of animal cognition based on associations rather than on concepts as a way of explaining the relationship that some animals appear to have to the future and the past.
5. On the Stoic conception of human perception in terms of propositional content, or *lekta*, see Steiner 2005:77ff. On the notion of the "as such," see Steiner 2008:76f.
6. See also Steiner 2008:20ff.
7. See also Steiner 2005:29–36.

References

Allen, C., and M. Hauser. 1991. "Concept Attribution in Nonhuman Animals: Theoretical and Methodological Problems in Ascribing Complex Mental Processes." *Philosophy of Science* 58:221–40.

Candland, D. K. 1993. *Feral Children and Clever Animals: Reflections on Human Nature.* New York: Oxford University Press.

Davidson, D. 1985. "Rational Animals." In E. Lepore and B. McLaughlin, eds., *Actions and Events: Perspectives on the Philosophy of Donald Davidson,* pp. 473–480. Oxford: Basil Blackwell.

Griffin, D. R. 1992. *Animal Minds.* Chicago: University of Chicago Press.

Nagel, T. 1974. "What Is It Like to Be a Bat?" *Philosophical Review* 82:435–456.

Pepperberg, I. M. 2005. "Number Comprehension by a Grey Parrot (*Psittacus erithacus*), Including a Zero-like Concept." *Journal of Comparative Psychology* 119:197–209.

Plutarch. 1995 [AD 70]. *Moralia.* Vol. 12. Cambridge: Harvard University Press.

Sextus Empiricus. 1996 [ca. AD 190]. *The Skeptic Way: Sextus Empiricus's "Outlines of Pyrrhonism."* New York: Oxford University Press.

Steiner, G. 2005. *Anthropocentrism and Its Discontents: The Moral Status of Animals in the History of Western Philosophy*. Pittsburgh: University of Pittsburgh Press.

——. 2008. *Animals and the Moral Community: Mental Life, Moral Status, and Kinship*. New York: Columbia University Press.

Vygotsky, L. 1986 [1934]. *Thought and Language.* Cambridge: MIT Press.

Wittgenstein, L. 1953. *Philosophical Investigations.* Trans. G. E. M. Anscombe. New York: Macmillan.

15

Assessing Evidence for Animal Consciousness

The Question of Episodic Memory

PAULA DROEGE

A squirrel bustles down the tree into the pachysandra to retrieve an acorn and then scurries back up to sit atop a knot in the bark while it shaves the shell and eats the nut meat. What is it like to be a squirrel? Is there any way to tell unless one is that very squirrel? Many people believe that the essentially private nature of consciousness closes off the possibility that science as an objective, third-person form of investigation could tell us anything about the subjective, first-person experience of an animal. This view is compelling. We are all familiar with the frustration of trying to describe our experiences to someone who has not shared those experiences. (What was Cairo like? There is no adequate answer.) Isn't the difficulty magnified to incomprehension when trying to understand the experience of an animal that does not even have the ability to express itself in language? Equally compelling is the opposite intuition in which we can indeed tell that the squirrel is conscious. There is no doubt in my mind that squirrels, dogs, bats, and babies are conscious, although I am less sure about fish, octopus, ants, and other creatures. How can anything as cute and lively as a squirrel not be conscious? Curiously, the same people who are adamant that consciousness is essentially subjective and inaccessible to third-person explanation are the most vociferous in their defense of animal consciousness.[1]

Why think that animal consciousness is indisputable, yet deny that it is explainable? A brief foray into a current controversy about the memory capacity of scrub jays suggests the limits of behavioral evidence for animal consciousness. While various forms of behavior serve as indicators of consciousness, more specific content is needed to be completely convinced that animals are conscious. Without the specifics, we fail to have a sense of what it is like for the animal to be conscious and consequently fail to have explained its consciousness. This failure then leads us to doubt the truth of the behavioral indicators, without yet being strong enough to overturn their intuitively persuasive evidence. Thus we are left with the tension between an unwavering conviction about animal consciousness and an inability to explain it. In this chapter I argue that a resolution of this tension can be found in representational theories of the mind. More detail about exactly how an animal represents its world—what the acorn means to the squirrel, how the scrub jay keeps track of its food cache—will give us a better sense of what it is like to be that animal. Where details are insufficient, as they are in the case of scrub jay episodic memory, we know what sort of evidence we need to gather. Though the description will never be adequate, even as a description of Cairo can never fully capture the richness of the city, representational detail is the bridge over the explanatory gap between animal and human consciousness.

What Is It Like?

Before we form this bridge, let's spend a minute examining the gap it must cross. Try to imagine what it is like to be a squirrel. You are very small and furry with a rapid metabolism that drives you to move quickly, chase your friends up and down the tree, and eat as many nuts, berries, and other edibles as possible. You have amazing balance and have no fear of leaping off limbs to land, clinging with your tiny nails, to the top of the swinging bird feeder. On the one hand, you probably can imagine what this is like. The Disney character Scratch is funny precisely because the cartoonists so vividly caricature the motivation and behavior of squirrels familiar to anyone who has ever watched them dash about. On the other hand, your imagination is limited by your own experience. You can only imagine a speedier version of your own thoughts and sensations, a kinesthetic ability more proficient, a hunger more difficult to sate. You are certain that squirrels must be conscious, because you can project yourself into their inner lives. Yet the very act of projection entails that you gain no information unique to the experience of the squirrel; you can know only what it would be like for *you* to be a squirrel.

This is the paradox of the privacy of experience. The resonance of great literature and art is often generated by the expression of intimate feelings and emotions, the public articulation of the most private states. Nonetheless, there remains the sense that no third-person account could convey the individual quality of experience. Only what is shared in common between my experience and the experience of artist or squirrel can be described. Because I am not a squirrel, I can only say what squirrel experience is like to the extent that it is like my experience in some respect. One reason for the difficulty in imagining the experience of an octopus or ant is that so little is shared in common between the life of the human and the life of these creatures.

Some interpret this structural limitation on our imaginative ability as an indication that an aspect of experience—when private or subjective—is inexplicable. The basic argument goes like this: only shared experiences can be described; some experiences (or aspects of experiences) are not shared; therefore some experiences cannot be described. The missing premise to constitute the explanatory gap is that only experiences that can be described can be explained. This is the premise I intend to challenge, and by overturning it I believe we can gain explanatory purchase on behavioral evidence for animal consciousness.

On one reading the premise is tautological. Explanation essentially involves description, and so anything to be explained must be described in some way or other. But this broad sense of description cannot be the sense used in the argument just presented because then it would not be true that only shared experiences can be described. Neuropsychologists examining fMRI data regularly describe experiences they may not have shared. So do primatologists, ornithologists, and entomologists, even though the animals they study have experiences that are unlike the researcher's experience in various ways. For the argument to be true and not equivocal, the sense of description at issue must be first-person description; to be explanatory, a description of experience must be from the perspective of the creature that has had the experience. Thus the explanatory gap to be bridged is this: humans offer explanations of animal consciousness, yet cannot describe animal consciousness from the animal's point of view. The question is whether animal experiences can be explained even if they cannot be described in the first-person sense (or the first-animal sense).[2]

My suggestion is that sufficient detail about the representational content of animal consciousness can bridge this gap to provide an adequate reason for attributing animal consciousness or, as in the case at issue, for not attributing consciousness to an animal.

In Search of the Cache

A spirited debate among animal researchers concerns whether any animals have the capacity for episodic memory, where *conscious re*experiencing of a past event is taken as a necessary component. The creature at the center of this debate is the western scrub jay, a furtive sort of bird that characteristically hides remnants of food in various locations, called caches, and then retrieves the food at a later point. This skill requires that the scrub jay be able to keep track of where the food is hidden, what kind of food is hidden, and, critically, *when* it was hidden in order to avoid degraded food. For example, if peanuts were hidden on Tuesday and worms hidden on Wednesday, the hungry scrub jay would go looking for worms on Friday because it likes worms better than peanuts and the worms will still be edible. By Sunday, however, the scrub jay will not bother with the worms but go straight to the peanuts, knowing that worms will have degraded by then.

What is it like for the scrub jay to remember its cache? The scrub jay cannot describe its experience from the first-person perspective, so we must see whether a third-person description can provide an adequate explanation of scrub jay experience. It is obvious that scrub jays do not keep track of their food caches by recording the days of the week on a calendar. There must be some other way that scrub jays represent the timing of the caches in order to retrieve the best food. How do scrub jays represent time, and does this tell us anything about what it is like for the scrub jays to remember?

One point at issue in the debate about scrub jay memory is the definition of "episodic memory." Unlike semantic memory, which involves the retention of knowledge, or procedural memory, which involves the retention of a skill, episodic memory involves the retention of a conscious experience. Memory researcher Endel Tulving (2005) takes the ability to consciously reexperience a past event to be a necessary condition of episodic memory. *Autonoetic consciousness* is the retrieval of experience from the past such that an individual now consciously remembers an event that happened at an earlier time.[3]

Thus the challenge: what sort of behavioral evidence could count as evidence for autonoetic consciousness, given the privacy of experience? Humans issue linguistic, first-person reports. Criteria for assessing episodic memory in children include the ability to use past-tense verb forms, to elaborate on the past experience with respect to place and time, and to detail sensory aspects of the experience, particularly visuospatial perspective (Clayton and Russell 2009; Nelson 2005; Wheeler, Stuss, and Tulving 1997). The name *episodic memory* refers to the distinctive remembrance of a particular episode in the person's life. Reports that include perceptual detail mark the event as the memory of a specific experience rather than the report of an event that could be made by anyone. My memory of the day I was born does not count as an episodic memory, even though the event is a particular episode in my life, because I cannot report any details of the event that count as evidence for autonoetic consciousness. The memory I have of the event is derived from knowledge imparted to me by others in the form of stories and artifacts like my birth certificate and photographs.

Since scrub jays cannot produce linguistic reports to provide evidence of recalled first-person experience, researchers have looked for other sorts of evidence that indicate memory for specific events of food caching, focusing on Tulving's original (1972) *www criterion* for episodic memory: behavioral evidence should identify the event in terms of *when* it occurred, *what* happened, and *where* it was located. An operative principle in assessing evidence for animal temporal representation is to ensure that behavior cannot be explained by an ability to keep track of *how long ago* an event happened rather than the ability to remember specific events of caching. According to William Roberts (2002), scrub jay caching behavior can be explained in two ways that do not involve memory for specific past events. First, cache behavior can be explained by some form of interval timing in terms of the number of ecological cycles elapsed, such as a diurnal cycle, lunar phases, or seasonal change. The memory of when an item was cached is represented in relation to the present, not as a representation of the past experience of caching. If scrub jays are able to associate the rate of food degradation at a cache site with a diurnal cycle, they can determine that after four days the worms have degraded but the peanuts have not. This ability does not require a conscious memory of the past event, since the past merely figures in the calculation of the present situation. Scrub jays simply represent how the world is *now* with respect to food freshness; how the world was in the past is a means to this information, yet it need not be explicitly represented in order to serve this function.

Is there behavioral evidence that might show a specific past time is utilized rather than an interval? Since accurate interval timing has only been demonstrated for short intervals of seconds or minutes, and in sync with the time of day, Clayton and colleagues (2002) timed cache and recovery intervals to be at four, twenty-eight, and one–hundred-hour gaps to prevent the reliance on internal or external biological cycles as a temporal cue. This is not to say that no form of interval timing is used to code elapsed time. Some kind of timing mechanism is needed to form an explicit temporal representation of the specific past event. At issue is the content of the representation—whether a specific past event is represented or general relations between events. Because Clayton's birds were able to differentially recover cached food at a variety of intervals, it is unlikely that interval timing alone can account for their performance.

In another experiment, de Kort, Dickinson, and Clayton (2005) sought to demonstrate representation of a particular past event rather than general knowledge of the world. Test conditions were designed to permit caching of particular sorts of food at different times in trays uniquely marked by Lego Duplo structures. The jays were allowed to retrieve their caches after a short interval on some trials and after a longer interval on other trials. In order to choose the best food, the birds had to remember what food was hidden in which particular tray. They also had to remember *when* the food was hidden so as to determine whether it was likely degraded or whether it was still fresh. What-where-when information about a specific episode of caching is necessary to retrieving the best cache.

Roberts's (2002) second explanation of scrub jay caching behavior that does not depend on memory for a specific event involves the strength of a memory trace.[4] Scrub jays may have learned to associate a strong memory trace with the general event of caching fresh worms and a weak memory trace with caching degraded worms. In this case the jays will only retrieve worms when the memory trace is sufficiently strong. Here again the scrub jays can rely on information about the present state of the world—strong or weak memory trace—to determine which cache has the best food. Information about the past is utilized generally but not specifically.

What about this alternative? Do scrub jays determine when food was cached by the strength of a memory trace? Apparently not, as studies have shown that scrub jays differentially retrieve food even when it is cached in the same tray, eliminating any cues for memory trace. Both worms and peanuts were cached in one tray at an earlier time, and a second batch was cached in a different tray at a later time. When the jays were allowed to recover their cache from both trays, they selectively retrieved the peanuts from the first tray and the worms from the second. The same memory trace strength applied to both forms of food, yet the jays differentially recovered worms and peanuts depending on the specific time and place of caching (Clayton et al. 2002). Again this demonstrates the what-where-when condition: the birds were able to keep track of what food was cached in which tray and when the food was hidden.

Still there may be some reluctance to ascribe autonoetic consciousness to the scrub jays on the basis of this evidence. Why think that behavioral evidence for the www criterion is truly an indicator that the jays are *reexperiencing* the event of caching in order to recover the appropriate form of food? After introducing the www criterion, Tulving himself had qualms about its sufficiency as evidence for autonoetic consciousness. In later work Tulving (2005) has focused on another feature of memory as an indicator of autonoetic consciousness. "Mental time travel" is the ability to consciously anticipate the future and to consciously relive the past. The hypothesis of a general purpose capacity for mental temporal projection is gaining support among both memory and animal cognition researchers (see reviews in Schacter, Addis, and Buckner 2008; Suddendorf and Corballis 2007). If one can imaginatively detach from current experiences to construct an image of a past event, the same sort of process should be available for constructing an image of a future event. The similarity in the two forms of temporal representation is even more compelling in light of the way memories are manufactured. Past experiences are not simply stored and retrieved intact in the way one might place a photograph in an album for later viewing. Instead the mind composes a memory based on retrieval cues so that the resulting experience is more or less like the event that took place. Certain elements may be perfectly accurate, while others are confabulated for various reasons. Likewise, in imagining a future event, the mind will collect a number of relevant features from past experiences that can more or less effectively anticipate the situation to come.

Relying on the similarity between prospective and retrospective imagination, Tulving (2005) proposes a test of the ability to "mentally time travel" into the future as a behavioral indicator of autonoetic consciousness. He calls it the "spoon test" and uses the following story to illustrate the required ability: "In an Estonian children's story with a moral, a young girl dreams about going to a friend's birthday party where the guests are served delicious chocolate pudding, her favorite. Alas, all she can do is watch other children eat it, because everybody has to have her own spoon, and she did not bring one. So the next evening, determined not to have the same disappointing experience again, she goes to bed clutching a spoon in her hand" (p. 44). The spoon test exhibits autonoetic consciousness, according to Tulving, because the girl's behavior shows she anticipates an event at a different place and time. By taking the spoon with her to bed, she will be equipped at the party later to eat some chocolate pudding.

The story is lovely and definitely shows something about temporal reasoning. However, two considerations suggest that the spoon test is still not adequate as a behavioral indicator for autonoetic consciousness. First, the test seems too strong. A key component in Tulving's description of the test is the ability to consider the future abstractly as unconnected to current needs and desires. This requirement is designed to eliminate the possibility that associative learning accounts for the behavior rather than an imaginative anticipation of a future event. While some such condition is needed to distinguish the case from simple association, it is not clear that Tulving's condition applies even in the exemplary story. Isn't the girl's current desire for pudding the motivation for her to carry the spoon? Similarly, the current hunger of an animal might well motivate an action that depends on the anticipation of a future situation. Consider an animal like yourself, for example. You are hungry for lunch but don't know what to eat. So you imagine various options—a sandwich, pizza, soup—to help you decide. The problem with this scenario from a methodological perspective is that there is no way to test when an animal utilizes imagination to plan what to eat rather than some other decision-making procedure.

This is also a problem for the spoon test and is the second reason to reject it as a test for autonoetic consciousness: it is too weak. Consciousness is not necessarily required for abstract thought, either in making inferences about future events or in any other form of inferential reasoning.[5] More to the point, the special form of autonoetic consciousness need not accompany thoughts about the future. In the spoon story the little girl must reason from her past dream experience to a future event, but she need not *consciously imagine* a future dream event to decide to carry the spoon to bed. Consequently, the spoon test is insufficient to indicate autonoetic consciousness.

Tellingly, scrub jays pass the spoon test by anticipating the possible future theft of their cache (Emery and Clayton 2001). When scrub jays have had the experience of stealing another bird's cache, they will recache their own food whenever another bird has observed the cache. Scrub jays that have never stolen food do not exhibit this behavior, even when they have watched other birds cache and when other birds have watched them. This case demonstrates quite sophisticated mental abilities. Not only do the birds reason from past experience as a thief to a possible future case of thievery, they also are able to reason from their own behavior to the possible future behavior of another bird. Very smart—but conscious? Without a clearer idea about what it is like for the scrub jay to think about the threat of theft, even the spoon test cannot provide a definitive answer.

All of which leaves us with the original question: what is it like for the scrub jay to remember its cache? Does it represent a specific event, and, more to the point, is the representation of a specific past event (or future event) sufficient to entail a conscious reexperiencing of that event? We just don't know. Which does not mean that we can't know, that the experience of scrub jays and other nonlinguistic animals is forever beyond our ability to explain. Despite the current impasse, the scrub jay debate shows important progress toward identifying the essential features of research design and interpretation.

First, researchers have identified the particular form of content required for episodic memory: a specific past event. Crucial to isolating this form of temporal content is the distinction between a representation of the event *as* past and representations that simply use knowledge of the past event in relation to present action. Scrub jay behavior clearly indicates *use* of information about a specific past event. On the basis of their knowledge about when various foods were hidden, these birds can accurately calculate the degradation of food in each cache. But evidence is insufficient to claim that scrub jays represent the event *as* an experience of the past.

Second, an important development in memory research is the recognition that representation of the past is related to representation of the future. Given that the mind is a representational system, representational abilities need to be understood in terms of how they relate to one another. In speculating about the adaptive value of memory for the past, Tulving (2005) suggests subjective awareness of future time is critical. "For anyone to take steps at one point in time that would make the unpredictable, frequently inhospitable natural environment more predictable at a future time, it is necessary to be able to be consciously aware of the existence of a future" (40). This thinking is the motivation for Tulving's spoon test and marks an important change in memory theory. The value of knowledge about the past is primarily its use in planning for the future. After all, the future can be changed while the past cannot.

Third, an understanding of the mind of a scrub jay involves entering the bird's world. How the scrub jay represents time depends on how it uses time. Apparently scrub jays are unique in their ability to discriminate specific past events according to the three criterial features: *when* an event occurred, *what* happened, and *where* it was located. Experimental results using the scrub jay research paradigm have not been reproduced in any other species (Hampton and Schwartz 2004). One explanation of this failure is the novelty of scrub jay cache behavior. While many other species hoard food, scrub jays cache in many different locations and conceal their food. Moreover, as de Kort, Dickinson, and Clayton (2005) note, "fast, flexible learning may be essential to the survival of corvids that cache perishable foods in an environment where the rate at which foods decay changes across the year, and from day to day, depending on the weather conditions between caching and recovery" (p. 172). The selective demands on the scrub jay biologically motivate the development of its ability to represent the time, location, and type of food it caches.

The next step in telling the story of scrub jay representation is to fit its abilities into a larger theory of the mind and consciousness. We now have a better idea about how scrub jays represent time; what we need is a way of reading this form of temporal representation in terms of what it might be like to be a scrub jay.

What It Is Like to Represent Time

The phenomenologist Edmund Husserl (1990 [1928]) offered a fundamental insight into conscious experience when he noticed that all experience is *about* something: Consciousness is essentially representational. Furthermore, the representation of *time* is always an aspect of conscious experience. Your current experience of reading this sentence represents the sentence as now being read, where the frame of "now" includes a brief temporal span from past into future. If your experience were limited to the momentary contents of a single saccade, for example, you would understand nothing of the sentence. Instead your experience of the present moment incorporates a sense of the earlier words as well as the anticipation of how the sentence could continue.

Following Husserl's insight, we can say that what it is like to be me right now is the sum total of the contents of this represented present moment. One of the reasons it is impossible to describe what it is like to be me right now is that there are just not enough words (or time) to capture the detail involved in a single moment of consciousness, especially when past associations and future anticipations are included in the representation of this moment. Still, the intuitive idea is that my consciousness is constituted by representations of the blue and the itch and the tap, tap, tap and the spinning of thought and . . . all that. *That* is what it is like for me right now.

To get a sense of what it is like for the scrub jay, then, is to learn what it represents, and more importantly, to learn whether it represents anything as "now." It is sometimes said that animals are "stuck in the present" to imply that they cannot represent the past or future explicitly (Roberts 2002). But, if this is the case, then animals are not really stuck in the present—they have no sense of time at all. Events simply occur. It makes no sense to say that events occur "now" without the ability to distinguish the present from either past or future. Evidence suggests that scrub jays are not in this situation; they do have a sense of time. In order to determine which cache to retrieve, scrub jays represent specific past events: what, where, and when the food was hidden. Scrub jays also exhibit evidence that they represent the future in their ability to anticipate potential theft and act to prevent that possible event.

On my view, the ability to represent the present moment, as distinct from the past and future, is sufficient for consciousness (Droege 2003, 2009). The selection and combination of sensory representations into a multimodal representation of the present moment is the elusive sine qua non of consciousness.[6] Following Husserl, I argue that temporal representation is the distinguishing mark of conscious as opposed to nonconscious states. Because scrub jays are able to distinguish past, present, and future events, scrub jays have conscious states.

But the crucial question regarding episodic memory is not whether scrub jays are conscious, it is whether they *reexperience* the past event when they represent it. Here again we should think about what distinctive sort of representational content is required for episodic memory on this description. Not only must a specific past event be represented, but the experience of the past event must be represented *as* occurring again in the present. While there is no need to reproduce an exact replica of the past experience, an essential phenomenological feature of episodic memory is the re-presenting of the past experience in the present. Consequently, the representational structure of episodic memory is higher order: a representation (present experience) of a representation (past experience).

To get a better sense of the phenomenological dimension of this structure, call up an episodic memory of your own. Try to find a vivid event, such as a favorite birthday or an exceptional vacation. You should find, in addition to your sensory representations of the world around you right now, visual and perhaps auditory representations of the past event (gustatory and olfactory representations are curiously rare in episodic memory, although these sensations often trigger an episodic memory). The sensory elements of the remembered event are represented as occurring now—this temporal currency is what makes them conscious—yet also represented as elements of an event in the past. The higher-order representational structure of memory eliminates any temporal contradiction by embedding the representation of the past within the representation of the present. What it is like to have an episodic memory, then, is to experience *right now* that I am experiencing a past event (Droege 2007).

In addition to evidence for consciousness, evidence for episodic memory includes the requirement that scrub jays have the capacity for higher-order representation. The development of higher-order thought seems to usher in a suite of abilities: a comprehension of deception, the capacity for linguistic self-reference, and an autobiographical framework for time. Investigations into scrub jay thievery and attendant preventative measures suggest that the birds have some comprehension of deception. The scrub jay who recaches food that was seen by another demonstrates an appreciation of vision as the source of representations about the world and a recognition that representations can be mistaken. After moving the food, the scrub jay has thwarted the potential thief by rendering the representations of the would-be thief false.

Of course scrub jays show no evidence for linguistic self-reference, since they are not linguistic creatures. The inability to issue first-person reports of experience is the reason we were forced to look for alternative behavioral criteria for episodic memory. However, in the absence of any reason to think language is essential to episodic memory, the general lack of linguistic ability should not count against the possibility that scrub jays are capable of reexperiencing the past. On the other hand, there *is* reason to think that an autobiographical framework of time may be essential to episodic memory. To see why, several pieces of the puzzle of episodic memory need to be assembled together. We are looking for evidence of autonoetic consciousness, reexperiencing of the past event, which has a higher-order representational structure. The representation of the present moment includes a representation of the past event *as* a past event. It may be that an autobiographical framework is necessary for the representation of a past event *as past*.

One of the reasons Roberts (2002) is skeptical that animals are capable of mental time travel is his conviction that the representation of specific past and future events requires their placement relative to other events along a sketchy yet definitive personal timeline. As he notes, episodic memory "implies a temporal structure for past time within which events can be located" (p. 486). This structure need not be elaborated in linear, temporally sequenced units; episodic memory is a rough and reconstructive affair. One's personal history tends to be structured by a selection of salient events that are used as markers to infer the temporal relations among other associated events. When the smell of the mock-orange bush evokes images of my childhood home, the memory may not come labeled with a specific date or even be identifiable as a single event of smelling the blossoms rather than an amalgam of events. Nonetheless, the mock-orange bush is locatable in my autobiographical timeline in relation to other clearly specifiable events: after the move to Indiana, before graduation from high school, etc. On this analysis, we would need evidence that the scrub jay organized its representations according to an autobiographical temporal framework to determine whether it has the higher-order temporal representations involved in episodic memory.

It is unclear how to design a research program that could gather this sort of evidence. Roberts wonders whether animals could be taught to use the calendars and clocks that facilitate an autobiographical sense of time in humans. This is an interesting question, which raises an even more interesting question. If animals are taught an autobiographical sense of time, will this new conceptual ability then generate episodic memories, complete with the requisite phenomenological *reexperiencing* of the past? There seems to be no reason, in principle, why not.

Notes

1. Thomas Nagel (1974), who famously argued that we cannot in principle know what it is like to be a bat, also maintained that it is simply obvious that they are conscious. Colin McGinn (1999) also insists on the certainty of consciousness in other human and animal minds, despite its essential privacy and consequent inexplicability.
2. To assume that an explanation of consciousness can only be described from the first-person perspective is a version of what William Lycan (1996:47) calls

the stereoscopic fallacy, since it relies on the difference between having an experience (first-person description) and watching a brain having an experience (third-person description).

3. The term *autonoetic consciousness* was introduced by Tulving to refer specifically to the type of consciousness that is characteristic of episodic memory. In addition to autonoetic consciousness, Tulving lists three other features as central to episodic memory: the function of remembering, the dependence on a remembering self, and a sense of subjective time (Tulving 2005:14).
4. A memory trace, or engram, is the neural change underlying associative learning.
5. Consider, for example, the inferential abilities of chess-playing computers or the more familiar situation where a vexing problem is solved only when one stops consciously thinking about it.
6. Conceptual representations are also usually included in a representation of the present moment. Since the capacity for animals to possess concepts is another matter of debate, I mention only sensory representations to avoid multiplying controversies. For similar reasons, I focus on sensory consciousness in *Caging the Beast* (Droege 2003).

References

Clayton, N. S., D. P. Griffiths, N. J. Emery, and A. Dickinson. 2002. "Elements of Episodic-like Memory in Animals." In A. Baddeley, M. Conway, and J. Aggleton, eds., *Episodic Memory: New Directions in Research*, pp. 232–249. Oxford: Oxford University Press.

Clayton, N., and J. Russell. 2009. "Looking for Episodic Memory in Animals and Young Children: Prospects for a New Minimalism." *Neuropsychologia* 47:2330–2340.

De Kort, S. R., A. Dickinson, and N. S. Clayton. 2005."Retrospective Cognition by Food-Caching Western Scrub-Jays." *Learning and Motivation* 36:159–176.

Droege, P. 2003. *Caging the Beast: A Theory of Sensory Consciousness*. Amsterdam: Benjamins.

——. 2007. "Memory and Consciousness." Paper presented at the Association for the Scientific Study of Consciousness. Retrievable at http://theassc.org/documents/memory_and_consciousness.

——. 2009. "Now or Never: How Consciousness Represents Time." *Consciousness and Cognition* 18:78–90.

Emery, N. J., and N. S. Clayton. 2001. "Effects of Experience and Social Context on Prospective Caching Strategies." *Nature* 414:443–446.

Hampton, R. R., and B. L. Schwartz. 2004. "Episodic Memory in Nonhumans: What, and Where, Is When?" *Current Opinion in Neurobiology* 14:192–197.

Husserl, E. 1990 [1928]. *On the Phenomenology of the Consciousness of Internal Time (1893–1917).* Trans. J. B. Brough. Dordrecht: Kluwer.

Lycan, W. 1996. *Consciousness and Experience.* Cambridge: MIT Press.

McGinn, C. 1999. *The Mysterious Flame: Conscious Minds in a Material World.* New York: Basic Books.

Nagel, T. 1974. "What Is It Like to Be a Bat?" *Philosophical Review* 82:435–456.

Nelson, K. 2005. "Emerging Levels of Consciousness in Early Human Development." In H. S. Terrace and J. Metcalf, eds., *The Missing Link in Cognition,* pp. 116–141. Oxford: Oxford University Press.

Roberts, W. 2002. "Are Animals Stuck in Time?" *Psychological Bulletin* 128:473–489.

Schacter, D. L., D. R. Addis, and R. L.Buckner. 2008. "Episodic Simulation of Future Events: Concepts, Data and Applications." *The Year in Cognitive Neuroscience 2008: Annals of the New York Academy of Sciences* 1124:39–60.

Suddendorf, T., and M. C. Corballis. 2007. "The Evolution of Foresight: What Is Mental Time Travel, and Is It Unique to Humans?" *Behavioral and Brain Sciences* 30:299–351.

Tulving, E. 1972. "Episodic and Semantic Memory." In E. Tulving and W. Donaldson, eds., *Organization of Memory,* pp. 381–403. New York: Academic.

——. 2005. "Episodic Memory and Autonoesis: Uniquely Human?" In H. S. Terrace and J. Metcalf, eds., *The Missing Link in Cognition,* pp. 3–56. Oxford: Oxford University Press.

Wheeler, M. A., D. T. Stuss, and E. Tulving. 1997. "Toward a Theory of Episodic Memory: The Frontal Lobes and Autonoetic Consciousness." *Psychological Bulletin* 121, no. 3: 331–354.

16

What Are Animals Conscious Of?

ALAIN MORIN

There is little doubt that animals are conscious. Animals hunt prey, escape predators, explore new environments, eat, mate, learn, feel, and so forth. If one defines consciousness as being aware of external events and experiencing mental states such as sensations and emotions (Natsoulas 1978), then gorillas, dogs, bears, horses, pigs, pheasants, cats, rabbits, snakes, magpies, wolves, elephants, and lions, to name a few creatures, clearly qualify. The contentious issue is, do these animals *know* that they are perceiving an external environment and experiencing internal events? Are animals *self*-conscious?

Recent attempts at understanding animal consciousness (e.g., Edelman and Seth 2009) agree that nonhuman animals most probably possess "primary" (or "minimal") consciousness. But these views also argue that, unlike humans, animals lack many (but not all) elements that make up higher-order consciousness—the capacity to reflect on the contents of primary consciousness. In this chapter I will aim at offering a more elaborate picture of this position. I will present detailed information on what is meant by "higher-order consciousness"—i.e., self-awareness. I will suggest that some dimensions of self-awareness (e.g., self-recognition, metacognition, mental time travel) may be observed in several animals, but that numerous additional aspects (e.g., self-rumination, emotion awareness) seem to be absent. Some other self-related processes, such as Theory-of-Mind, have been identified in animals, but not as the full-fledged versions found in humans. I will postulate that these differences in levels of self-awareness between humans and animals may be attributable to one distinctive feature of human experience: the ability to engage in inner speech.

Definitions, Measures, and Effects of Self-Awareness

Mead (1934) established a classic distinction between focusing attention outward toward the environment (consciousness), and inward toward the self (self-awareness). This framework was recaptured and expanded by Duval and Wicklund (1972). It became very popular in experimental social and personality psychology, where it has been guiding empirical research for more than four decades (see Carver 2003 for a review).

Unconsciousness refers to the absence of processing of information, either from the environment or the self. As previously stated, consciousness constitutes the processing of environmental information or responding to external stimuli. Cabanac, Cabanac, and Parent (2009) suggest that this kind of consciousness (the nonreflective type) could very well be present in reptiles, including tortoises, turtles, lizards, snakes and crocodiles, and in birds and mammals. These animals possess comparable brain volume (as measured by the ratio of brain to body mass), structure, and neurochemistry and, like conscious humans, exhibit emotions, feel pleasure, play, and dream.

Self-awareness is usually defined as becoming the object of one's own attention. It represents a state in which one actively identifies, processes, and stores information about the self (Morin 2004). Self-awareness constitutes a complex multidimensional phenomenon that comprises various self-domains (e.g., thinking about one's past and future, emotions, thoughts, personality traits, preferences, intentions; sense of agency) and corollaries (e.g., making inferences about others' mental states, self-description, self-evaluation, self-esteem, self-regulation, self-efficacy, death awareness, self-conscious emotions, self-recognition, self-talk). Self-awareness also entails knowing that one shows some continuity as a person or entity across time and that one distinguishes oneself from the rest of the environment (Kircher and David 2003).

Initial empirical work focused on short-term effects and long-term consequences of being self-aware. Much of this research involved exposing participants to self-focusing stimuli known to remind the person of his or her object status to others—e.g., mirrors, cameras, an audience, recordings of one's voice; such stimuli reliably produce heightened self-awareness (Carver and Scheier 1978). Other manipulations and measures of self-awareness include 1. questionnaires that assess dispositional self-focus (Fenigstein, Scheier, and Buss 1975) or spontaneously occurring fluctuations in self-awareness (Govern and Marsch 2001); 2. first-person singular pronoun use (Davis and Brock 1975); 3. self-novelty manipulation (Silvia and Eichstaedt 2004), where participants are invited to write about ways in which they differ from others; 4. word-recognition measures (Eichstaedt and Silvia 2003), where subjects are asked to identify self-relevant or self-irrelevant words as quickly as possible; and 5. match between self- and other-ratings on cognitive or personality measures to evaluate self-knowledge in healthy people (Hoerold et al. 2008) and self-awareness of deficits in patients with traumatic brain injury (Cocchini et al. 2009). Note that all these measurement techniques are inappropriate for animals because they require language; to my knowledge the most used nonverbal measure of self-awareness is self-recognition, which will be discussed in another section.

Inducing self-awareness with self-focusing stimuli produces self-evaluation (Duval and Wicklund 1972), whereby the person compares any given salient self-aspect to an ideal representation of it. Self-criticism is then likely to occur, leading to an avoidance of the state of self-awareness or a reduction of the intraself discrepancy by either modifying the target self-aspect or by changing the ideal itself. Another effect of self-awareness is emotional intensity—the proposal that focusing on one's emotions or physiological responses amplifies one's subjective experience (Gibbons 1983). To illustrate, empirical evidence suggests that angry self-aware individuals behave more aggressively than nonself-aware participants. Self-awareness also increases accurate access to one's self-concept; for instance, self-reports of self-aware individuals are more accurate. Other effects or consequences of self-awareness are heightened consistency between one's behavior and attitudes, increased self-disclosure in intimate relationships, stronger reaction to social rejection, greater social conformity, and lower antinormative behavior (see Franzoi 1986 for references). While it is ultimately pointless to ponder if animals exposed to self-focusing stimuli would, like humans, engage in self-evaluation, "self-disclose" more, or act more ethically, it is surely intriguing to wonder if they would become more aggressive if angered or if they would react more strongly to social rejection—these are potentially observable events. Remarkably, this has never been done—i.e., trying to replicate some effects and consequences of self-awareness in animals using self-focusing stimuli. Positive results would suggest the presence of the aforementioned forms of self-awareness in tested creatures. Of course this assumes that the presence of a camera or a mirror would successfully induce self-focus in animals.

Past research also shows that self-awareness increases the likelihood of more effective self-regulation (e.g., Carver and Scheier 1981). Self-regulation includes altering one's behavior, resisting temptation, changing one's mood, selecting a response from various options, and filtering irrelevant information (Baumeister and Vohs 2003). Do animals self-regulate? They must be able to monitor their ongoing behavior and compare it to set goals or else they would not survive. But, given the demonstrated importance of speech-for-self in self-regulation, I would suggest that animal self-regulation is much more primal than that of humans. Vygotsky (1962 [1934]) pioneered the view that language can be used as a verbal self-guidance device, and decades of work on private speech use in problem solving, planning, and decision making support this view (Winsler 2009). In short, and depending on the content of speech-for-self, people who rely on self-talk while engaged in self-regulatory activities (as previously defined) perform significantly better than those who do not. Thus nonverbal animal and verbal human self-regulation should probably not be equated.

Although self-awareness clearly represents an evolutionary advantage, it has its setbacks as well (Leary 2004). In humans excessive self-focus creates worry, guilt, shame, jealousy, insomnia, etc., and may contribute to social anxiety (Buss 1980), depression (Pyszczynsky and Greenberg 1987), and even suicide (Baumeister 1991); unhealthy people are also known to self-ruminate (Smith and Alloy 2009). Do animals experience these psychological ailments? To my knowledge wild animals have never been observed worrying and do not seem to experience sleeping difficulties as a result. Note that some lab animals (e.g., white rats) show evidence of anxiety when asked to perform extremely difficult discriminatory tasks (Cook 1939). Maybe they can feel guilty or jealous—it's hard to say. There is no evidence of suicide in animals (Ramsden and Wilson 2010), suggesting that nonhuman creatures may not be able to mentally represent their own death. Do animals possess a more general awareness of death? Here the evidence is mixed. On one hand, we have African elephants known to pick up and scatter the bones of deceased elephants (McComb, Baker, and Moss 2006). Perhaps these elephants are aware of death, but the bone scattering could also be seen as simple survival behavior that hides their migration routes or feeding patterns. On the other hand, the way rabbits react to the death of their companion is puzzling. While some cautiously smell the body of the deceased, others take turns lying nested against the dead rabbit, and all eventually abandon the body permanently. J. A. Smith (2005) suggests that these behaviors indicate an understanding that a partner has undergone a permanent and catastrophic change.

Animals apparently do not blush (Darwin 1872)—a physiological response typically associated with social anxiety. Rats have been shown to experience negative emotional states, or "depression," as suggested by increased sensitivity to the unanticipated loss of food reward (Burman et al. 2008). Humans are more sensitive to reward loss than gain, but depressed individuals tend to be even more responsive to reward loss. Although pet owners may describe their cat or dog as being "depressed," clear scientific data to this effect is still lacking. In addition, such depressive episodes, if they do exist, may have little to do with heightened self-awareness or with the animal being aware of experiencing them.

In humans, self-awareness does not represent a uniform construct and is made up of two different tendencies (Trapnell and Campbell 1999): self-reflection, which constitutes an authentic curiosity about the self, and self-rumination, which represents anxious attention paid to the self. While it is clearly impossible to know if self-aware animals are genuinely curious about their selves, one can state with some confidence that self-rumination in animals is unlikely. Self-rumination consists in recurrent, intrusive, and disruptive thoughts. These thoughts are most probably articulated with inner speech, and, indeed, measures of self-rumination strongly correlate with the Automatic Thoughts Questionnaire (Conway et al. 2000)—in essence an inner speech scale. Animals lack inner speech, but perhaps they could "ruminate" with mental images? Self-ruminators are more likely to suffer from depression, anxiety, stress, and social phobia or withdrawal; they also experience problem-solving and concentration difficulties (Smith and Alloy 2009). Animals do not seem to present any of these problems, at least not as self-induced conditions.

Agency and Mental Time Travel

As indicated previously, self-awareness consists of various dimensions, among which are sense of agency and mental time travel. Most animals probably have a sense of agency based on representations of the relation between their action and the subsequent effects that develop through operant conditioning. This capacity allows organisms to interact with and control their own environment (Engbert, Wohlschläger, and Haggard 2008).

Mental time travel (MTT), also known as autonoetic consciousness, represents the ability to remember personally experienced events that occurred in the past and to imagine personal happenings in the subjectively felt future (Tulving and Kim 2009). It includes a "what, where, and when" (www) of events, as well as the conscious reexperiencing of oneself in the remembered event. The "remembering one's past" portion of MTT has been assessed in various animals; in primates a typical experiment goes as follows. The animal first witnesses a unique event—e.g., seeing a familiar person doing something odd, such as stealing a cell phone. This is followed by a fifteen-minute retention interval, and then the subject is shown three photographs (two distractors and the witnessed event) and is asked to select the correct photograph depicting what he/she had seen before. A good answer is rewarded with food. Chimpanzees and gorillas perform significantly higher than chance on that type of trial (Menzel 2005). Variations of this task have been created to test MTT in other creatures such as scrub jays, magpies, and rats. They also remember the "www" of personal events, suggesting rudiments of an MTT system (Roberts and Feeney 2009). It remains impossible to determine if these animals actually mentally relive the events.

Can animals anticipate and plan for the future? The evidence is very limited and is open to alternate interpretations. Scrub jays can anticipate the need for specific food at breakfast the following day by storing seeds in novel locations where they have not encountered food before, and captive chimpanzees will cache stones they will later hurl at human visitors (Roberts and Feeney 2009). However, the possibility remains that in both cases these animals rely on semantic knowledge (generalized knowledge that does not involve anticipation of a specific event—e.g., human visitors periodically appear) rather than on genuine planning (human visitors will appear tomorrow morning). In short, it is still unclear if animals are "stuck in time" or not.

Private and Public Self-Awareness

One fundamental distinction that has been proposed early on is the difference between private and public self-focus (Fenigstein, Scheier, and Buss 1975). Self-awareness includes a knowledge of one's own mental states (private self-aspects) such as thoughts, emotions, preferences, personality traits, opinions, goals, sensations, attitudes, etc., and visible characteristics (public self-aspects) such as one's body, physical appearance, mannerisms, and behaviors.

Humans routinely focus on private and public self-dimensions. Do animals also focus on both private and public self-aspects? All animals must possess some rudimentary form of body self-awareness (Bekoff and Sherman 2004). They position their body parts in space so that they do not collide with nearby conspecifics and they travel as a coordinated hunting unit or flock. Also, some animals are capable of self-recognition, which requires a mental representation of one's body (Mitchell 2002). Note that one's body can be apprehended both as a private self-aspect (kinesthetic experience) and public self-aspect (the image of one's body seen in a mirror). Do animals reflect on their physical appearance? Unlike humans, they do not wear body adornments such as bracelets and beads to be more attractive (Mitchell 2002). This suggests a different form of self-presentation in humans than in other animals.

Do animals reflect on their unobservable mental states? It is virtually impossible to determine animals' awareness of their own sensations, motives, opinions, or attitudes. Animals, including primates, dogs, cats, and birds, certainly experience emotions. Physiological, behavioral, and cognitive changes that spontaneously accompany affective states are remarkably similar in humans and animals (Paul, Harding, and Mendl 2005). Chimpanzees and orangutans display emotional reactions of pride, shame, and embarrassment (Tracy and Robins 2004). However, an organism may experience emotions, including self-conscious emotions, without being aware of them (Salzen 1998), and, to my knowledge, evidence for emotional awareness in animals is nonexistent. This remark also applies to animals' awareness of preferences (e.g., food). Animals do have preferences, but there is no known way of determining if they know about their preferences. Animals also exhibit individual differences, but awareness of one's personality characteristics in terms of traits entails linguistic representation that nonhuman creatures lack.

Metacognition

What about awareness of thoughts? Metacognition consists in thinking about thinking or cognition about cognition (Nelson and Narens 1994). Examples of metacognition in humans are becoming aware of a thought one just had or a solution to a problem one just discovered. One other case of metacognition is when we feel uncertain about some information we might possess or not—for instance: can I recall this phone number or do I need to look it up in the directory? Uncertainty responses during perceptual tasks (e.g., tone discrimination) or memory tasks (e.g., item recall) are often used in animals as an indicator of metacognition (Smith 2009). In addition to discrimination responses per se, subjects are given the possibility to decline completion of any trials they want. Doing so suggests uncertainty and probably knowledge that information is missing to adequately perform the task. Available evidence indicates that dolphins and monkeys, but not rats, make uncertainty responses in a variety of tasks (J. D. Smith 2005). Hampton (2001) reports studies on "memory awareness" showing that rhesus monkeys, but not pigeons, turn down trials when they are uncertain they will pass a memory test.

Another metacognition test consists in asking subjects to make metaconfidence judgments in which they evaluate up to what point they are confident that a previously made response is correct (Son and Kornell 2005). This requires metacognition because one has to think about one's own knowledge (or lack thereof) when assessing one's confidence in an answer. In a representative experiment, monkeys are asked to identify the longest of nine lines. The "metaconfidence judgment" part of the task consists in subjects making a high bet (for high reinforcement) or a low bet (for lower reinforcement) on the correctness of their previous answer. Making a high bet means that the animal is very confident in the previously given answer, and vice versa. Monkeys are indeed able to make accurate confidence judgments.

Not only do some animals decline a task because of lack of knowledge—some will also seek information when it is incomplete (Call 2005), suggesting that they are aware of not knowing. Subjects in a typical study have to choose one of two containers to obtain a reward. The experimenter places food in one of the two containers. In one situation subjects can see in which container the food is put before choosing. In another situation they do not have direct visual access to that information, but if they bend down and look under the containers they can see where the food is. Chimpanzees, gorillas, and orangutans actively seek additional information in that type of experiment by looking under the containers before choosing; dogs, however, do not.

Various experimental paradigms consequently suggest that some animals could be aware of their thoughts, although alternative nonmentalistic (first-order, as opposed to second-order) explanations are available (Carruthers 2008). Also, animals show functional parallels to human conscious metacognition, but they may not experience everything that can accompany conscious metacognitive experience in humans.

Self-Recognition and Theory-of-Mind

Most organisms that are confronted with a reflective surface react as if they were seeing another conspecific creature by engaging in a variety of social responses such as bobbing, vocalizing, and threatening. Only humans, chimpanzees, orangutans, and bonobos, elephants, dolphins, and magpies have been shown to exhibit spontaneous mirror-guided self-exploration, e.g., self-directed behaviors such as examining body parts only visible in the mirror (see Morin 2011 for a review). The aforementioned animals also pass the more formal "mark test" and will touch a red dot that has been inconspicuously applied to their brow or forehead (or throat feathers in magpies' case). Emitting self-directed responses in front of a mirror and passing the mark test indicate self-recognition. In humans this developmental landmark is achieved between eighteen and twenty-four months of age (Amsterdam 1972).

How does self-recognition relate to self-awareness? According to Gallup (1982), emitting self-directed behaviors in front of a mirror indicates that the organism can take itself as the object of its own attention. In addition, *re*-cognizing oneself in front of a mirror presupposes preexisting "self-cognition" (i.e., self-knowledge, a self-concept) and therefore self-awareness. There is little doubt that self-recognition implies some form of self-awareness; rather, the question should be: *What type*, or what *level*, of self-awareness is involved? Mitchell (2002) and others suggest that self-recognition only requires knowledge of one's body. The organism matches the kinesthetic representation of the body with the image seen in the mirror and infers that "it's me." This interpretation implies that an awareness of one's own thoughts (or any other more private dimensions of the self) is not needed for self-recognition to take place. An awareness of private self-aspects does not seem relevant for self-recognition, whereas an awareness of the body is critical for self-identification in front of a mirror. Note that perhaps self-recognizing creatures do have access to their thoughts (e.g., metacognition), but passing the self-recognition test does not demonstrate this.

Theory-of-Mind (ToM) consists in attributing mental states such as goals, intentions, beliefs, desires, thoughts, and feelings to other social agents (Gallagher and Frith 2003). The social cognitive and evolutionary benefits of ToM are the ability to predict others' behavior and to help, avoid, or deceive others as the situation dictates. Do animals engage in ToM? This represents a highly controversial question, with some claiming that primates, and possibly birds, (e.g., ravens) do (Gallup 1982; Premack and Woodruff 1978), and others denying ToM in animals altogether (e.g., Heyes 1998). A more likely scenario is that primates are capable of some forms of ToM but do not possess the fully developed human version. While early experiments on chimpanzees seemed to imply an ability to understand human goals, much subsequent work increasingly suggested that they do not appreciate human goals or visual perception, as exemplified by Povinelli and Eddy's study (1996) in which chimpanzees begged from humans facing them, but also solicited the attention of other humans who had buckets over their heads. According to Call and Tomasello (2008), more recent evidence instead shows an understanding of goals, intentions, perceptions, and knowledge in others—but not of others' beliefs. In a typical experiment on intention understanding, the animal observes the human experimenter trying to turn on a light with his head because his hands are occupied holding a blanket. The subject reacts to this not by imitating the experimenter's behavior (miming turning on a light with its head) but rather by imitating the *intention* behind the physical constraint—by turning on the light with its hands.

Povinelli and Vonk (2003) remain skeptical and propose that chimpanzees form mental concepts of visible, concrete objects in their environments (e.g., apples, facial expressions, leopards), but not about inherently unobservable things (e.g., God, gravity, love). In ToM experiments, chimpanzees would reason solely about the abstracted statistical regularities that exist among certain events and the behavior, postures, and head movements of others (behavioral abstractions), but not about others' unobservable mental states.

In this chapter I raised the question What are animals conscious of? Table 16.1 summarizes my analysis. I suggest that some animals are conscious of their body (as measured by self-recognition) and of being the agent behind their actions (sense of agency). Animals seem to be unaware of various private and public self aspects (e.g., traits, physical appearance, attitudes), and some might know about their thoughts (metacognition)

TABLE 16.1. What Self-Aspects and Processes Are Animals Aware Of?

Yes	No	Less sophisticated version	Perhaps	Unlikely	Unknown
Body awareness	Suicide	Self-regulation	Jealousy	Rumination	Social anxiety
Sense of agency	Insomnia	Theory of mind	Depression	Ethical judgments	Self-evaluation
Self-recognition	Awareness of		Mental time travel	Worry	Emotional amplification
Guilt	• personality traits		Metacognition	Awareness of	Self-disclosure
Shame	• physical appearance		Awareness of	• attitudes	Sensitivity to social rejection
			• goals	• opinions	Awareness of
			• death		• sensations
					• emotions
					• preferences

as well as their past and future (MTT). Unlike humans, who can experience negative states resulting from excessive self-focus, animals do not appear to worry, ruminate, or self-destruct. However, like humans, animals seem to engage in self-regulation and ToM, but these represent less refined versions. Take note that the last "unknown" column contains quite a few self-related processes, indicating that our knowledge of animal self-awareness is still precarious. This analysis supports the widespread view that self-awareness differences in humans and animals are not radical and come in degrees.

Why is self-awareness less sophisticated in animals? The lack of language in animals is often cited, but I propose more specifically that it is the absence of inner speech in animals that should be credited. Inner speech is known to contribute to the development of self-awareness in humans (e.g., DeSouza, DaSilveira, and Gomes 2008). Self-directed speech allows us to verbally label our internal experiences and characteristics; as a result, these become more salient—more conscious. A significant positive correlation exists between diverse validated scales measuring the frequency of private self-focus and use of inner speech (see Morin 2005 for a review); accidental loss of inner speech following brain injury impedes self-awareness (Morin 2009). In light of this evidence, I would predict that linguistically tutored apes, such as those trained by Savage-Rumbaugh, Fields, and Taglialatela (2000), should exhibit heightened self-awareness, assuming that speech-for-self automatically follows social speech.

Note

I would like to thank Breanne Hamper, Petra Kamstra, and Jack Robertson for their helpful editorial comments on previous versions of this chapter.

References

Amsterdam, B. 1972. "Mirror Self-Image Reactions Before Age Two." *Developmental Psychobiology* 5:297–305.

Baumeister, R. F. 1991. *Escaping the Self.* New York: Basic Books.

Baumeister, R. F., and K. D. Vohs. 2003. "Self-Regulation and the Executive Function of the Self." In M. R. Leary and J. P. Tangney, eds., *Handbook of Self and Identity,* pp. 197–217. New York: Guilford.

Bekoff, M., and P. W. Sherman. 2004. "Reflections on Animal Selves." *Trends in Ecology and Evolution* 19:176–180.

Burman, O. P., R. M. A. Parker, E. S. Paul, and M. Mendl. 2008. "Sensitivity to Reward Loss as an Indicator of Animal Emotion and Welfare." *Biology Letters* 4:330–333.

Buss, A. H. 1980. *Self-Consciousness and Social Anxiety*. San Franscisco: Freeman.

Cabanac, M., A. J. Cabanac, and A. Parent. 2009. "The Emergence of Consciousness in Phylogeny." *Behavioural Brain Research* 198:267–272.

Call, J. 2005. "The Self and Others: A Missing Link in Comparative Social Cognition." In H. S. Terrace and J. Metcalfe, eds., *The Missing Link in Cognition: Origins of Self-Knowing Consciousness,* pp. 321–342. New York: Oxford University Press.

Call, J., and M. Tomasello. 2008. "Does the Chimpanzee Have a Theory of Mind? Thirty Years Later." *Trends in Cognitive Sciences* 12:187–192.

Carruthers, P. 2008. "Meta-cognition in Animals: A Skeptical Look." *Mind and Language* 23:58–89.

Carver, C. S. 2003. "Self-Awareness." In M. R. Leary and J. P. Tangney, eds., *Handbook of Self and Identity,* pp. 179–196. New York: Guilford.

Carver, C. S., and M. F. Scheier. 1978. "Self-Focusing Effects of Dispositional Self-Consciousness, Mirror Presence, and Audience Presence." *Journal of Personality and Social Psychology* 36:324–332.

——. 1981. *Attention and Self-Regulation: A Control Theory Approach to Human Behavior*. New York: Springer.

Cocchini, G., A. Cameron, N. Beschin, and A. Fotopoulou. 2009. "Anosognosia for Motor Impairment Following Left Brain Damage." *Neuropsychologia* 23, no. 2: 223–230.

Conway, M., P. A. R. Csank, S. L. Holm, and C. K. Blake. 2000. "On Assessing Individual Differences in Rumination on Sadness." *Journal of Personality Assessment* 75:404–425.

Cook, S. W. 1939. "The Production of 'Experimental Neurosis' in the White Rat." *Psychosomatic Medicine* 1:293–308.

Darwin, C. 1872. *The Expression of the Emotions in Man and Animals.* London: John Murray.

Davis, D., and T. C. Brock. 1975. "Use of First Person Pronouns as a Function of Increased Objective Self-Awareness and Prior Feedback." *Journal of Experimental Social Psychology* 11:381–388.

DeSouza, M. L., A. DaSilveira, and W. B. Gomes. 2008. "Verbalized Inner Speech and the Expressiveness of Self-Consciousness." *Qualitative Research in Psychology* 5, no. 2: 154–170.

Duval, S., and R. A. Wicklund. 1972. *A Theory of Objective Self-Awareness.* New York: Academic.

Edelman, D. B., and A. K. Seth. 2009. "Animal Consciousness: A Synthetic Approach." *Trends in Neurosciences* 32, no. 9: 476–484.

Eichstaedt, J., and P. J. Silvia. 2003. "Noticing the Self: Implicit Assessment of Self-Focused Attention Using Word Recognition Latencies." *Social Cognition* 21: 349–361.

Engbert, K., A. Wohlschläger, and P. Haggard. 2008. "Who Is Causing What? The Sense of Agency Is Relational and Efferent-Triggered." *Cognition* 107, no. 2: 693–704.

Fenigstein, A., M. F. Scheier, and A. H. Buss. 1975. "Public and Private Self-Consciousness: Assessment and Theory." *Journal of Consulting and Clinical Psychology* 36:1241–1250.

Franzoi, S. L. 1986. "Self-Consciousness and Self-Awareness: An Annotated Bibliography of Theory and Research." *Social and Behavioral Sciences Documents* 16, no. 1: listing 2744, 1–53.

Gallagher, H. L., and C. D. Frith. 2003. "Functional Imaging of 'Theory of Mind.'" *Trends in Cognitive Science* 7, no. 2: 77–83.

Gallup, G. G., Jr. 1982. "Self-Awareness and the Emergence of Mind in Primates." *American Journal of Primatology* 2:237–248.

Gibbons, F. X. 1983. "Self-Attention and Self-Report: The 'Veridicality' Hypothesis." *Journal of Personality* 51:517–542.

Govern, J. M., and L. A. Marsch. 2001. "Development and Validation of the Situational Self-Awareness Scale." *Consciousness and Cognition* 10, no. 3: 366–378.

Hampton, R. R. 2001. "Rhesus Monkeys Know When They Remember." *Proceedings of the National Academy of Sciences* 98:5359–5362.

Heyes, C. M. 1998. "Theory of Mind in Nonhuman Primates." *Behavioral and Brain Sciences* 21, no. 1: 101–134.

Hoerold, D., P. M. Dockree, F. M. O'Keeffe, H. Bates, M. Pertl, and I. H. Robertson. 2008. "Neuropsychology of Self-Awareness in Young Adults." *Experimental Brain Research* 186, no. 3: 509–515.

Kircher, T., and A. S. David. 2003. "Self-Consciousness: An Integrative Approach from Philosophy, Psychopathology, and the Neurosciences." In T. Kircher and

A. S. David, eds., *The Self in Neuroscience and Psychiatry*, pp. 445-473. Cambridge: Cambridge University Press.

Leary, M. R. 2004. *The Curse of the Self: Self-Awareness, Egotism, and the Quality of Human Life*. Oxford: Oxford University Press.

McComb, K., L. Baker, and C. Moss. 2006. "African Elephants Show High Levels of Interest in the Skulls and Ivory of Their Own Species." *Biology Letters* 2:26–28.

Mead, G. H. 1934. *Mind, Self, and Society*. Chicago: University of Chicago Press.

Menzel, C. 2005. "Progress in the Study of Chimpanzee Recall and Episodic Memory." In H. S. Terrace and J. Metcalfe, eds., *The Missing Link in Cognition: Origins of Self-Knowing Consciousness*, pp. 188–224. New York: Oxford University Press.

Mitchell, R. W. 2002. "Subjectivity and Self-Recognition in Animals." In M. R. Leary and J. P. Tangney, eds., *Handbook of Self and Identity*, pp. 3–15. New York: Guilford.

Morin, A. 2004. "A Neurocognitive and Socioecological Model of Self-Awareness." *Genetic, Social, and General Psychology Monographs* 130, no. 3: 197–222.

——. 2005. "Possible Links Between Self-Awareness and Inner Speech: Theoretical Background, Underlying Mechanisms, and Empirical Evidence." *Journal of Consciousness Studies* 12, no. 4–5: 115–134.

——. 2009. "Self-Awareness Deficits Following Loss of Inner Speech: Dr. Jill Bolte Taylor's Case Study." *Consciousness and Cognition* 18, no. 2: 524–529.

——. 2011. "Self-Recognition, Theory-of-Mind, and Self-Awareness: On What Side Are You On?" *Laterality* 16, no. 3: 367–383.

Natsoulas, T. 1978. "Consciousness." *American Psychologist* 33, no. 10: 906–914.

Nelson, T. O., and L. Narens. 1994. "Why Investigate Metacognition?" In J. Metcalfe and A. P. Shimamura, eds., *Metacognition: Knowing About Knowing*, pp. 1–25. Cambridge: MIT Press.

Paul, E. S., E. J. Harding, and M. Mendl. 2005. "Measuring Emotional Processes in Animals: The Utility of a Cognitive Approach." *Neuroscience and Biobehavioral Reviews* 29, no. 3: 469–491.

Povinelli, D. J., and T. J. Eddy. 1996. "What Young Chimpanzees Know About Seeing." *Monographs of Social Research in Child Development* 61:1–152.

Povinelli, D. J., and J. Vonk. 2003. "Chimpanzee Minds: Suspiciously Human?" *Trends in Cognitive Science* 7:157–160.

Premack, D., and G. Woodruff. 1978. "Does the Chimpanzee Have a Theory of Mind?" *Behavioral and Brain Sciences* 1, no. 4: 515–526.

Pyszczynsky, T., and J. Greenberg. 1987. "Self-Regulatory Perseveration and the Depressive Self-Focusing Style: A Self-Awareness Theory of Reactive Depression." *Psychological Bulletin* 10:122–138.

Ramsden, E., and D. Wilson. 2010. "The Nature of Suicide: Science and the Self-Destructive Animal." *Endeavour* 34, no. 1: 21–24.

Roberts, W. A., and M. C. Feeney. 2009. "The Comparative Study of Mental Time Travel." *Trends in Cognitive Sciences* 13:271–277.

Salzen, E. A. 1998. "Emotion and Self-Awareness." *Applied Animal Behaviour Science* 57:299–313.

Savage-Rumbaugh, S., W. M. Fields, and J. Taglialatela. 2000. "Ape Consciousness—Human Consciousness: A Perspective Informed by Language and Culture." *American Zoologist* 40:910–921.

Silvia, P. J., and J. Eichstaedt. 2004. "A Self-Novelty Manipulation of Self-Focused Attention for Internet and Laboratory Experiments." *Behavior Research Methods, Instruments, and Computers* 36:325–330.

Smith, J. A. 2005. "'Viewing' the Body: Toward a Discourse of Rabbit Death." *Worldviews* 9, no. 2: 184–202.

Smith, J. D. 2005. "Studies of Uncertainty Monitoring and Metacognition in Animals and Humans." In H. S. Terrace and J. Metcalfe, eds., *The Missing Link in Cognition: Origins of Self-Knowing Consciousness,* pp. 247–271. New York: Oxford University Press.

——. 2009. "The Study of Animal Metacognition." *Trends in Cognitive Sciences* 13, no. 9: 389–396.

Smith, J. M., and L. B. Alloy. 2009. "A Roadmap to Rumination: A Review of the Definition, Assessment, and Conceptualization of This Multifaceted Construct." *Clinical Psychology Review* 29:116–128.

Son, L. K., and N. Kornell. 2005. "Metaconfidence Judgments in Rhesus Macaques: Explicit Versus Implicit Mechanisms." In H. S. Terrace and J. Metcalfe, eds., *The Missing Link in Cognition: Origins of Self-Reflective Consciousness,* pp. 296–320. New York: Oxford University Press.

Tracy, J. L., and R. W. Robins. 2004. "Show Your Pride: Evidence for a Discrete Emotion Expression." *Psychological Science* 15:194–197.

Trapnell, P. D., and J. D. Campbell. 1999. "Private Self-Consciousness and the Five-Factor Model of Personality: Distinguishing Rumination from Reflection." *Journal of Personality and Social Psychology* 76:284–304.

Tulving, E., and A. S. N. Kim. 2009. "Autonoetic Consciousness." In P. Wilken, T. Bayne, A. Cleeremans, eds., *The Oxford Companion to Consciousness,* pp. 96–98. Oxford: Oxford University Press.

Vygotsky, L.S. 1962 [1934]. *Thought and Language.* Cambridge: MIT Press.

Winsler, A. 2009. "Still Talking to Ourselves After All These Years: A Review of Current Research on Private Speech." In A. Winsler, C. Fernyhough, and I. Montero, eds., *Private Speech, Executive Functioning, and the Development of Verbal Self-Regulation,* pp. 3–41. New York: Cambridge University Press.

PART V

Tailoring Representations to Audiences

17

Chimpanzees Attribute Beliefs?

A New Approach to Answering an Old Nettled Question

ROBERT W. LURZ

The capacity to attribute mental states to others ("mindreading" or "theory-of-mind") was once thought by most empirical researchers and philosophers to be a distinctively human talent. The tide has changed, and in the past eight or nine years there has been a growing consensus in comparative psychology and philosophy that humans are not alone in being mindreaders. It is now generally believed that a number of highly intelligent social animals, from apes to scrub jays, are capable of attributing simple perceptual states, such as seeing and hearing, as well as goals and intentional actions (e.g., Buttelmann et al. 2007; Emery and Clayton 2008; Hare et al. 2000; Hare, Call, and Tomasello 2001; Santos, Flombaum, and Webb 2007; Uller 2004; Wood and Hauser 2008).[1]

Yet, despite this largely accepted opinion, there is considerable debate in the field over whether animals are capable of attributing beliefs. On the one hand, there are some philosophical arguments (Bermúdez 2003, 2009; Davidson 2001) that purport to show the a priori impossibility of belief attribution in animals and a few empirical studies with chimpanzees that appear to confirm this conclusion (Call and Tomasello 1999; Kaminski, Call, and Tomasello 2008; Krachun et al. 2010). On the other hand, there are at least two empirical studies with chimpanzees that have produced what appear to be opposite findings (Krachun et al. 2008; O'Connell and Dunbar 2003).

I do not believe that the philosophical arguments or the empirical studies on either side of this debate are particularly persuasive. Since I have argued elsewhere (Lurz 2007, 2008, 2011) that the main philosophical arguments against belief attribution in animals are invalid, my focus here will be on the empirical side of the debate. What is needed to move the debate and the empirical field forward, I propose, is a fundamentally new experimental approach to testing belief attribution in chimpanzees, one that is capable of distinguishing genuine belief-attributing subjects from their perceptual-state and behavior-reading counterparts. I present just such an experimental protocol in the last section of this chapter.

Belief Attribution Studies with Chimpanzees

Two types of experimental protocols have been used to test for belief attribution in chimpanzees and other animals. The first type, commonly called the cooperative knowledge-ignorance (or bait-and-switch) paradigm, is designed to test an animal's ability to use a human experimenter's communicative gestures to choose between two containers, one of which covers food underneath. Call and Tomasello (1999) ran one of the first of such experiments, but found no evidence that their apes could discriminate between the containers on the basis of whether the experimenter was knowledgeable or ignorant of the food's true location.

It was thought by some that the chimpanzees failed the cooperative knowledge-ignorance test because it involved cooperative communication regarding the location of hidden food, a type of communicative act that is not part of the natural behavioral repertoire of chimpanzees (see Hare and Tomasello 2004). This, in turn, prompted Kaminski, Call, and Tomasello (2008) to run a competitive version of the knowledge-ignorance experiment. In their experiment two chimpanzees (a subject and a competitor) participated in a competitive game over a highly desirable piece of food (e.g., a grape) hidden beneath one of three cups on a sliding table in a middle area (see figure 17.1).

At the start of the game, the chimps watched while an experimenter baited one of three cups on the table. After the baiting, a screen was positioned (or lowered) in front of the competitor, blocking its view of the middle area. Once the screen was in place, the experimenter performed

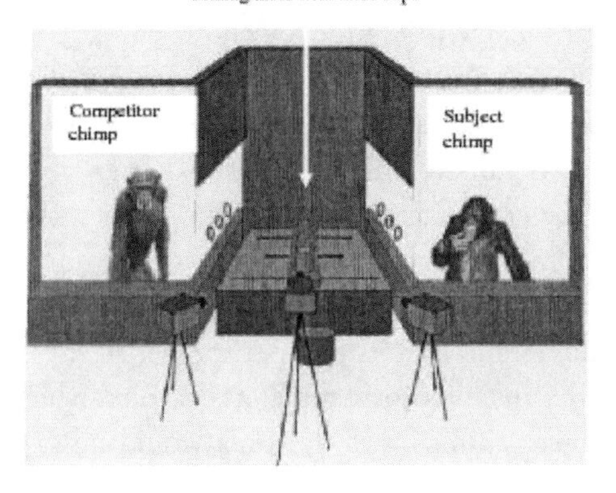

FIGURE 17.1. Experimental setup from Kaminski et al. (2008). Reprinted from *Cognition,* vol. 109, J. Kaminski, J. Call, and M. Tomasello, "Chimpanzees Know What Others Know, But Not What They Believe," pp. 224–234, 2008, with permission from Elsevier.

in front of the subject chimp one of two types of manipulations on the cups and bait. In the *unknown-lift* trials the experimenter removed the bait from underneath the cup and then replaced it underneath the same cup. In the *unknown-shift* trials the experimenter removed the bait from underneath the cup and placed it under a different cup. After this, a screen was positioned in front of the subject chimp, blocking its view of the middle area. The screen before the competitor was then removed, and the table with the cups was slid to the competitor's side. The competitor was allowed to choose a cup by pointing at it. If the baited cup was chosen, the food underneath it was removed and given to the animal; the empty cup was then returned to the table. If an unbaited cup was chosen, the animal was shown that the cup was empty and received no reward; the empty cup was then returned to the table. The subject chimp, of course, could not see which cup the competitor had chosen; although from its prior training in the experiment it knew that the competitor had selected a cup at this point and had received the hidden bait if it had chosen correctly. After the competitor had made its choice and the cups were returned to their original locations, the screen before the subject chimp was removed. The table was then slid to the subject chimp's side. The subject chimp was allowed to make one of two choices at this point: It was allowed to choose a cup on the table that (depending upon the competitor's choice) may or may not have the highly desirable piece of food underneath or it was allowed to choose a cup, which was placed inside its booth, that had a less preferred but guaranteed food reward inside (e.g., monkey chow).

Kaminski, Call, and Tomasello (2008) hypothesized that if the subject chimp was capable of attributing beliefs, then it should understand that the competitor in the unknown-shift trials was likely to choose the empty cup on the table, given its false belief regarding which cup was baited, and thus the subject chimp in such trials would be expected to choose the high-quality baited cup on the table. But in the unknown-lift tests the subject chimp should understand that its competitor was likely to choose the baited cup on the table, given its correct belief regarding which cup was baited, and thus the subject chimp in these trials would be expected to choose the low-quality but guaranteed baited cup inside its booth. Kaminski and colleagues found no such preferential choosing on the part of the subject chimp. In fact, the subject chimps were just as likely to choose the high-quality baited cup on the table in the unknown-shift tests as they were in the unknown-lift tests. The authors took these results, plus the findings from Call and Tomasello (1999), as positive support for the hypothesis that chimpanzees are unable to attribute beliefs.

Negative results are always tricky to interpret. And although Kaminski, Call, and Tomasello's interpretation of the findings is plausible, it is not the only plausible interpretation. It is quite possible that the knowledge-ignorant paradigm (either competitive or cooperative) is simply poorly suited to elicit belief attribution in chimpanzees. Arguably if animals are capable of mindreading, then this is for the express purpose of producing reliable expectations of others' behaviors. So it stands to reason that mindreading animals should be expected to excel at realistic (perhaps competitive) mindreading tasks that explicitly test their ability to anticipate others' behavior. And, as a matter of fact, that is what we do seem to find with perceptual-state attribution tests (see Hare 2001; Hare, Call, and Tomasello 2001; Santos, Flombaum, and Webb 2007). However, the knowledge-ignorance paradigm is not designed to test an animal's ability to anticipate another's behavior; it is principally a discrimination task, designed explicitly to test an animal's ability to choose between different containers.[2] Thus it is quite plausible that in Kaminski, Call, and Tomasello's study the subject chimp's capacity to attribute beliefs was not piqued by the discriminatory nature of the task problem, forcing the animal to employ a nonmindreading strategy to make its selection of cups—perhaps some "high-risk strategy in favor of the high-quality reward in preference to the safe, low-quality reward," as Kaminski, Call, and Tomasello suggest.

Most recently, Krachun et al. (2010) received negative results from a knowledge-ignorance (bait-and-switch) protocol that was purposively designed to be *neither* a communicative-cooperative *nor* competitive paradigm. In the training phase of the experiment, a chimp observed while an experimenter (the baiter) placed a grape or a piece of banana inside a small yellow box with a lid. The baiter closed the lid on the box and placed it underneath one of two colored cups according to the following rule: if the food in the box was a grape, the baiter placed the box under the blue cup; if the food in the box was a piece of banana, the baiter placed the box under the white cup. After the baiter placed the box under the correct cup according to the protocol, the chimp was then allowed to select one of the cups on the table. Five of the six chimps tested on this phase of the experiment eventually learned to select the white cup if they had observed the baiter place a banana piece in the yellow box and the blue cup if they had observed the baiter place a grape in the yellow box.

Once the chimps reached criterion on this phase of the experiment, they were then given a false-belief test and a true-belief test. In the former the chimp observed while the baiter placed a grape or a banana piece inside the yellow box (as in the training phase), closed the lid of the box, and then left the testing room. While the baiter was absent, the chimp then observed another experimenter (the switcher) replace the contents of the box with the alternative food item (e.g., if the box contained a grape, the switcher removed the grape and replaced it with a banana piece). After the switcher switched the contents of the box and closed its lid, the baiter returned to the room. A screen was then raised in front of the chimp, blocking its view while the baiter placed the yellow box under the appropriate cup according to what item the baiter (not the switcher) had placed inside it. The screen was then removed and the chimp was allowed to choose one of the cups. The design of the true-belief test was the same except that the baiter remained in the room and watched while the switcher switched the contents of the box.

The researchers reasoned that if the chimps understood that the baiter in the false-belief trials had a mistaken belief regarding the contents of the yellow box, then they should choose the cup that corresponded to the yellow box's *original* contents and not to its actual contents. Likewise, if the chimps understood that the baiter in the true-belief trials had a correct belief regarding the contents of the yellow box (having witnessed the switch by the switcher), then they would be expected to choose the cup that corresponded to the box's *actual* contents, not its original contents.

What Krachun and colleagues discovered, however, was that, across the different test trials, their chimps chose the cup according to the yellow box's *actual* contents. One rather plausible explanation of these results, the researchers suggest, is that the chimps "might have unthinkingly carried the strategy they learned during the training trials over to the true and false belief test trials (e.g., to choose based on the current contents of the yellow container)" (2010:162).[3] That is to say, the chimps may have failed to comprehend the task presented to them in the training trials as one requiring the *prediction* of the baiter's future behavior in light of what the baiter currently *believes* about the contents of the box, but rather one requiring them to discriminate between colored cups according to a conditional rule that had nothing directly to do with the mental states of the baiter. Hence it is not implausible to suppose that the failure of the chimpanzees on the Krachun et al. (2010) test is due not to their inability to attribute beliefs but to their inability to understand the tasks at hand as requiring or involving mindreading as opposed to requiring or involving discrimination according to a nonmindreading conditional rule.

The conclusion I wish to draw here is that a more feasible belief-attribution test for chimpanzees, one that is more likely to trigger their belief-attributing capacities, is one in which the test animal is required to anticipate a target's behavior. This is the method used in the experimental protocol described in the last section.

Theoretical Background: The ARM Theory

The experimental protocol I will outline is based upon an appearance-reality mindreading (ARM) theory, various versions of which have been defended over the years by Humphreys (1980), Gallup (1982), and Carruthers (1996). According to theorists of ARM, mental-state attribution in animals evolved for the express purpose of anticipating a target's (e.g., a conspecific's, predator's, or prey's) behavior in those situations in which the animal's behavior-reading counterparts (i.e., conspecifics that lack mental-state concepts) could not. In many cases the way things perceptually *appear* to a target is a better predictor of its behavior than the way things objectively are. Behavior-reading animals—animals that are unable to attribute mental states to targets to predict their behaviors—can appeal only to the latter sorts of objective, observable, mind-independent facts (e.g., facts about a target's past behaviors or its current line of eye gaze to an object in the environment) to predict targets' future behaviors. Mindreading animals, however, are able to appeal to the subjective way things perceptually appear to a target to predict its behavior. It is hypothesized that mental-state attribution in animals evolved as a result of their coming to introspect their own ability to distinguish appearance from reality, using this introspective distinction for the purpose of anticipating others' behaviors.

Of course, targets can sometimes see through deceptive appearances and act on what they *believe* to be really occurring in the environment. And so, on the ARM theory, mental-state attribution in animals should reflect two levels of development. The first level involves *perceptual-state attribution*, wherein the animal is able to anticipate a target's behavior in terms of how it thinks the world perceptually appears to the target. The second level of development is *belief attribution*, wherein the animal is able to anticipate a target's behavior in terms of how it thinks the target *believes* things are, the way things perceptually *appear* to the target notwithstanding.

To test ARM empirically, we need an experimental method capable of distinguishing (a) those animals capable of attributing perceptual states (e.g., seeing or visually appearing) from (b) those capable of attributing perceptual states *and* beliefs. All the belief-attribution studies just surveyed, unfortunately, rest upon a particular view of the difference between belief attribution and perceptual-state attribution that prevents them from experimentally distinguishing (a) and (b). The view itself, quite correctly, is based upon an undeniable fact about beliefs and perceptions—namely, that past veridical perceptions (e.g., having seen food placed in location *x*) can sometimes lead a subject to later act on a false belief at a time when the veridical perception no longer exists (e.g., after the deceptive switching of the bait to location *y*, the competitor/cooperator now incorrectly believes that the bait is still in location *x*). The studies, therefore, assume that if chimpanzees are capable of attributing beliefs, and not just current perceptual states, then they should be able to predict a target's behavior on the basis of his/her currently held false belief in the absence of the veridical perception that caused it. The chief problem with this approach is that a target's currently held false belief will always be confounded with his/her *having had* a past veridical perception (as well as his/her having had a direct-line-of-gaze to the perceived object), and, as a result, the animal's successful prediction of the target's action can just as well be explained in terms of it attributing a past veridical perception (or a past direct-line-of-gaze to an object) to the target as it can in terms of it attributing a currently held false belief.

FIGURE 17.2. Müller-Lyer diagram.

Luckily there are two other ways of understanding the difference between belief attribution and perceptual-state attribution that escape this problem. Both these approaches reflect a distinctive feature of belief in relation to perception—namely, revisability and abstractness. A distinctive feature of beliefs (at least as a kind of mental state) is that they are revisable in light of countervailing evidence in a way that perceptual states (as a kind) largely are not.[4] As a result of measuring the lines in a Müller-Lyer diagram (see figure 17.2), for example, one's belief about the lines' unequal length changes despite their continuing to look unequal, and this change in one's belief naturally leads to a change in one's behavior.

Thus an animal capable of attributing beliefs, and not just perceptual states, is expected to be sensitive to the behavioral difference between 1. a target that currently perceives an object *as F* but has countervailing evidence indicating that the object is really not *F*, and 2. a target that currently perceives an object *as F* but has no such countervailing evidence. In Lurz (2009, 2011) I have described a number of innovative experimental protocols for testing chimpanzees' and other animals' sensitivity to this epistemic difference; so I shall not discuss those here.

The other distinctive feature of belief (and thought in general) is its ability to allow subjects to represent facts or properties in the world that cannot be represented by means of perception alone. Such facts and properties are routinely called "abstract." Of course, abstractness comes in degree, and so it is unlikely that there will be any sharp distinction between perception and cognition on this feature. Be this as it may, the point here is merely that the more abstract a fact or property a mental state represents, the more the state belongs to cognition. And this is surely correct. Although I do not have a metric of abstractness to offer, it is quite clearly the case that representing an object as having a second-order (or determinable) property (i.e., a property that is defined in terms of other properties) is more abstract than representing an object's first-order (determinate) properties (i.e., properties that are not defined in terms of other properties). Consider, for example, the second-order (determinable) property of having shape in general. Intuitively, representing an object as having shape in general is more abstract than representing it as square or circular, for example. And, for similar reasons, representing a pair of objects as being same/different in shape in general is more abstract than representing them as having the same/different determinate shape (e.g., as both being square).

Following Premack (1983), cognitive scientists have come to call relations of same/difference in second-order (determinable) properties of color or shape *second-order relations* and relations of same/difference in determinate colors or shape *first-order relations,* taking mental representations of the former (abstract) relations as a mark of genuine cognition. It would be ideal, then, in developing a test for *belief* attribution in chimpanzees, to run an experiment that tested the apes' capacity to attribute representations of second-order relations to targets. Just such a protocol can be developed, I believe. Chimpanzees and other great apes, after all, are known for their distinctive talent in solving and responding to relational match-to-sample (RMTS) tasks that arguably require representing abstract, second-order relations or properties (see Premack 1983; Smith et al. 1975; Thompson and Oden 2000; Thompson, Oden, and Boysen 1997; Vonk 2003; Vonk and MacDonald 2002, 2004).[5] The idea behind the experimental protocol I am proposing here is to exploit this unique cognitive talent in chimpanzees and see whether they can attribute such abstract representations (i.e., beliefs) to others.

Testing the Attribution of Abstract Representations by Chimpanzees

The following belief-attribution test employs a violation-of-expectancy (VOE) paradigm with video-animated stimuli using looking time as a measure—a methodology that is commonly used in developmental psychology and becoming more so in animal studies (see e.g., Csibra 2008; Gergely et al. 1995; Heider and Simmel 1944; Kuhlmeier, Wynn, and Bloom 2003; Premack and Premack 1997; Premack and Woodruff 1978; Surian, Caldi, and Sperber 2007; Uller 2004). However, unlike any VOE experiment to date, the following experiment requires the test animal to be able to make an appearance-reality distinction involving occluded objects. Therefore, the belief-attribution experiment will include a series of pretests designed to probe the chimpanzees' ability to discriminate appearances from reality involving partially occluded objects.

APPEARANCE-REALITY (A/R) DISCRIMINATION TEST USING COMPUTER ANIMATED STIMULI

Abundant empirical evidence shows that chimpanzees and monkeys are susceptible to some of the same perceptual illusions as humans (for review see Fujita 2006). One well-known example of this is the case of seeing occluded objects as unified, typically referred to as amodal completion.[6] Using a two-choice delayed matched-to-sample task, Sato, Kanazawa, and Fujita (1997), for example, showed that an adult chimpanzee reliably matched a unified straight bar when presented with an occluded figure (see figure 17.3a) whose top and bottom portions were aligned and moving in a congruent manner, but matched a broken bar when presented with an occluded figure (see figure 17.3b) whose top and bottom portions were misaligned and moving in an incongruent manner.

This and other amodal completion experiments have been taken as evidence that chimpanzees see occluded figures as having certain features (e.g., being straight or unified) even though the figures may not in fact have such features. Chimpanzees appear to be subject to a type of amodal completion *illusion*. What researchers have not investigated, however, is whether chimpanzees (or any other animal) can distinguish the apparent from the real shape of an occluded object in those cases where they are given clear evidence that appearances are deceptive.[7] The appearance-reality (A/R) task I will describe asks whether chimpanzees, when given

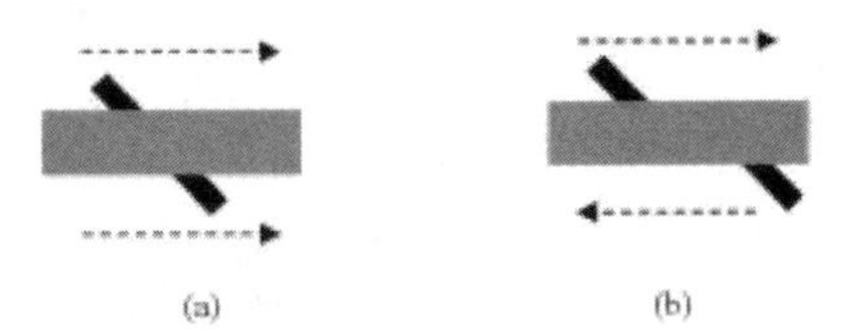

FIGURE 17.3. Amodal completion stimuli.

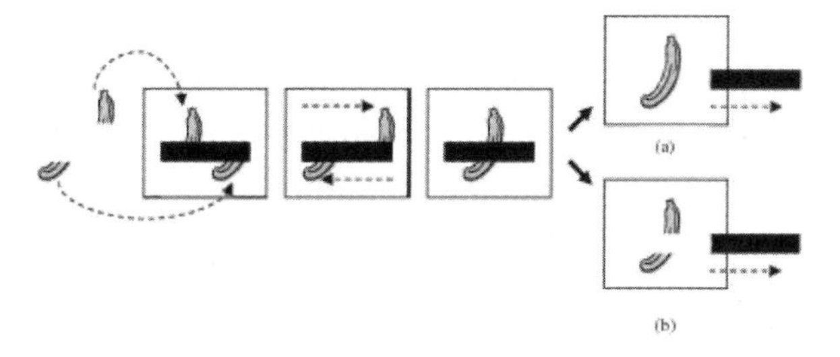

Figure 17.4. Simple A/R test using computer animation.

evidence that an occluded figure is actually bent or disjointed, despite it looking straight or whole, take the figure to be as it really is (i.e., as bent or disjointed) or as it appears to be (i.e., as straight or whole). If chimpanzees are capable of distinguishing appearances from reality in such cases, then they should take the occluded figure to be bent or disjointed, appearances to the contrary. If they are incapable of making the distinction, then they should take the occluded figure to be as it finally appears to be (i.e., as a straight or whole object behind an occluder).

Although different methods can be used to answer these questions (e.g., match-to-sample or delay match-to-sample tasks, such as used in Sato, Kanazawa, and Fujita 1997), I present here a habituation/dishabituation procedure using computer-animated stimuli. Figure 17.4 shows a series of still frames from the videos that could be used in the experiment. In the habituation phase of the experiment, the animal is shown a video of two pieces of a disjointed object (e.g., two pieces of banana or two pieces of a rod) being placed on top of an occluder. (Note that the width of the rectangular occluder is the same size as the missing section between the two banana pieces. Thus, when the banana pieces align themselves on top of the occluder in the third frame, there will still be a missing portion between them that is the width of the occluder.) The object's continued disjointedness can be further demonstrated by moving the pieces on top of the occluder in opposite directions. The pieces are then brought into alignment, giving the illusion of a whole occluded object, as illustrated by the third frame in figure 17.4. This ends the habituation phase of the experiment. The testing phase occurs immediately afterward. In this phase the occluding bar is removed to reveal either the occluded object (17.4a) as whole or (17.4b) as disjointed.[8]

If the animal is able to make an appearance-reality distinction in the case of occluded objects, then it should be able to comprehend that, from the countervailing evidence given in the habituation phase, there really is not a whole object (banana) behind the occluder, despite it coming to look that way once the pieces are aligned. Such an animal, then, should be surprised (look longer) to see a whole object (17.4a) emerge from behind the occluder once the occluder is removed but unsurprised (look less) to see a disjointed object (17.4b) emerge. On the other hand, if the animal is unable to make an appearance-reality distinction and simply thinks "if it looks like a whole object (banana) behind the occluder, then it *is* a whole object (banana) behind the occluder," then it should be surprised (look longer) to see a disjointed object (17.4b) emerge from behind the occluder but unsurprised (look less) to see a whole object (17.4a) emerge.

Follow-up tests can be run using a set of different types of objects, both familiar and unfamiliar to the animal, as well as different types of occluders and settings (e.g., vertical as opposed to horizontal occluders, three-dimensional backgrounds as opposed to two-dimensional or blank backgrounds). Those animals that pass this A/R test will be allowed to participate in the belief-attribution test.

BELIEF ATTRIBUTION TEST

Those animals that test positive on the A/R task will participate in the following belief-attribution test using a VOE paradigm with computer ani-

Figure 17.5. Habituation phase with same-shape stimuli. While the CGI figure is on screen and facing *away* from the trees (not shown here), 1. the same-shaped stimuli (red and blue rectangles) come from offscreen and place themselves behind the trees. 2. The CGI figure then turns around to face the occluded stimuli (as shown here). After looking toward each occluded shape, the CGI figure knuckle walks down the right path (white arrow).

mation. In the familiarization phase of the experiment, the test animal is habituated to the movements of a computer-generated image (CGI) of a conspecific (perhaps one with which the test animal is familiar). Figures 17.5 and 17.6 show a still frame from the habituation videos. The chimpanzee in the still frames is the CGI figure that the test animal is watching, not the test animal viewing the video. The video sequence opens with the CGI figure standing in front of two trees. The CGI figure at this point is facing *away* from the trees and toward the test subject watching the video. While the CGI figure's head is turned away from the trees, two sample stimuli (e.g., a red and blue rectangle) from offscreen silently move onto the screen and place themselves behind the trees, as shown in step 1 in Figures 17.5 and 17.6. After the stimuli are in place, the CGI figure turns around to face the occluded rectangles, as shown in step 2 in figures 17.5 and 17.6. The CGI figure now has the same view of the occluded stimuli as the test subject. After the CGI figure has turned around and has looked at each of the occluded stimuli, it knuckle walks down one of the two paths. Thus there will be two types of habituation videos shown to the test animal. In those videos where the sample stimuli are the same in shape (as illustrated in figure 17.5), the CGI figure knuckle walks down the right path.

In those videos where the sample stimuli are different in shape (as illustrated in figure 17.6 below), the CGI figure knuckle walks down the left path.[9]

The test animal is shown these two types of habituation videos—those in which the sample stimuli are the same in shape and the CGI figures takes

Figure 17.6. Habituation phase with different-shaped stimuli. While the CGI figure is on screen and facing *away* from the trees (not shown here), 1. the different-shaped stimuli (red rectangle and yellow crescent) come from offscreen and place themselves behind the trees. 2. The CGI figure then turns around to face the occluded stimuli (as shown here). After looking toward each occluded shape, the CGI figure knuckle walks down the left path (white arrow).

FIGURE 17.7. "Look-same-but-are-different" test video. While the CGI figure is on screen and facing *away* from the trees (not shown here), 1. the different-shaped stimuli come from offscreen and place themselves behind the trees. 2. The CGI figure then turns around to face the occluded stimuli (as shown here).

the right path and those in which the sample stimuli are different in shape and the CGI figure takes the left path—a number of times to induce habituation (i.e., the expectation of the CGI figure's actions in the presence of same- or different-shaped stimuli). Various shapes (e.g., circles and triangles) can be used as sample stimuli in the habituation videos.

In the testing phase of the experiment, the test animal watches two types of similar videos in which a *new* set of occluding stimuli is used. In

FIGURE 17.8. "Look-different-but-are-same" test video. While the CGI figure is onscreen and facing *away* from the trees (not shown here), 1. the same-shaped stimuli come from offscreen and place themselves behind the trees. 2. The CGI figure then turns around to face the occluded stimuli (as shown here).

the "look-same-but-are-different" test video, two sample stimuli (e.g., an amputated orange oval and a whole blue oval) come on the scene. The pieces of the amputated orange oval align themselves to the sides of the left tree while the whole blue oval moves behind the right tree (see figure 17.7).

To reinforce to the viewing chimpanzee (test animal) that the disjointed pieces that are attached to the left tree do not constitute an occluded whole oval, the pieces themselves can be moved in opposite directions before aligning, much in the way that the banana pieces were moved in the A/R task present earlier. The final placement of the stimuli gives the illusion that there are two similarly shaped objects behind the trees (see figure 17.7). After the placement of the different sets of shapes, the CGI figure turns around to look at the sample stimuli, just as in the habituation videos. At this point the test animal is shown one of two different endings to the video. In the "surprise" ending the CGI figure is shown knuckle walking down the left path.[10] In the "expected" ending the CGI figure is shown knuckle walking down the right path. For each ending, the test animal's looking time is measured.

The second test video is the "look-different-but-are-same" video, in which two sample stimuli of the same shape come onto the screen and move behind the trees (see figure 17.8).

After the placement of the stimuli, the CGI figure turns around and looks at them, as in the habituation videos. The test animal is then shown one of two endings. In the "surprise" ending the CGI figure is shown knuckle walking down the right path.[11] In the "expected" ending the CGI figure is shown knuckle walking down the left path. Again, for each of these endings, the test animal's looking time is measured.

It goes without saying that different types of "deceptive" shapes can be used in the test videos (see figure 17.9) and that the quality of the moving images in the videos themselves will be more detailed than the schematic representations given here.

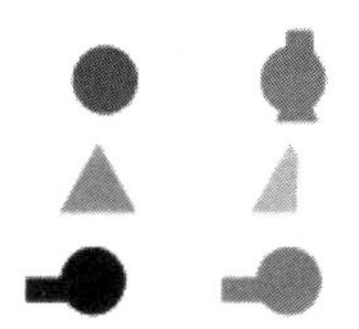

FIGURE 17.9. Various deceptive shapes.

The beauty behind the belief-attribution protocol is that it is capable of discriminating between belief-attributing chimps and their perceptual-state and behavior-reading counterparts. If those chimps that pass the A/R task are also capable of attributing beliefs involving second-order relations, then they should be able to understand that the CGI figure takes the right path when it *believes* the occluded objects are the same in shape (irrespective of their determinate shapes and colors) and the left path when it *believes* the occluded objects are different in shape (again, irrespective of their different determinate shapes or colors). Armed with this understanding of the CGI figure's behavior in the habituation videos, the chimps should be able to understand that in the look-same-but-are-different video the CGI figure likely *believes* that the occluded objects are the same in shape, since they look to be the same in shape and the CGI figure did not see the objects go behind or align themselves to the trees. And so, on the basis of this second-order belief attribution, the chimps should anticipate the CGI figure taking the right path and look longer at the surprise ending than the expected one.

Likewise, in the look-different-but-are-same video, the chimps should understand that the CGI figure likely believes that the occluded objects are different in shape—again, since they look different in shape and the CGI figure did not see them go behind the trees. The chimps should thus predict that the CGI figure will move down the left path and should look longer at the surprise ending than at the expected one.

The chimpanzees would *not* have such expectations of the CGI figure's actions, however, were they capable of attributing only first-order representations (i.e., perceptual states) to the CGI figure. Although such chimpanzees would be capable of understanding that the CGI figure mistakenly sees the occluded objects as having the same determinate shape (both oval) in the first test video and different determinate shapes (circle and packman) in the second, they would not be able to predict the CGI figure's actions on the basis of either of these first-order perceptual-state representations. Since the determinate shapes of the occluded objects on these test trials are novel, the CGI figure was never shown to the chimps as ever having such first-order perceptual representations in the habituation videos. Thus first-order perceptual-state-attributing chimps would have no way of knowing how the CGI figure might behave on the basis of such novel first-order perceptual states. The point of introducing novel shapes (real and apparent) into the test videos is precisely to test the chimpanzees' ability to rise to the more abstract second-order level of mental-state attribution in their understanding of the CGI figure's actions.

The last prediction we need to consider is that of the behavior-reading chimpanzee that passed the A/R task. These chimps cannot attribute first-order or second-order representations to targets (such as the CGI figure), but they can themselves represent first-order and second-order relations in the world. Hence these chimps, through their viewing of the habituation videos, could come to understand that the CGI figure takes the right path when the occluded objects actually have the second-order relation of being same in shape, taking the left path when they have the second-order relation of being different in shape. The question, then, is how these chimps would represent the second-order relation of the occluded objects in the test videos. Do they represent them as they actually *are* or as they *appear* to be? If they represent them as they are, then these behavior-reading chimps would make the exact opposite predictions of the CGI figure's behavior in the test videos from their belief-attributing counterparts. They would expect the CGI figure in the look-same-but-are-different test video to knuckle walk down the *left* fork, since the sample stimuli are in fact incongruent shapes, and this is how the CGI figure responded to incongruent shapes in the habituation videos, and they would expect the CGI figure in the look-different-but-are-same test video to knuckle walk down the *right* fork, since the sample stimuli are in fact congruent shapes, and this is how the CGI figure responded to congruent shapes in the habituation videos.

On the other hand, if the behavior-reading chimps represent the sample stimuli as having the second-order relations that they *appear* to have, then these behavior-reading chimps *would* make the same predictions of the CGI figure's actions as their belief-attributing counterparts. However, it needs to be remembered that the point of the prescreening A/R test was to screen out such chimps in the first place. If the chimps take the occluded objects as having the second-order relation that they *appear* to have, despite their being shown that the objects really do not instantiate these second-order relations, then these chimps are not able to make an appearance-reality distinction involving occluded objects, and so they would not be included as subjects in the belief-attribution test. Thus the only behavior-reading chimps that could plausibly squeeze through the A/R test are those that would be expected to represent the occluded objects in the test videos as having the second-order relations that they actually have, appearances to the contrary. But these chimps, as noted, would be expected to react differently to the different endings in the test videos from their belief-attributing counterparts.

Thus, unlike any test used in the past or currently being used, this belief-attribution test has the power to distinguish genuine belief-attributing chimpanzees (as well as other great apes) from their behavior-reading and perceptual-state attributing counterparts. It has been the challenge of designing an experimental approach capable of making this distinction that has severely retarded progress in animal mindreading research (see Fitzpatrick 2009; Heyes 1998; Lurz 2009, 2011; Penn and Povinelli 2007; Povinelli and Vonk 2006). The experimental protocol proposed here has the potential to move the field forward. Finally, it is relevant to note that this experimental approach can be used to test belief-attribution in infants and severely aphasic subjects, which are areas of research in need of more sensitive diagnostic measures of belief attribution (see Song and Baillargeon 2008; Varley 2001).

Notes

I wish to thank Marietta Dindo, Bob Mitchell, and Julie Smith for their very helpful comments on an earlier draft of this essay and Whitney Pillsbury for creating the images for figures 17.5–17.8.

1. There is a vocal dissenting minority, however, that maintains that these animals are at best merely gifted behavior readers (Heyes 1998; Penn and Povinelli 2007; Povinelli and Vonk 2006).
2. Harman (1978) once made a similar point regarding Premack and Woodruff's (1978) original discriminatory mind-reading tests.
3. Krachun and colleagues (2010) reject this possibility, however, on the grounds that "chimpanzees almost always looked back and forth between the containers [on the test trials] before making a choice, suggesting that they were not making an automatic, impulsive decision" (p. 162). But there is no reason to suppose that following the rule of choosing the colored cup based on the current contents of the yellow box would necessarily result in the chimps making an automatic, impulsive decision. After all, the chimp would still need to (a) remember what the current content of the yellow box is and then (b) apply the discrimination rule to this remembered bit of information. It is surely not implausible to suppose that such an act of deliberation might manifest itself in the chimp looking back and forth between the cups, because perhaps doing so aids the animal in calling up the particular discrimination rule for each of the colored cups.

4. The notion of "revisability" here is similar to Bennett's (1976) notion of "educability," which he also took to be a mark of beliefs (as a kind) that distinguished them from perceptions (as a kind).
5. Thompson and Oden (2000) go further and interpret the success of chimpanzees on such tasks as evidence that the animals judge the sameness/difference of the abstract relations exemplified by the different pairs. All that is being assumed, however, is that the chimpanzees represent the abstract, second-order relations exemplified by each of the different pairs, not that they understand that the abstract relation exemplified in one pair of objects is numerically identical with/different from the abstract relation exemplified in the second pair of objects. On this point, it is relevant to note that Penn, Holyoak, and Povinelli (2008) have argued that chimpanzees could pass RMTS tests by simply representing the level of variability (entropy) existing between the items in a pair and by learning the conditional rule that if the between-item variability of the sample pair is low/high, then they should select the pair in the choice display that has a between-item variability that is low/high. The chimps, on such a view, are certainly not credited with understanding relations about relations—that is, that the between-item variability relation exemplified in the sample pair is the very same relation exemplified in the choice pair. Nevertheless, they are still credited with representing the between-item level of variability of the sample and choice pairs, which Penn, Holyoak, and Povinelli acknowledge is a representation of an abstract, second-order relation. Thus Penn, Holyoak, and Povinelli's interpretation of the RMTS data is quite consistent with the argument made here.
6. The term is from Kanizsa (1979). Kanizsa distinguished two kinds of filling-in illusions. In some instances the visual system fills in the missing shape of an occluded object without filling in other properties characteristic of the sense modality of vision, such as a color and brightness. The represented part of the occluded shape, in such cases, is amodally filled in. The black bar in figure 17.3a, for example, is seen as complete, but the visual system does not fill in the color or brightness of the occluded section of the bar. In other cases of filling in, however, the color and brightness of the illusory shape are filled in as well, as in a Kanizsa triangle illusion (not shown here). In such cases, the illusory object is represented with all the properties (e.g., color, brightness, shape, etc.) associated with the visual modality. Thus Kanizsa described the illusory object in such cases as being modally completed by the visual system.
7. Krachun, Call, and Tomasello (2009) received positive results with an appearance-reality (A/R) task on chimpanzees using magnifying/minimizing lenses. Their results do not bear directly on the A/R task here, however, since they do not involve the use of occluded objects as stimuli.
8. Such methods are routinely used by magicians to evoke surprise in their audience, as in Goldin's box-sawing trick. So one can think of the A/R test that I am

proposing here as a kind of magic trick for animals. It is not entirely surprising, then, that when I showed my three-year-old son the actual video sequence illustrated in figure 17.4a he exclaimed, "Daddy, that's magic!"

9. Since right and left are sometimes difficult for animals to distinguish, the difference between the two paths can be made more salient to the test animal by placing a distinct type of object at the end of each, such as a man standing at one end and a woman at another. These distinguishing figures, moreover, could be pictures of human subjects, such as trainers or experimenters, that the test animal knows.
10. This is "surprising" since the CGI figure did not see the placement of the stimuli and thus has no reason to believe that they are other than they appear—namely, occluded stimuli that have the same shape. Thus, given the CGI's behavior in the habituation videos in response to occluded stimuli that looked to be (and were) the same shape, the CGI figure in this situation is expected to take the right path—at least, that is what a test animal should expect of the CGI figure if it is capable of understanding that the CGI figure, in the situation, mistakenly believes that the occluded stimuli are the same in shape.
11. This is "surprising" since the CGI figure did not see the placement of the stimuli and thus has no reason to believe that they are other than they appear—namely, occluded stimuli that have different shapes. Thus, given the CGI's behavior in the habituation videos in response to occluded stimuli that looked to be (and were) different in shape, the CGI figure in this situation is expected to take the left path—at least, that is what a test animal should expect of the CGI figure if it is capable of understanding that the CGI figure, in the situation, mistakenly believes that the occluded stimuli are different in shape.

References

Bennett, J. 1976. *Linguistic Behaviour*. Indianapolis, IN: Hackett.

Bermúdez, J. L. 2003. *Thinking Without Words*. Oxford: Oxford University Press.

——. 2009. "Mindreading in the Animal Kingdom." In R. Lurz, ed., *The Philosophy of Animal Minds*, pp. 145–164. Cambridge: Cambridge University Press.

Burdyn, L., and R. Thomas. 1984. "Conditional Discrimination with Conceptual Simultaneous and Successive Cues in the Squirrel Monkey (*Saimiri sciureus*)." *Journal of Comparative Psychology* 4:405–413.

Buttelmann, D., M. Carpenter, J. Call, and M. Tomasello. 2007. "Encultured Chimpanzees Imitate Rationally." *Developmental Science* 10:F31–F38.

Call, J., and M. Tomasello. 1999. "A Nonverbal False Belief Task: The Performance of Children and Great Apes." *Child Development* 70:381–395.

Carruthers, P. 1996. *Language, Thought and Consciousness*. Cambridge: Cambridge University Press.

Csibra, G. 2008. "Goal Attribution to Inanimate Agents by 6.5-Month-Old Infants." *Cognition* 107:705–717.

Davidson, D. 2001. *Subjective, Intersubjective, and Objective.* Oxford: Oxford University Press.

Emery, N. J., and N. S. Clayton. 2008. "How to Build a Scrub-Jay That Reads Minds." In S. Itakura and K. Fujita, eds., *Origins of the Social Mind,* pp. 65–98. Tokyo: Springer.

Fagot, J., E. Wasserman, and M. Young. 2001. "Discriminating the Relation Between Relations: The Role of Entropy in Abstract Conceptualization by Baboons (*Papio papio*) and Humans (*Homo sapiens*)." *Journal of Experimental Psychology* 27:316–328.

Fitzpatrick, S. 2009. "The Primate Mindreading Controversy: A Case Study in Simplicity and Methodology in Animal Psychology." In R. Lurz, ed., *The Philosophy of Animal Minds,* pp. 258–277. Cambridge: Cambridge University Press.

Fujita, K. 2006. "Seeing What Is Not There: Illusion, Completion, and Spatiotemporal Boundary Formation in Comparative Perspective." In E. A. Wasserman and T. R. Zentall, eds., *Comparative Cognition,* pp. 29–52. Oxford: Oxford University Press.

Gallup, G. 1982. "Self-Awareness and the Emergence of Mind in Primates." *American Journal of Primatology* 2:237–248.

Gergely, G., Z. Nadasdy, G. Csibra, and S. Biro. 1995. "Taking the Intentional Stance at Twelve Months of Age." *Cognition* 56:165–193.

Hare, B. 2001. "Can Competitive Paradigms Increase the Validity of Experiments on Primate Social Cognition?" *Animal Cognition* 4:269–280.

Hare, B., J. Call, B. Agnetta, and M. Tomasello. 2000. "Chimpanzees Know What Conspecifics Do and Do Not See." *Animal Behaviour* 59:771–785.

Hare, B., J. Call, and M. Tomasello. 2001. "Do Chimpanzees Know What Conspecifics Know?" *Animal Behaviour* 61:139–151.

Hare, B., and M. Tomasello. 2004. "Chimpanzees Are More Skillful in Competitive Than in Cooperative Cognitive Tasks." *Animal Behaviour* 68:571–581.

Harman, G. 1978. "Studying the Chimpanzee's Theory of Mind." *Behavioral and Brain Sciences* 4:576–577.

Heider, F., and M. Simmel. 1944. "An Experimental Study of Apparent Behavior." *American Journal of Psychology* 57:243–259.

Heyes, C. 1998. "Theory of Mind in Nonhuman Primates." *Behavioral and Brain Sciences* 21:101–148.

Humphrey, N. 1980. "Nature's Psychologists." In B. Josephson and V. Ramachandran, eds., *Consciousness and the Physical World,* pp. 57–75. Oxford: Pergamon.

Kaminski, J., J. Call, and M. Tomasello. 2008. "Chimpanzees Know What Others Know, But Not What They Believe." *Cognition* 109:224–234.

Kanizsa, G. 1979. *Organization in Vision: Essays on Gestalt Perception.* New York: Praeger.

Krachun, C., J. Call, and M. Tomasello. 2009. "Can Chimpanzees Discriminate Appearances from Reality?" *Cognition* 112:435–450.

Krachun, C., M. Carpenter, J. Call, and M. Tomasello. 2008. "A Competitive Nonverbal False Belief Task for Children and Apes." *Developmental Science* 12:521–535.

——. 2010. "A New Change-of-Contents False Belief Test: Children and Chimpanzees Compared." *International Journal of Comparative Psychology* 23:145–165.

Kuhlmeier, V., K. Wynn, and P. Bloom. 2003. "Attribution of Dispositional States by Twelve-Month-Old." *Psychological Science* 14:402–408.

Lurz, R. 2007. "In Defense of Wordless Thoughts About Thoughts." *Mind and Language* 22:270–296.

——. 2008. "Animal Minds." *Internet Encyclopedia of Philosophy*. Retrieved from http://www.iep.utm.edu/ani-mind/.

——. 2009. "If Chimpanzees Are Mindreaders, Could Behavioral Science Tell? Toward a Solution to the Logical Problem." *Philosophical Psychology* 22:305–328.

——. 2011. *Mindreading Animals*. Cambridge: MIT Press.

O'Connell, S., and R. Dunbar. 2003. "A Test for Comprehension of False Belief in Chimpanzees." *Evolution and Cognition* 9:131–140.

Penn, D., and D. Povinelli. 2007. "On the Lack of Evidence That Non-human Animals Possess Anything Remotely Resembling a 'Theory of Mind.'" *Philosophical Transactions of the Royal Society B* 362:731–744.

Penn, D., K. Holyoak, and D. Povinelli. 2008. "Darwin's Mistake: Explaining the Discontinuity Between Human and Nonhuman Minds." *Behavioral and Brain Sciences* 31:109–178.

Povinelli, D., and J. Vonk. 2006. "We Don't Need a Microscope to Explore the Chimpanzee's Mind." In S. Hurley and M. Nudds, eds., *Rational Animals*, pp. 385–412. Oxford: Oxford University Press.

Premack, D. 1983. "The Codes of Man and Beast." *Behavioral and Brain Sciences* 6:125–167.

Premack, D., and A. J. Premack. 1997. "Motor Competence as Integral to Attribution of Goal." *Cognition* 63:235–242.

Premack, D., and G. Woodruff. 1978. "Does the Chimpanzee Have a Theory of Mind?" *Behavioral and Brain Sciences* 1:515–526.

Santos, L., J. I. Flombaum, and P. Webb. 2007. "The Evolution of Human Mindreading: How Nonhuman Primates Can Inform Social Cognitive Neuroscience." In S. M. Platek, J. P. Keenan, and T. K. Shackelford, eds., *Evolutionary Cognitive Neuroscience*, pp. 433–456. Cambridge: MIT Press.

Sato, A., S. Kanazawa, and K. Fujita. 1997. "Perception of Object Unity in a Chimpanzee (*Pan troglodytes*)." *Japanese Psychological Research* 39:191–199.

Smith, H., J. King, E. Witt, and J. Rickel. 1975. "Sameness-Difference Matching from Sample by Chimpanzees." *Bulletin of the Psychonomic Society* 6:469–471.

Song, H., and R. Baillargeon. 2008. "Infants' Reasoning About Others' False Perceptions." *Developmental Psychology* 44:1789–1795.

Surian, L., S. Caldi, and D. Sperber. 2007. "Attribution of Beliefs by Thirteen-Month-Old Infants." *Psychological Science* 18:580–586.

Thompson, R., and D. Oden. 2000. "Categorical Perception and Conceptual Judgments by Primates: The Paleological Monkey and the Analogical Ape." *Behavioural Processes* 35:149–161.

Thompson, R., D. Oden, and S. Boysen. 1997. "Language-Naive Chimpanzees (*Pan troglodytes*) Judge Relations Between Relations in a Conceptual Matching-to-Sample Task." *Journal of Experimental Psychology* 23:31–43.

Uller, C. 2004. "Disposition to Recognize Goals in Infant Chimpanzees." *Animal Cognition* 7:154–161.

Varley, R. 2001. "Severe Impairment in Grammar Does Not Preclude Theory of Mind." *Neurocase* 7:489–493.

Vonk, J. 2003. "Gorilla (*Gorilla gorilla*) and Orangutan (*Pongo abelii*) Understanding of First- and Second-Order Relations." *Animal Cognition* 6:77–86.

Vonk, J., and S. MacDonald. 2002. "Natural Concept Formation by a Juvenile Gorilla (*Gorilla gorilla gorilla*) at Three Levels of Abstraction." *Journal of the Experimental Analysis of Behaviour* 78:315–332.

——. 2004. "Levels of Abstraction in Orangutan (*Pongo abelii*) Categorization." *Journal of Comparative Psychology* 118:3–13.

Wood, J., and M. Hauser. 2008. "Action Comprehension in Non-human Primates: Motor Simulation or Inferential Reasoning?" *Trends in Cognitive Sciences* 12: 461–465.

18

Minding the Animal in Contemporary Art

JESSICA ULLRICH

At the Venice Biennale in 2003 the Swiss artist duo Peter Fischli and David Weiss projected onto a black wall hundreds of questions, from the in-depth to the banal, including "Will happiness find me?" "Is everything drifting apart?" and "What is my dog thinking?" (see Fischli and Weiss 2003 for all the questions presented). The questions are such that they cannot be answered and remind us of the limits of human understanding and influence. It is probably not possible to really know what my dog thinks—possibly the same applies to human beings—but this is no reason for showing any less concern or respect to an animal. It is the otherness of animals that appeals most to some contemporary artists. They are curious about "the absolute alterity of [the] neighbor," as Derrida (2002:380) put it.[1]

In this chapter I present works by six artists from different countries to show different modes of investigating the minds of animals in art. Artists, just as other human beings, have to invent ways to interpret living animals (see Ullrich, Weltzien, and Fuhlbrügge 2008). The artists discussed here have developed methods to engage in art with animal agents. In order to select these six viewpoints, I asked, "What ways of approaching live animals can be identified in contemporary art?" I found three basic strategies that are useful to work with: observing an animal, imitating an animal, and adopting the point of view of an animal.

Observing an Animal

The first example is the video *Vigia* (Guard) by Brazilian artist Hugo Fortes (2005; see Ullrich and Weltzien 2009). It shows the face of a young dachshund occupying the entire screen, filling it with the animal's presence (see figure 18.1). The dog is loosely wrapped in white sheets and thereby framed like a picture. The artwork claims this dog is worth the effort of attentive observation: it is forcing the viewer to attend to the face of this little dog, to an unspectacular moment of its life doing nothing. The artist calls the video a "dog performance by Brioche" and thereby not only informs us about the name of his dog but also assigns some agency to the dog, whose actions or rather whose inactivity determines the narration and the length of the video. She barely moves, so that the video almost seems like a film

FIGURE 18.1. Hugo Fortes, *Vigia* (Guard), 2005. Video, 25 min. Courtesy of the artist.

still. Almost nothing happens in the next twenty-five minutes: the dog falls asleep once for a few seconds, wakes up again, and finally leaves her bed, thereby ending the performance.

The video enacts a fascinating dynamic. The face and especially the tired eyes of the dog that never meet the eyes of the beholder almost cast a spell on the viewer and thus involve him or her in the work. The viewer has ample time to meditate on his or her own way of looking and the different way of looking of the animal. He or she can notice the subtle movements of the pupils, the delicate vibration of the ears, and the gentle twitching of the nose. Brioche's eyes are never fully closed, not even when she falls asleep. It is as if her eyes mock our eagerness to read something in the eyes of the other. In one segment of the film, she starts dreaming, evidenced in the wild fluttering of the eyelids and jerking of the nostrils. The longer we look, the more alien the dog becomes. We are excluded not only from her sensory experience—we do not even see what she sees—but also from her state of mind, her feelings or dreams. As John Berger (1980) has shown, to become involved with looking is the starting point of relating to the animal, to address the animal no matter if the animal itself offers some kind of interaction or not. Just by closely looking at the dog, we experience the supposedly familiar pet animal as strange and unfamiliar. The functions that are normally attributed to dogs, such as alertness, a quality implied in the title of the work, lose their significance: Brioche's sleepy, indifferent gaze rules out any use of these functions and even any artistic appropriation of them. The video makes clear that Brioche may possibly see in a physiological way similar to the way a human being sees. Nevertheless, it demonstrates that Brioche sees the world through her own eyes and that we, as humans, will never be able to experience Brioche's particular view of the world.

The philosopher Jacques Derrida was shocked when he realized that animals have their own perception of humans. He tells us about this moment of epiphany in his long essay "The Animal That Therefore I Am (More to Follow)" (Derrida 2002).[2] By seeing his cat looking at him, he recognizes himself as somebody who is being looked at and realizes the equal meaning of the gaze for human as well as nonhuman animals. At this moment he becomes aware of his responsibility toward animals. Hugo Fortes's video, in my view, also presents Brioche as a "significant other" with her own point of view, one who cannot simply be used as a symbol or metaphor reflecting human desires, interests, needs, or fears (see Ullrich 2008). Fortes's dispassionate documentary of the gaze of his pet makes us aware of the fact that dogs have their own inner worlds, and it thereby limits any claim to the absoluteness of the anthropocentric gaze.

Albanian video artist Anri Sala (2003) has a different approach suggesting the same thing. In his five-minute film *Time After Time,* we see the black silhouetted image of a solitary horse surrounded by busy traffic on a highway in front of an undefined cityscape. We can only guess what happened or how the horse was trapped in this situation. The city noise sounds like a choppy sea and imparts a dynamic rhythm to the scene. Just like Fortes, Sala is working with a near total absence of camera movement, which has the effect of freezing scenes into paintings. Both artists are giving their full attention to seemingly marginal scenes. They both mainly observe, refusing to embed their images in a clear flowing narration. The static images enhance the perceptions and sensations of the viewer. This strategy transcends the purely physical being of the observed animal and opens a door to its inner state, even though we might not understand it or understand it correctly. The scene illumination changes constantly, and the camera alternates between a sharp and blurred focus so that the horse keeps losing its clear outline while two light sources in the background grow bigger and seem to come closer until they almost fill a big part of the screen. It is as if this camera method represents our own omnipotent stare—the omnipotent stare of a human being. By forcing us to keep looking, Sala might also imply that we owe a form of attentiveness, that we are too blinded by our own interests to see this animal.

Making this film, Anri Sala might also have asked himself heterophenomenological questions, such as "What is the perception of an approaching car like for an animal with a well-developed flight instinct?" In this video we are obviously not in the position of the horse, but we may see and hear the environment as if we are the horse. The headlights of cars, for example, become the menacing eyes of a predator. And this threat achieves such a central meaning that everything else loses importance. When our attention is drawn back to the horse, everything looks the same as before. The images are in a state of constant becoming and decomposition, creating a nightmarish atmosphere of transition. The fading focus of the camera might serve to simulate the weakening gaze of the horse, the slow attenuation of its visual senses, possibly foreshadowing its death.

Critics see the video as a "metaphor for a clash of nature and progress, and our inability to focus on the needs of our fellow creatures" (Linga 2007). Anri Sala forces the viewer to watch a scene in which nothing really happens, one that he or she might otherwise disregard. No car stops, nobody tries to help, not even the artist and certainly not the viewer of this film. Our inability to act makes us compassionate offenders. The horse fades out of focus again as if it tries to draw itself back from attention or to withdraw into its inner being by closing its eyes. It does not seem to be able to actively flee the situation. Everything becomes diffused and blurred: The horse becomes part of the night. Just before everything turns into blackness, a swelling noise from approaching trucks brings the animal back into focus. The headlights illuminate the scene abruptly and unfold the extent of the horse's misery. It is famished, barely able to stand; its posture suggests resigned pain. It lifts its hind leg, possibly because it hurts or in a vain attempt at self-defense. As soon as the light, the noise, and thus the threat have passed, the setting of the camera changes again, and the night returns. The headlights save the horse from disappearing, but they also seem to harm the animal. The maltreated horse lifts its leg for a painfully long time, most likely because it is injured. By representing the horse in this way, Sala teaches the viewer a lesson in empathy. He or she can empathize with the horse's bodily pain and its hopelessness.

Anri Sala's camerawork gives a brief insight into the inner life of the horse by simulating the visual and auditory perceptions that it might experience. The video posits a heterophenomenological question: how does this horse feel (or how do we imagine it feels) when we consider the perceptions it probably experiences, its apparent pain and dead-end fate? We could give an answer by thinking about how helpless and hopeless we have felt when we were in pain and alone in a more or less hostile environment. Perhaps we can answer the question affectively by watching the video with empathy and analyzing the emotions we have while watching it. Artworks like these may generate concern or respect for the animal and by doing so could possibly lead to a sustained ethical relationship toward animals. (For discussion of heterophenomenology, see Radner 1994).

Imitating an Animal

Other artists try to come closer to the minds of animals not simply by observing animals but also by imitating them. The Austrian Arnulf Rainer, for example, wished to attain the state of mind of an animal in order to become a better artist (see Lenain 1997). He chose two chimpanzees, Lady and Jimmy, to be his teachers (see figure 18.2). Rainer declared the apes to be the coauthors of their common painting activities in 1979.

Figure 18.2. Arnulf Rainer, *Parallel malaktion mit schimpansen* (Parallel painting action with chimpanzee), 1979. Mixed media on photograph. Courtesy Jablonka Galerie, Cologne.

Rainer tried to imitate Jimmy's and Lady's every movements and brush strokes as they painted together. He hoped thereby to attain the same clarity and intensity of abstraction as the chimps. The experiment could have been a success considering the belief that "the capacity to recognize the bodily feelings of another is related to one's own imitative capacities" (see "Empathic" n.d.). If Rainer had observed closely the bodily movements and facial expressions of the chimpanzees, he probably would have gotten a sense of how the animal felt and would have been able to produce corresponding movements or expressions. His prejudices, however, about the way chimps are likely to behave obstructed this process: He assumes that they are wild animals that cannot control their emotions, so he tried to behave in this assumed manner and exaggerated his own "wild" behavior without paying close attention to his fellow painters. In fact, as Thierry Lenain has shown, apes can concentrate and control their actions and have to exert much less physical effort to fill a canvas with rhythm and energy than humans do.[3]

Film footage of the simultaneous painting performances reveal that the actions were irritating for the chimp (see Brödl 1979). At first one can see Jimmy painting—calm and absorbed—until Rainer's agitated movements start to irritate him. Jimmy stops painting, infected with the nervous aggression of the human, and chases Rainer away from the canvas. Rainer soon abandoned the experiment. One reason he gave was that he never managed to attain the immediacy, unselfconscious spontaneity, and directness of his animal partners—in his mind, attributes of the ideal artist he admired in the chimps.

This artwork does not show much about the animal's mind, even though it is intended to do so. This project reveals almost nothing about the creative animal, but a great deal more about Rainer's view of himself and his idea of an artist. By focusing on the animal, he paradoxically delivered a commentary on human attitudes. Even if Rainer claimed that the chimp provided a model for the artist, and even if he considered his own capacities to be inferior, he nonetheless deployed artistic categories and methods that are distinctly human. He made the chimp an involuntary participant in a game that only humans understand, with the result that Jimmy basically functioned merely as an extra in Rainer's meditative conversation with himself. Despite this fact, it is an interesting artwork because it challenges the dominant idea of artistic creativity as the result of individual inspiration, human genius, and intention. In other words, the work can be read as an attempt to unmask the illusory character of notions of artistic genius.

Other artists also believe that concepts of individual creativity are outdated. Collective creativity is the new paradigm, apparently applicable to the following work but ultimately not. Beginning in the 1980s, Berlin artist Katharina Meldner has worked on art in some ways collaboratively with ants (see Ullrich and Weltzien 2009). Ants, like some other insects, are often said to have a collective mind rather than individual minds. That makes it even more difficult to relate to them and imitate their behavior. But the imitation of individual ants was the generator of meaning making for Meldner. She traced in pencil with great precision the daily routes covered by ants, thus imitating their traces with her pencil (see figure 18.3), and wrote an accompanying report in which she recorded the weather and the insects' behavior. To encourage them to enter the paper, she put sugary water on one spot on the paper. Sometimes she had to tease the ants out of their nests by spilling some sweet water. Then she operated in the same,

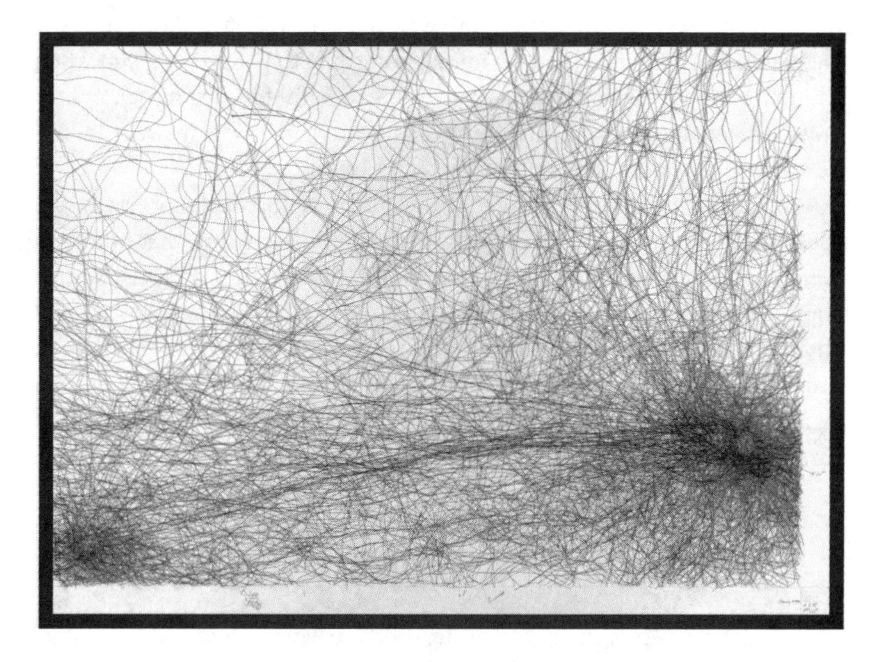

FIGURE 18.3. Katharina Meldner, *Wege der ameisen (17.8.1998, 17:03–20:11 Uhr)* (Ways/paths of the ants (08/17/1998, 5:03 pm–8:11 pm), 1998. Drawing 1 out of 3, pencil on paper. Courtesy of the artist.

almost pedantic, scientific way as an ethologist, even drawing up an ethogram (a category of all types of behavior of a species). However, she did not try to connect pencil drawings and notes but instead left everything open for a more poetic interpretation. She sought and found something in the traces of the insects and made it visible—made it "come into the picture"—without assigning any meaning to this "something." She accepted that the traces have a meaning but did not propose what it might be. In her book *When Species Meet*, Donna Haraway (2007) reminds us that "there is no general answer to the question of animals' agential engagement in meanings, any more than there is a general account of human meaning making" (p. 262). The traces of Meldner's ants are not merely proofs of authenticity or extraneous by-products of the creative process; they are in fact the signature of the animal coauthors and constitute the work itself. This kind of attending to an animal and its alleged intentions and desires for its own sake paints a utopian picture of a world in which one can acknowledge animals without considering what use they might have for humans.

Meldner meets the animal in a polite way, to borrow the words of Vinciane Despret (2008). Politeness to an animal according to Despret is given when a human-animal asks questions that are meaningful for the animal. That is how one gets more interesting, innovative, and surprising answers than if he or she does not even try to interact with the animal.

Both imitating approaches result in fascinating artworks precisely because they keep the secret of animal minds. But it also seems that imitation is not the best strategy to really understand an animal. If an artist simply copies movements without insight into the motivations or the inner state of an animal, a real dialogue cannot develop. Haraway (2007) suggests that, in order to communicate with the animals, "we have to learn who they are in all their nonunitary otherness in order to have a conversation on the bias of carefully constructed, multisensory, compounded languages" (p. 263).

Adopting the Point of View of an Animal

The last strategy artists pursue in order to come close to the animal mind is adopting the point of view of an animal. Bruce Nauman uses the setting and the practices of behaviorism for his work *Learned Helplessness in Rats (Rock and Roll Drummer)* of 1988. The title of the work comes from a *Scientific American* article about experiments on rats and auditory stress ("Stressed Out" 1987). The term *learned helplessness* in the scientific literature describes a "psychological condition in which a human being or an animal has learned to believe that it is helpless in a particular situation. It has come to believe that it has no control over its situation and that whatever it does is futile. As a result the human being or the animal will stay passive in the face of an unpleasant or harmful situation, even when one does actually have the power to change its circumstances" (see "Learned Helplessness" 2010). Nauman wanted to bring his audience into the position of rats in a learned helplessness experiment. The installation is a closed system that can only be entered and left through one door. There is no daylight, and the whole setting is difficult to grasp at first sight. The audience is confronted with an empty yellow Plexiglas maze and a video projector showing three alternating video tracks of rats negotiating the maze. The maze is scanned in real time by a surveillance camera and a punk rock drummer is playing furiously and badly. The different, simultaneously changing, visual impressions in a rather dark environment and the almost unbearable noise disorient viewers. Because of the noise, they cannot hear themselves think anymore, as one visitor has put it. The viewer moves around in the system and experiences himself or herself as part of it. A viewer may lose his or her external point of view and gain a more internalized one. The causal connection between projection and monitors is not apparent from first visual impressions. The functioning of the system is not obvious. Nauman strategically irritates the viewer to keep him or her uncertain and uninformed. The viewer loses control over the experience, which generates feelings of subjection and anxiety. He or she experiences entrapment, isolation, intensified bodily awareness, excessive sensitivity, and feelings of being "overwhelmed." The viewer becomes helpless because he or she cannot predict or anticipate the next stimulus—but, unlike a lab animal, the viewer is free to leave the gallery. Nauman's exhibit supports research on learned helplessness proving that stress reactions can be similar in humans and nonhuman animals.

This artwork, however, does not really make one aware of states of animals' minds in laboratory situations. It rather seems to me that, in situations of pure stress, confusion, and helplessness, animals, like humans, would lose the connection to the self ("I cannot hear myself think anymore") and also the connection to other living beings, even to those in similar situations. The ability to empathize weakens as one's own pain makes one less receptive to the pain of others. But, just by letting us experience stress and making us confused (an intention suggested by the title), this artwork reminds us of our own animal state and of the fact that we share not only physiological but also psychological and mental traits with animals.[4]

There have also been other attempts to explore the way an animal views the world. My last example is thus a composition involving nonhuman animal, human artist, and a series of cameras: American artist Sam Easterson, based in Toronto, has been collecting footage from the perspective of animals and also of plants since 1998. He equipped individuals from different species—birds, mammals, reptiles, insects—with cameras mounted on a lightweight helmet that he designed (see Easterson 2007, 2010). As the animals moved around, they filmed their world from their own perspectives.[5] The idea was for each animal to determine the camera's point of view and for the artist to be guided by a way of seeing that is not his own. Even though we always see the world only through our own eyes, the viewer nonetheless is allowed to imagine, for a few minutes, that his or her view of things is not the only one. One can ask if these animals are "just objects for the data-gathering subjects called people" or rather "symmetrical actors" with a similarity to humans (Haraway 2007:262).

Certainly these video recordings are not intentional as are those of a human camera operator, but they are not pure objects either. But how close can one get to the animal's mind by seeing life through a camera attached to its head? One is at least forced to confront a question posed by Daisie Radner (1994): "What would my experiences be like if I were in the subject's circumstances and I had certain features in common with it?" (p. 392). Thus when seeing Easterson's videos one might ask, "What would a human being feel like while digging the earth like a mole or crawling on the grass like an armadillo? Because of the forward-pointing motion of the wobbly footage, the viewer even has the chance to identify, just as when one adopts the persona of a videogame avatar (for examples of his videos, see Easterson 2007, 2010). The animals in Easterson's videos claim their involvement by means of vertigo-inducing, often confusing pictures and odd-angle shots of their environment.

The artist hopes that his work "will help expand the public's capacity to understand the natural environment—in empathetic terms" (quoted in Easterson 2010). For example, the viewer can experience some kind of exhaustion after watching the buffaloes run or the curious, restless point of view of a wandering wolf. Simply viewing the nervous and jumping camera movements may produce exhaustion as well. The viewer may gain some kind of bodily empathy with the animal, even though the more important senses of the animal are still inaccessible. The digital imagery sometimes even evokes other senses, although in an impoverished state. For example, as a predator, the wolf manifests qualities quite different from those of a human: He is always on the move, responding to scent stimuli a human would not even register, the rhythm of his movements determined by his sense of smell. The work does not, therefore, truly embrace the essential differentness of animals, since it disregards their specific senses, particularly ones that are not visual.

Sam Easterson also wanted to learn more about the animals' habitats. His work enables the viewer to learn about animals' lives in remote areas without human interference, giving the viewer a glimpse into the animals' interaction with mates and other species and their encounters with rough landscape. He also hopes to increase the viewer's sensibility to endangered species' habitats and therefore lead viewers to make an impact on the animals that live there. Whereas in art animals are often treated as mere objects that are seen but do not themselves see, in this work the animals not only become the documentarians of their environment but also the subjects of their own gaze and actions. The very attempt to adopt the perspective of an animal contains an emancipatory function, because it means accepting that the animal has a point of view.

Both Nauman and Easterson aim at influencing the feelings of viewers. Their work encourages viewers to be involved with their senses. What matters, though, besides the fact that viewers experience interesting artworks, are the ethical obligations to animals that are the consequences of works such as those discussed. When we are put in the position of an animal, we might be more concerned about their well-being and we might feel empathy or respect for them. Using animals in art might thus have impact on animals in real life.

All the artists that have been presented engage with animals not as mythological figures nor symbols but as real, individual, living entities. Each in his or her own way demonstrates an interest in the mind of the animal. All the works help imagine possible nonhuman states of mind, with their alternative realities, and allow us to see that animals also interpret and invent their own worlds. But in none of them are the animals shown as subjects with whom one can communicate. Even though the visual sense is important in all the works, in none of them does the beholder ever meet the gaze of an animal. Even though the works affirm or empower the animal-other and accept animal agency, they do not consider animals to be relevant partners in real interactions. Artists still tend to talk about animals rather than try to talk with them, and, if they do, they do not risk the intersecting gaze. Two examples illustrate this common approach to animals that avoids every real bodily or mental engagement with a nonhuman being. In the video *This Little Piggy . . . Fades to Pink* (2003), by Australian artist Catherine Bell, Bell is caring for a piglet while wearing chain armour. And German artist Joseph Beuys, in the much earlier, famous performance *How to Explain Pictures to a Dead Hare* (1965), appeared to be explaining artworks to a dead animal.

It would be interesting to see more artworks that refrain from classifying animal experience with human categories and conceptions. Daisie Radner (1994) put it this way: "The question 'What is the animal like?' has to be answered in light of the animal's own physiology and behavior. The key to making progress via Innenwelt heterophenomenology is to recognize that animal experiences are not just pale imitations of our own" (p. 403). Indeed, the investigation of color vision in birds and bees has shown that it is necessary to invent new terms and color categories for these animals, like, for example, *bee violet* or *bird purple.* Still, artworks that train one's ability to attribute mental states to others and to understand that others have desires, knowledge, and intentions that are different from one's own cultivate empathy. The display of animals' difference and the implicit acceptance that animals' experience is of equal value, even while they remain "other," pave the way toward a new understanding and appreciation of animals.

Notes

1. In the Judeo-Christian tradition animals are constructed as the primary other from which man differentiates himself. In front of his cat Derrida (2002) writes, "nothing will have ever done more to make me think through this absolute alterity [otherness] of the neighbor than these moments when I see myself seen naked under the gaze of a cat" (p. 380). On Derrida's encounter with his cat and Haraway's reading of it, see Khang (2008).
2. In this essay, Derrida critiques the stance toward animals of the entire tradition of Western philosophy, including Plato, Descartes, Kant, Heidegger, Lacan, and Lévinas.
3. Outside the art world, there have been many experiments with apes drawing pictures. Nadezhda Nikolaevna Ladygina-Kohts (2002 [1935]) worked with a chimpanzee named Joni as early as 1913. There were further experiments in the 1940s such as those by Paul Schiller (1951) at the Yerkes Laboratories of Primate Biology. Zoos use drawing or painting as a kind of occupational therapy for primates and elephants (Komar and Melamid 2001; Lenain 1997). In the 1970s a number of artists painted with apes or made them coauthors of their work. One of these artists discussed by Lenain was the French painter Lucien Tessarolo, who worked side by side with the chimpanzee Kunda. He signed their canvases by writing his name while she did so with a handprint. Kunda did not always like the figurative elements Tessarolo added to their work and sometimes painted over them or waited for him to add some abstract markings. Kunda is said to have had a sense of balance and a preference for regular patterns and to have known when a work was finished.
4. Such anthropomorphic attributions of mental states to animals are often believed to be speculative or metaphysical and therefore not legitimate. Bernard Rollin (2003), however, argues that anthropomorphic locutions "based on ordinary empathetic experiences of animals' lives are needed to make meaningful claims about what animals experience" just as "we use our own individual experiences as a guide to understanding that of other humans" (p. 67).
5. Apart from Easterson, other artists who have also used cameras attached to living animals are Nobuhira Narumi and Jana Sterbak, who both work with dogs.

The idea of a dog cam is not new: David Letterman first fitted out the golden retriever Travis with a camera in 1981. It transmitted live footage into the studio. Later Letterman introduced the monkey cam.

References

Bell, C. 2003. *This Little Piggy . . . Fades to Pink.* Performance at Sutton Gallery, Fitzroy, Australia. Video available from http://www.videoartchive.org.au/cbell/.

Berger, J. 1980. "Why Look at Animals?" *About Looking,* pp. 1–26. New York: Pantheon

Beuys, J. 1965. *How to Explain Pictures to a Dead Hare.* Performance at Galerie Schmela, Düsseldorf, November 26.

Brödl, H. 1979. *Arnulf Rainer: Sternsucher* [Arnulf Rainer: Star Seeker]. Germany: Baumhaus Film Brödl.

Derrida, J. 2002. "The Animal That Therefore I Am (More to Follow)." Trans. D. Wills. *Critical Inquiry* 28:369–418.

Despret, V. 2008. Paper presented at the workshop Animal Subjects Under Observation, July 10. Berlin: Max-Planck-Institut für Wissenschaftsgeschichte.

Easterson, S. 2007. *Nature Holds My Camera: The Video Art of Sam Easterson.* Retrieved from http://www.imamuseum.org/exhibitions/natureholdsmycamera/videos.php.

———. 2010. *Sam Easterson.* Retrieved from http://greenmuseum.org/content/artist_index/artist_id-106__nosplit-z.html.

"Empathic." N.d. Retrieved from http://empathic.askdefine.com/.jh.

Fischli, P., and D. Weiss. 2003. *Will Happiness Find Me?* London: Alberta.

Fortes, H. 2005. *Vigia* [Guard]. Video presented at Festival Prog:Me, Rio de Janeiro, July 18. Retrieved from http://www.progme.org/.

Haraway, D. 2007. *When Species Meet.* Minneapolis: University of Minnesota Press.

Khang, D. 2008. *Animot: Reflections on the Works of Erin Perry and Amy Thompson.* Retrieved from http://www.vaarc.ca/wp-content/downloads/Animot_pdf.

Komar, V., and A. Melamid. 2001. *When Elephants Paint: The Quest of Two Russian Artists to Save the Elephants of Thailand.* New York: Perennial.

Ladygina-Kohts, N. N. 2002 [1935]. *Infant Chimpanzee and Human Child.* Oxford: Oxford University Press.

"Learned Helplessness." 2010. Retreived from http://wikidoc.org/index.php/Learned_helplessness.

Lenain, T. 1997. *Monkey Painting.* London: Reaktion.

Linga. M. 2007. Press release [*Time After Time* by Anri Sala]. Cambridge: MIT List Visual Arts Center. Retrieved from http://listart.mit.edu/node/169.

Morris, D. 1966. *The Biology of Art: A Study of the Picture-Making Behaviour of the Great Apes and Its Relationship to Human Art.* London: Methuen.

Nauman, B. 1988. *Learned Helplessness in Rats (Rock and Roll Drummer)*. Art installation at the Museum of Modern Art, New York. http://www.moma.org/collection/browse_results.php?object_id=81490.

Radner, D. 1994. "Heterophenomenology: Learning About the Birds and the Bees." *Journal of Philosophy* 91:389–403.

Rollin, B. E. 2003. "Scientific Ideology, Anthropomorphism, Anecdote, and Ethics." In S. J. Armstrong and R. G. Botzler, eds., *The Animal Ethics Reader*, pp. 67–74. New York: Routledge.

Sala, A. 2003. *Time After Time, Lumen Eclipse*. Zurich: Gallery Hauser and Wirth. Podcast retrieved from http://www.lumeneclipse.com/gallery/04/sala/index.html.

Schiller, P. 1951. "Figural Preferences in the Drawings of a Chimpanzee." *Journal of Comparative and Physiological Psychology* 44:101–111.

"Stressed Out." 1987. *Scientific American* 257, no. 5 (November): 30–32.

Ullrich, J. 2008. "Der Blick der Wächterin: Die Hündin Brioche in Hugo Fortes' Video 'Vigia'" [The View of the Guard: The Dog Brioche in Hugo Fortes' Video "Vigia"]. In J. Ullrich, F. Weltzien, and H. Fuhlbrügge, eds., *Ich, das tier: Tiere als Persönlichkeiten in der kulturgeschichte* [I, the Animal: Animals as Personalities in the History of Civilization], pp. 253–268. Berlin: Reimer

Ullrich, J., and F. Weltzien. 2009. *Tierperspektiven* [Animal perspectives]. Berlin: Georg-Kolbe-Museum.

Ullrich, J., F. Weltzien, and H. Fuhlbrügge, eds. 2008. *Ich, das Tier: Tiere als Persönlichkeiten in der Kulturgeschichte* [I, the Animal: Animals as Personalities in Cultural History]. Berlin: Reimer.

19

Popular Beliefs and Understanding of the Dolphin Mind

JESSICA SICKLER, JOHN FRASER, AND DIANA REISS

Dolphins have long captured the interest and imagination of the public and scientists alike, whether they are observed in the wild, onscreen, or in aquariums. This fascination has led aquariums, zoos, and similar institutions to create live dolphin exhibits and demonstrations aimed at educating the public about these very intelligent animals and their conservation. At the same time, scientists have intensified their efforts to better understand dolphin cognitive abilities. As the results of this scientific research became publicized, highlighting the mental abilities of dolphins, the Wildlife Conservation Society developed a study to explore whether an exhibition focused on dolphin cognition, communication, and scientific research could successfully engage the public in thinking about this species and wildlife in general. The goal was to focus visitor attention on the whole animal, to provide visitors with new information about the social lives and intelligence of dolphins, and to relate this information to humans' own learning and behavior. The rationale was to develop in the public a sense of connection, understanding, and empathy for these animals. The creation of the exhibition about dolphin intelligence also offered the team a unique opportunity to study public conceptions about animal minds and advance knowledge about how the public conceives and characterizes dolphins' social and cognitive abilities. With the goal of creating recommendations for a new exhibition at the Wildlife Conservation Society's New York Aquarium, we undertook an extensive, 2.5-year social science research program to understand public perceptions of dolphin cognition to inform and guide this design. This research program led to a suite of findings about public characterizations of dolphin intelligence, which we present in this chapter.

Scientific Evidence of Dolphin Intelligence

Scientists have documented many aspects of dolphins' rich social and cognitive lives through systematic study of these animals' behavior in the field, research laboratories, and aquariums. This research has demonstrated that dolphins are intelligent mammals. For example, longitudinal field studies indicate that dolphins have a complex fission-fusion social structure rivaling that of chimpanzees (e.g., Connor, Smolker, and Richards 1992; Wells, Scott, and Irvine 1987; Würsig 1978). These studies indicate that social relationships and coalitions are critical to the social lives and survival of these creatures, and they contribute to a growing literature supporting the hypothesis that dolphin and other cetacean social groups have "culture" that includes tool use and the transmission of information across individuals (Connor, Mann, Tyack, and Whitehead 1998; Krutzen et al. 2005). Like chimpanzees, dolphins also show advanced capabilities for classifying, remembering, and discovering relationships between events, for forming response rules of general utility, and for manipulating symbols (Herman 1983 for a review). Dolphins, and possibly other cetaceans, are also vocal learners (e.g., Payne, Tyack, and Payne 1983; Richards, Woltz, and Herman 1984), a rare trait in the animal world. Further, research has indicated that dolphins show stages in their early vocal learning, such as imitation, overproduction, babbling, and attrition (e.g., McCowan and Reiss 1995), that parallel those previously described for humans and avian species (Marler and Peters 1982). The evidence of such parallels in phylogenetically distinct species is striking and suggests a convergence in strategies of vocal learning. Research has also shown that dolphins are capable of mirror self-recognition (Reiss and Marino 2001), a rare capacity in the animal kingdom demonstrated only in humans, the great apes, Asian elephants (Plotnik, de Waal, and Reiss 2006), and magpies (Prior, Schwarz, and Güntürkün 2008). The demonstration of mirror self-recognition in dolphins and other nonprimates implies that it is not specific to large-brained primates and represents a striking case of cognitive convergence.

Public Knowledge of Dolphin Intelligence

The social science literature surrounding public perceptions and attitudes about animal cognition, and about dolphins in particular, is remarkably sparse considering the popularity of these animals. Studies of public perceptions have found strongly positive and supportive opinions about marine mammals, their conservation, and their welfare in captivity (Kellert 1999), as well as common characterizations of dolphins in terms of physical characteristics, ocean environment, intelligence, and use for research and entertainment (Barney, Mintzes, and Yen 2005). Some researchers have revealed how the public distinguishes the mental abilities of humans and animals, with certain types of complex thinking reserved exclusively for humans (Rasmussen, Rajecki, and Craft 1993), whereas others have found that the attribution of mental states strongly follows perceived phylogenetic lines, with large-brained mammals (such as primates and dolphins) and pets (especially cats and dogs) being placed at the top of the hierarchy of mental abilities (Driscoll 1995; Herzog and Galvin 1997; Phillips and McCulloch 2005). Two of these studies (Driscoll 1995; Herzog and Galvin 1997) specifically found that people attribute extremely high, even human-like, cognitive abilities to dolphins. The conflict between Rasmussen, Rajecki, and Craft's work and the latter findings suggests that dolphins may be a unique case in the public's perception of animals' mental abilities and confirms the premise that this species offers great potential for engaging public thinking about animal cognition.

Popular Beliefs and Understanding of the Dolphin Mind

Our first exploratory study with zoo and aquarium visitors used a two-step process to confirm previous research about perceptions of dolphins. A majority of visitors responding to open-ended interview questions characterized dolphins using personal and positive adjectives. Other frequently mentioned attributes focused on intelligence and communication abilities, primarily through use of the words *intelligent, smart,* or *communicate*, but not indicating to what extent the public understands these aspects or the depth of dolphin cognition. A semantic differential study confirmed that these descriptions supported a social stereotype about the species as positive, affective, and intelligent (Fraser, Bicknell, and Sickler 2006). This clarified that a large portion of the aquarium-going public is aware that dolphins are intelligent and communicative animals, but there was little evidence that the public understands the extent of that intelligence.

Based on these findings and the literature review, the team embarked on an in-depth study to understand how the public understands and characterizes dolphins' cognitive abilities. To do so, the team examined popular beliefs about the dolphin mind through two different lenses: characterizations of dolphins in popular media and social narratives held by the public about dolphins' cognitive abilities.

Popular Media and Public Perceptions

We recognized that dolphins are prevalent characters and symbols in the media and that this prevalence may influence public perceptions or attitudes about dolphins. The team was aware that many accounts of experiences with dolphins have been classified as nonfiction and present "facts" not validated through scientific research. Of particular note was dolphin researcher John C. Lilly's (1961, 1967) work, which initially brought the idea of an intelligent and communicative dolphin to public attention, but whose speculation (Lilly 1978) that dolphins communicate telepathically and have a high level of awareness and culture led to misconceptions.

To explore these factors, the research team reviewed a broad sample of dolphin portrayals in popular media that had some likelihood of influencing the aquarium-going public, including representations in film, television, music, and literature published since the 1950s. Sources were grouped by their general portrayal of dolphins' intelligence and communication abilities, examining how intelligence was demonstrated through knowledge, complex learning skills, and ability to plan. We categorized whether dolphins communicated (or "talked"), how they communicated, and the context of communication. We also noted when values or traits were attached to dolphins via descriptors such as peaceful, helpful, friendly, innocent, or wise, as well as when dolphins were used as literary devices. Four overarching portraits emerged from this review (Fraser et al. 2006):

1. dolphin as peer to humans, of equal intelligence or at least capable of communicating with or helping humans;
2. dolphin as symbol of freedom, peace, love, or an ideal existence;
3. dolphin as naive or innocent, in which they are subordinate to humans and vulnerable;
4. dolphin as superior to humans, associated with a higher power or intelligence.

DOLPHIN AS PEER TO HUMANS

The most common theme in popular media is dolphins as equals to humans, acting as friends and helpers. In these works dolphins welcome and desire human companionship, act as loyal assistants, or serve as protectors. For example, the dolphin Flipper, a well-known character in American film and television (e.g., Clark 1963; Shapiro 1996; Tors 1964–1968), is closely bonded with a human family with whom he communicates through vocalization and body movements. He is smart, playful, and friendly, has direct agency over his actions and a high level of intelligence, and chooses to maintain a peerlike relationship with humans. Other examples of this theme include the novel and film *Day of the Dolphin* (Merle 1969; Nichols 1973), other nonfiction accounts (e.g., *Between Species: Celebrating the Dolphin-Human Bond,* Frohoff and Peterson 2003), and a great deal of children's literature where dolphins are protectors that help child characters stay safe against a variety of threats, for example, *The Music of the Dolphins* (Hesse 1996).

DOLPHIN AS SYMBOL OF FREEDOM, PEACE, OR LOVE

In some cases dolphins are used not as a character, but as a literary device to represent a theme of idealized peace, freedom, or unconditional love. This theme most commonly surfaces in song lyrics, infrequently in literature. For example, the Red Hot Chili Peppers ("Green Heaven") and Prince ("Dolphin") use dolphins as the embodiment of peace, freedom, and love. They describe dolphins' lifestyle, behavior, and connection to the ocean to symbolize freedom and independence. In other cases dolphins are symbols of unconditional love (e.g., Olivia Newton-John's "The Promise" and Live's "The Dolphin's Cry") or represent peace, as in John Denver's song "Ancient Rhymes" and the children's novel *A Ring of Endless Light* (L'Engle, 1980).

DOLPHIN AS NAIVE, VULNERABLE, OR SUBORDINATE

In contrast to works that characterize dolphins as human peers, some children's literature places the dolphin in subordinate roles by representing them as naive, innocent, and vulnerable creatures. These depictions imply human superiority, giving humans responsibility over the care, protection, and well-being of dolphins. The characterizations idealize dolphins as innocent, peaceful, and playful, with an underlying subtext indicating that humans have agency and decision-making ability, while dolphins are subordinate as creatures that obey or can be trained. In the book *Story of a Dolphin* (Orr 1993), a dolphin is portrayed as lonely, playful, and naive, causing harm to people because it does not know any better. Orr's dolphin requires understanding, guidance, and training from humans to learn acceptable social behavior. Other books supporting this theme focus on human intervention to protect dolphins in peril. Most of these dolphin characters accept all human control with gratitude.

DOLPHINS AS SUPERIOR TO HUMANS

Interestingly, another theme portrays dolphins as superior to humans in intelligence, communication abilities, and/or spiritual purpose. While there are diverse abilities attributed to dolphins in these portrayals, each establishes competencies well beyond those of humans, including interspecies communication, telepathy (see the earlier discussion of Lilly), extraterrestrial communication, healing powers, and spiritual powers. For example, the 1990s science fiction television series *SeaQuest: DSV* (O'Bannon 1993–1996) portrayed a dolphin onboard a submarine who assists the crew, speaks with humans through a technological language interface, and even communicates with extraterrestrials. The creators of these portrayals imaginatively extend the abilities of dolphins based on some factual information from the scientific literature. We noted that some nonfiction books also make many claims about the superhuman abilities of dolphins, depicting them as spiritual intelligences, agents of a higher power, or communicators with extraterrestrial beings (e.g., Robbins 1997; Wyllie 2001).

DOLPHINS IN HUMOR

These four themes were also found in cartoons and satire, but the humorists mocked or used one or more of these characterizations to make a joke. This humor implies an acknowledgment by the humorist and reader of the existence and pervasiveness of the theme being mocked. The most well-known example is Douglas Adams's four-part "trilogy" *The Hitchhiker's Guide to the Galaxy* (1979, 1980, 1982, 1984), which satirizes human understanding of intelligence and the idea that dolphins are connected to a higher extraterrestrial power. Humorists often used the dolphin-as-peer theme, referencing the Flipper characterization but spoofing the idea that dolphins are mental equals to humans; some mocked themes of dolphin-human friendships. We even found themes of the peaceful dolphin mocked through depictions of them as tough and mean, and as concealing their aggressive nature beneath a cute image.

Our study of dolphins in popular culture defined these four themes, which seemed relatively cohesive and distinct. Each theme might influence how the public understands (or misunderstands) scientific findings about dolphin cognition. To further inform the development of the exhibit, we next studied the perspectives of the visiting public, beginning with this fou-part thematic framework, in order to delve further into how people think about dolphin intelligence and how these narratives from media might support public viewpoints about dolphin cognition.

Social Narratives About Dolphin Intelligence: Q-Methodology Study

We sought to determine public beliefs about the capacity of the dolphin mind by documenting socially held perspectives about dolphin intelligence. We began with the premise that people are likely to have varying perspectives on this topic based on different experience and worldviews, but we also felt that these views would tend to group together into overarching sets of more commonly held social perspectives. To reveal these, we used Q-methodology (Brown 1993; Stephenson 1952). Q-methodology collects a set of distinctly different individual perspectives based upon respondents' sorting of a sample of predetermined statements that represent the full range of discourse on a topic to reflect that individual's overall thinking about the issue. Inverted factor analysis is used to find patterns across the

Q sorts. These factor analyses help the researchers to determine the defining qualities of each socially held perspective.

We conducted two parallel studies, one of adults' perspectives and one of children's perspectives (Sickler et al. 2006). The analyses of the adults' and children's individual Q sorts each resulted in three social perspectives, labeled A, B, and C, for adults, and X, Y, and Z, for children. Among the adults there was consensus across all three narratives that dolphins are highly intelligent animals, with particular capacity in the area of communication; there was even a willingness to call the communication "language." Among the children there was consensus about the physical and acrobatic abilities of dolphins, which seemed to be prominent in their understanding. Beyond these similarities, the perspectives were quite distinct.

ADULT PERSPECTIVE A

Perspective A conceptualizes all animals, including humans, in an intelligence hierarchy. To this way of thinking, dolphins are extremely intelligent, particularly in their capacities for learning and communication, but remain firmly below humans in the hierarchy. Although this perspective is comfortable classifying humans as animals, it is very resistant to depictions of dolphins as superior to humans in any way, especially any portrayals of dolphins as spiritual or mystical. This perspective is far more comfortable classifying the dolphin mind as more like that of other animals than that of humans.

ADULT PERSPECTIVE B

Perspective B holds a highly favorable view of dolphins, attributing to them a range of cognitive abilities. It delineates the superiority and uniqueness of dolphins in comparison to other animals, specifically in terms of learning and emotional capacity. The context in which cognitive capacities are demonstrated is also important to this perspective. Intention and emotion are credited when behavior is presented in a favorable or altruistic light, but not supported if behavior is portrayed negatively. This group is also the most open-minded of the three to portrayals of dolphins representing spiritual or healing qualities, although members of the group do not wholeheartedly subscribe to such ideas.

ADULT PERSPECTIVE C

Perspective C attributes far more humanlike mental capacities to dolphins than perspective A, particularly regarding self-awareness, which it views as similar to that of humans. Although it does not explicitly equate dolphins to humans or indicate that their intelligence surpasses our own, it does recognize many similar cognitive skills, including reasoning, planning, and having a sense of humor, that place humans and dolphins at a similar level of intelligence. This is a far more pragmatic view of dolphins than the idealistic perspective of B; it does not adhere to a perception of dolphins as unerringly kind creatures. And, like perspective A, it rejects all concepts of healing ability and spirituality in dolphins.

CHILD PERSPECTIVE X

Perspective X focuses on positive relationships and interactions between dolphins and humans, asserting that dolphins choose, and even enjoy, being with people. To children in this perspective, dolphins have emotional and intellectual capacities that are strong, though not quite equivalent to or in excess of those of humans. This perspective believes dolphins are self-aware and make choices, yet it identifies these abilities as more like those of dogs than those of humans. It also resists attributing to dolphins certain humanlike qualities and disagrees with claims that dolphins are more intelligent than humans.

CHILD PERSPECTIVE Y

Although all of the perspectives believe that dolphins are aware of themselves, perspective Y is strongest in this belief. Perspective Y depicts dolphins as having a complex sense of self and believes strongly that dolphins think different things than humans do. This perspective clearly asserts that dolphins can talk with people, but does not grant dolphins a strong emotional capacity, learning capacity, or intentionality. This perspective stands out from the others for being open to some magical portrayals of dolphins, but it does not believe them to be smarter than people. Another distinguishing characteristic is the idea that dolphins are primarily instinctual beings.

CHILD PERSPECTIVE Z

Perspective Z views dolphins as extremely smart, self-aware, rational creatures with highly developed capacities for learning, communication, and emotion. Children in this perspective also felt that dolphins make intentional choices for their own benefit. Its most unique assertion is that dolphins are more intelligent than humans, which was ranked near the top of this perspective. At the same time, this perspective indicates that dolphins are qualitatively different from people. Although they have emotional capacity and language, those emotions are different than those of humans, and their language is not to communicate with people but to serve their own purposes.

Discussion

Our study suggests that adults believe that dolphins are highly intelligent, consistent with findings from the literature (Driscoll 1995; Herzog and Galvin 1997). The study reveals that people believe dolphins have an advanced communication system that people feel comfortable calling a language. Although such an attribution related to dolphin communication is still unclear and controversial within the scientific community, it is useful to know that the public will comfortably assume and use the term *language* in relation to dolphin communication. This study also suggests that popular media representations that present dolphins as having superhuman intelligence or spiritual qualities are generally unsupported among the public or are far less strongly believed than are other aspects of intelligence.

Two of the three adult perspectives reveal a willingness to believe that dolphins might demonstrate humanlike abilities regarding learning capacity and personal agency (planning and self-direction). Although this study does not indicate the distribution of these perspectives among the population, it suggests that individuals whose beliefs tend toward perspectives B or C are more open to demonstrations of such abilities. Additionally, although research shows that dolphins have cognitive abilities and learning skills comparable to those of humans, this study suggests that visitors whose beliefs tend toward perspective A will find some of these capacities more difficult to accept without compelling evidence. Common across all views, however, was the belief that dolphins have a complex communication system, a belief that can be an entry point for discussion across all perspectives.

Children's beliefs about dolphin intelligence differ from those of adults, with children demonstrating a greater openness to the beliefs that dolphins learn through observation, engage in interspecies communication, and are self-aware. This openness contrasts with the typically less generous adult narratives, and yet the three children's perspectives are distinct from one another. The profile of perspective X, for instance, takes an anthropocentric and anthropomorphic view in defining beliefs about dolphin abilities, attributing to them humanlike emotional abilities and focusing on statements indicating a friendly and helpful relationship between humans and dolphins. This pattern seems to echo the "dolphin as peer to human" theme from much of children's literature. Perspective Y, on the other hand, represents a unique mix of beliefs that is at once tolerant of magical portrayals of dolphins' abilities and skeptical of the animals' intentionality in behavior. Children holding perspective Y are, however, receptive to the idea that dolphins exhibit complex communication and self-awareness, highlighting that these two topics might be more successful entry points for children in this perspective to begin encountering and learning about research findings regarding dolphin intelligence. Finally, perspective Z demonstrates a more complete comprehension of animal lives as distinct from human life, which supports characteristics of psychosocial thinking (Myers, Saunders, and Garrett 2003). Animals are understood to have distinct needs not necessarily mirroring those of humans. This suggests that these children will be open to presentations that consider dolphin abilities from the dolphin's perspective.

Communication with the Public

These findings became the springboard for considering how the public might perceive, understand, and interact with the key themes in an exhibition. To test the validity of these themes, a temporary experimental pilot exhibit was developed under the operating title *Aquarium Think Tank*, organized around five supracontent topics selected for their relationship to cognitive tasks that would be familiar and concrete to things visitors do in everyday life. Each of these content topics was presented in a separate zone of the exhibit, highlighted by a one-word title stating a familiar cognitive task in common terms (see table 19.1). Within each zone the exhibit gave visitors an activity in which to engage, so as to encourage them to reflect upon what they (as humans) mentally do in that task, and showed evidence

Table 19.1 Exhibit Zones and Content Presented Within *Aquarium Think Tank* Pilot Exhibit

Exhibit Zone Title	Dolphin Research Content Presented
Interact	The social lives, group structure, and cooperative behaviors of dolphins in the wild
Think	Echolocation abilities, cross-modal perception, delayed match-to-sample, and mirror self-recognition
Communicate	Communication ability, complex understanding of symbol-based and gestural communication, flexibility of understanding and use of symbols in new contexts
Learn	Different ways of learning used by dolphins, including trial and error, observational, imitative, training, and flexible learning
Play	Use of self-directed, creative, and inventive play

and explanation from scientific research on how dolphins engage in similar, equally complex cognitive tasks.

We used a standard formative testing process for honing the messages and structure of exhibits to communicate these themes, first testing exhibit mock-ups with aquarium visitors, incorporating changes from those mock-up tests, and eventually constructing a full scale pilot exhibit for public engagement. In each case the team explored how social perspectives might inform the way visitors explored themes of cognitive science. Following the opening of the *Aquarium Think Tank* pilot exhibit at the New York Aquarium, we did a final assessment to determine how individuals representing the different social perspectives responded to and learned about the content of the exhibit.

The findings of these comparisons were striking. In terms of understanding the overarching theme of the exhibit, there were no differences between those who identified with each of the perspectives. However, there were differences in the specific topics that were of greatest interest. Those in perspective A focused most on the topics in the communication area, as was anticipated. Those in perspective B, however, were drawn to the play area that highlighted dolphins' creativity and self-directed experimentation with bubble rings, made by expelling air from their blowhole while underwater. Again, this seemed a natural fit for those who were drawn to personal and affective characterizations of dolphins. Similarly, we found differences in how these perspectives expressed concern for dolphins. Perspective B visitors who expressed concern often framed responses in terms of emotional affinity for dolphins, consistent with the tendencies in their social narrative. Those in perspectives A and C, whose entry narratives focused on scientifically based concepts of intelligence, tended to express awe at the human-dolphin parallels that emerged from these exhibits. For both, the exhibit's articulation of these capacities by highlighting scientific evidence of similarities with humans led to their connection with dolphins.

This project helped staff at the Wildlife Conservation Society to advance their thinking about how people understand dolphins and to appreciate how the public understands animal minds in general. From our perspective in the museum field, we were interested in not just conducting research but considering and testing how to apply the results of research to make the topics of cognitive science more accessible and relevant and create learning opportunities for visitors to engage with the fascinating world of the animal mind. This project highlighted the need to develop communication strategies that are in tune with the perspectives, beliefs, and biases of any audience.

Note

The authors wish to thank Paul Boyle, currently the senior vice president of conservation and education at the Association of Zoos and Aquariums, for his vision to pursue this research while he was director at Wildlife Conservation Society's New York Aquarium; a generous anonymous donor; and the Institute for Museum and Library Services, which supported the research (grant #LG-20–03–0197–03).

References

Adams, D. 1979. *The Hitchhiker's Guide to the Galaxy.* New York: Harmony.

——. 1980. *The Restaurant at the End of the Universe.* New York: Harmony.

——. 1982. *Life, the Universe, and Everything.* New York: Harmony.

——. 1984. *So Long, and Thanks for All the Fish.* New York: Harmony.

Barney, E. C., J. J. Mintzes, and C. Yen. 2005. "Assessing Knowledge, Attitudes, and Behavior Toward Charismatic Megafauna: The Case of Dolphins." *Journal of Environmental Education* 36, no. 2: 41–55.

Brown, S. R. 1993. "A Primer on Q Methodology." *Operant Subjectivity* 16:91–138.

Clark, J. B. dir. 1963. *Flipper*. Produced by I. Tors. Hollywood: Metro-Goldwyn-Mayer.

Connor, R. C., and R. A. Smolker. 1985. "Habituated Dolphins (*Tursiops* spp.) in Western Australia." *Journal of Mammology* 66:398–400.

Connor, R. C., R. A. Smolker, and A. F. Richards. 1992. "Dolphin Alliances and Coalitions." In A. H. Harcourt, and F. B. M. de Waal, eds., *Coalitions and Alliances in Humans and Other Animals*, pp. 415–443. Oxford: Oxford University Press.

Connor, R. C., J. Mann, P. L. Tyack, and H. Whitehead. 1998. "Social Evolution in Toothed Whales." *Trends in Ecology and Evolution* 13, no. 6: 228–232.

Driscoll, J. W. 1995. "Attitudes Toward Animals: Species Ratings." *Society and Animals* 3, no. 2: 139–150.

Fraser, J., J. Bicknell, and J. Sickler. 2006. "Assessing the Connotative Meaning of Animals Using Semantic Differential Techniques to Aid in Zoo Exhibit Development." *Visitor Studies Today* 9, no. 3: 1–9.

Fraser, J., D. Reiss, P. Boyle, K. Lemcke, J. Sickler, E. Elliott, B. Newman, and S. Gruber. 2006. "Dolphins in Popular Literature and Media." *Society and Animals* 14, no. 4: 321–350.

Frohoff, T., and B. Peterson, eds. 2003. *Between Species: Celebrating the Dolphin-Human Bond*. San Francisco: Sierra Club.

Herman, L. M. 1983. "Cognitive Characteristics of Dolphins." In L. M. Herman, ed., *Cetacean Behavior: Mechanisms and Functions*, pp. 363–429. New York: Wiley Interscience.

Herzog, H. A., and S. Galvin. 1997. "Common Sense and the Mental Lives of Animals: An Empirical Approach." In R. W. Mitchell, N. S. Thompson, and H. L. Miles, eds., *Anthropomorphism, Ancedotes, and Animals*, pp. 237–253. Albany: SUNY Press.

Hesse, K. 1996. *The Music of Dolphins*. New York: Scholastic.

Hills, A. M. 1995. "Empathy and Belief in the Mental Experience of Animals." *Anthrozoös* 8, no. 3: 132–142.

Kellert, S. R. 1999. *American Perceptions of Marine Mammals and Their Management*. Washington, DC: Humane Society of the United States.

Krutzen, M., J. Mann, M. R. Heithaus, R. C. Connor, L. Bejder, and W. B. Sherwin. 2005. "Cultural Transmission of Tool Use in Bottlenose Dolphins." *Proceedings of the National Academy of Sciences* 102, no. 25: 8939–8943.

L'Engle, M. 1980. *A Ring of Endless Light*. New York: Farrar, Straus, Giroux.

Lilly, J. C. 1961. *Man and Dolphin*. Garden City, NY: Doubleday.

——. 1967. *The Mind of the Dolphin: A Nonhuman Intelligence*. Garden City, NY: Doubleday.

——. 1978. *Communication Between Man and Dolphin: The Possibilities of Talking with Other Species*. New York: Crown.

McCowan, B., and D. Reiss. 1995. "Whistle Contour Development in Captive-born Infant Bottlenose Dolphins: A Role for Learning?" *Journal of Comparative Psychology* 109, no. 3: 242–260.

Marler, P., and S. Peters. 1982. "Subsong and Plastic Song: Their Role in the Vocal Learning Process." In D. E. Kroodsma and E. H. Miller, eds., *Acoustic Communication in Birds: Song Learning and Its Consequences*, 2:25–50. New York: Academic.

Merle, R. 1969. *Day of the Dolphin*. Trans. H. Weaver. New York: Simon and Schuster.

Myers, O. E. Jr., C. D. Saunders, and E. Garrett. 2003. "What Do Children Think Animals Need? Aesthetic and Psycho-social Conceptions." *Environmental Education Research* 9, no. 3: 305–325.

Nichols, M., dir. 1973. *Day of the Dolphin*. Produced by J. E. Levine. Wilmington, DE: AVCO Embassy Pictures.

O'Bannon, R. S., creator. (1993–1996). *SeaQuest DSV*. Produced by D. J. Burke, P. Hasburgh, R. S. O'Bannon, S. Spielberg, and T. Thompson. New York: NBC and Universal TV.

Orr, K. S. 1993. *Story of a Dolphin*. Minneapolis: Carolrhoda.

Payne, K., P. Tyack, and R. Payne. 1983. "Progressive Changes in the Songs of Humpback Whales (*Megaptera novaeangliae*): A Detailed Analysis of Two Seasons in Hawaii." In R. Payne, ed., *Communication and Behavior of Whales*, pp. 9–57. Boulder: Westview.

Phillips, C. J. C., and S. McCulloch. 2005. "Student Attitudes on Animal Sentience and Use of Animals in Society." *Journal of Biological Education* 40, no. 1: 17–24.

Plotnik, J. M., F. de Waal, and D. Reiss. 2006. "Self-Recognition in an Asian Elephant." *Proceedings of the National Academy of Sciences* 103, no. 45: 17053–17057.

Prior, H., A. Schwarz, and O. Güntürkün. 2008. "Mirror-Induced Behavior in the Magpie (*Pica pica*): Evidence of Self-Recognition." *PLoS Biology* 6, no. 8: e202.

Rasmussen, J. L., D. W. Rajecki, and H. D. Craft, 1993. "Humans' Perceptions of Animal Mentality: Ascriptions of Thinking." *Journal of Comparative Psychology* 107, no. 3: 283–290.

Reiss, D., and L. Marino. 2001. "Mirror Self-recognition in the Bottlenose Dolphin: A Case of Cognitive Convergence." *Proceedings of the National Academy of Sciences* 98, no. 10: 5937–5942.

Richards, D. G., J. P. Woltz, and L. M. Herman. 1984. "Vocal Mimicry of Computer-Generated Sounds and Labeling of Objects by a Bottlenose Dolphin, *Tursiops truncatus*." *Journal of Comparative Psychology* 98, no. 1: 10–28.

Robbins, D. 1997. *The Call Goes Out: Messages from the Earth's Cetaceans, Interspecies Communication*. Livermore, CA: Inner Eye.

Shapiro, A., dir. 1996. *Flipper*. Produced by L. Hool. Universal City, CA: Universal Pictures.

Sickler, J., J. Fraser, T. Webler, D. Reiss, P. Boyle, H. Lyn, K. Lemcke, and S. Gruber. 2006. "Social Narratives Surrounding Dolphins: Q-Method Study." *Society and Animals* 14, no. 4: 351–382.

Stephenson, W. 1952. "Q-Methodology and the Projective Techniques." *Journal of Clinical Psychology* 107, no. 3: 301–312.

Tors, I., producer. 1964–1968. *Flipper*. New York: NBC.

Wells, R. S., Scott, M. D., and A. B. Irvine. 1987. "The Social Structure of Free-ranging Bottlenose Dolphins." In H. Genoways, ed., *Current Mammalogy,* 1:247–305. New York: Plenum.

Würsig, B. 1978. "Occurrence and Group Organization of Atlantic Bottlenose Porpoises (*Tursiops truncatus*) in an Argentine Bay." *Biology Bulletin* 154:348–359.

Wyllie, T. 2001. *Adventures Among Spiritual Intelligences: Angels, Aliens, Dolphins, and Shamans*. Novato, CA: Wisdom.

20

Perceiving the Minds of Animals

Sociological Warfare, the Social Imaginary, and Mediated Representations of Animals Shaping Human Understandings of Animals

BRIAN M. LOWE

Mediated Spectacles Impact Human Perceptions of the Minds of Animals

Across cultures, human understandings of the minds of most nonhuman animals are derived primarily from mediated representations rather than direct interactions with nonhuman animals (hereafter termed *animals*). It is through such mediated representations that postindustrial postmodern humans typically encounter many species of animals (Kalof and Fitzgerald 2007). Because these representations are created, they necessarily reflect certain interests of their creators. They also function as sites of interaction where the public is entertained and informed about animals, even as their perceptions of the animals' minds are shaped.

The contemporary animal rights movement is increasingly using mediated encounters between postindustrial humans and animals. Jasper and Nelkin (1992) assert that the contemporary animal rights movement began with the publication of Peter Singer's *Animal Liberation* (1975) and was bolstered by Tom Regan's *The Case for Animal Rights* (1983). Singer's utilitarian and Regan's rights-based works employed linear and sustained arguments for the ethical treatment of animals supported by scientific evidence for the capacity of (some) animals to suffer. These works inspired widely noted animal rights protests, which often involved visual representations of the animals in question. In this sense, both the contemporary animal rights movement and animal advocacy movement are emblematic of the Enlightenment tradition of using evidence alongside persuasion with the goal of creating social and political change.

Case in Point: "Chimps Aren't Chumps"

A *New York Times* Op-Ed piece "Chimps Aren't Chumps," written by Steve Ross (2008), supervisor of behavioral and cognitive research at the Lester Fisher Center for the Study and Conservation of Apes at the Lincoln Park Zoo, emphasizes the connections between public perceptions of animals and their mediated representations: "A survey that I and several colleagues conducted in 2005 found that one in three visitors to the Lincoln Park Zoo assumed that chimpanzees are not endangered. Yet more than 90 percent of these same visitors understood that gorillas and orangutans face serious threats to their survival. And many of those who imagined chimpanzees to be safe reported that they based their thinking on the prevalence of chimps in advertisements, on television and in the movies" (p. 19). Thus, according to a professional zoologist, mediated representations of animals in popular entertainment and advertising influence public understandings (accurate or otherwise) of chimpanzees. How animals are represented is critical to the ways in which the public perceives them and impacts how people experience animals in their moral imagination.

This finding is in keeping with other observations of claims making and public perceptions of any issue's social significance. Bob (2005) contends that a "global morality market" exists (p. 4) within which nongovernmental organizations and others compete for public attention regarding the significance of a specific humanitarian crisis or oppressed population. For example, the Uyghurs occupy a similar position to that of the Tibetans in terms of their occupation by China, the documented human rights abuses against them, and the significant possibility of eventual cultural extermination. And, yet, despite their objective similarities in terms of their human grievances, the Uyghurs have not received equivalent popular or political receptions. In postindustrial countries like Canada and the United States (especially around colleges and universities), "Free Tibet" bumper stickers, T-shirts, and other objects promoting some aspect of Tibetan Buddhism, culture, and/or independence may be observed, film actors such as Richard Gere speak of their religious faith in Tibetan Buddhism, and public intellectuals like Jeffrey Hopkins and Robert Thurman generate scholarship, translations, and positive attention regarding Tibetan Buddhism to English-speaking audiences. These factors may be significant in explaining why, for example, in 2007 His Holiness the Dali Lama, the religious leader of the Tibetan people, after lecturing at Cornell University and Ithaca College, was received by Speaker Pelosi and President Bush in Washington, DC (over the expressed concerns of the People's Republic of China). The Uyghurs have none of the aforementioned means of gaining public attention. The ability to maintain a presence within popular mediated culture contributes to explaining why Tibet, and not other nations and peoples enduring similar crises, receives public and political attention. This case also suggests that the ability to position certain understandings of animals may significantly impact societal perceptions of them.

Social Imaginary and Public Moral Imagination

The perceptions humans hold of the minds and other aspects of animals are derived largely from a highly mediated popular culture, which informs what Taylor (2004) terms the *social imaginary*. The social imaginary consists of the norms, expectations, and practices that are taken for granted (i.e., imagined to be just the way things are) by the majority in a culture. The *social imaginary* includes "the deeper normative notions and images that underlie these expectations" and the "common understanding that makes possible common practices and a widely shared sense of legitimacy" (pp. 23–24). The social imaginary can encourage or inhibit forms of social, cultural, and economic change. For example, Steger (2008) argues that the macro-ideologies of modernity—liberalism and conservatism—emerged as their host societies were evolving a national imaginary that encouraged concepts of citizenship (instead of subjects beholden to monarchs) and national identity. Similarly, how the minds of animals are understood within the social imaginary informs the social positions and statuses of animals. Taylor argues that social life is an amalgamation of different social practices and their related images, ideas, and understandings. In this case, it strongly influences the legitimacy of human practices toward particular animals, such as which animals may be eaten (pp. 23–24). Animal portrayals from animal advocates that depict animals as worthy of ethical consideration may influence the status of the animals within the social imaginary.

Social Imaginary Impacted Through Spectacles

Contemporary animal advocates who focus on logical argumentation face a formidable problem in engaging the public regarding their understanding of the minds of animals: the public fascination with mediated spectacles. In mediated spectacles data may become obscured or distorted in favor of the visually dramatic. This mediated milieu has been termed "the spectacular" by Debord (1995, 2002), Kellner (2003), Edelman (1988), Duncombe (2007), and others (also see Baudrillard 1994). The vast majority of the information and claims (Best 2007) are disseminated, amplified, and/or distorted through the spectacular. Duncombe (2007) contends that claims makers must recognize and adapt to today's spectacular milieu rather than be swayed by an appealing myth about the efficaciousness of rational persuasion leading to agreement. Rather than believing that "if reasoning people have access to the Truth, the scales will fall from their eyes and they will see reality as it is and, of course, agree with us," claims makers must embrace the spectacular as an essential vehicle for persuasion and communication: "Spectacle is our way of making sense of the world. Truth and power belong to those who tell the better story" (pp. 7–8).

Put another way, my point is that movements that emerge out of an empirical and philosophical tradition tend to overemphasize the weight that data and evidence alone play in influencing third parties. As Duncombe argues, many social and political movements will have unrealistic expectations about facilitating social and political change if they draw on the Enlightenment assumption that reasoned arguments and empirical evidence *alone* are overwhelming weapons in their persuasive arsenal. Like the contemporary animal rights movement, many progressive political and social movements have emphasized data and theory as essential tools for building consensus while ignoring the roles played by emotional and/or visual appeals, thereby potentially alienating supporters (or unintentionally ceding these tools to their opponents): "Appeals to truth and reality, and faith in rational thought and action, are based in fantasy of the past, or rather past fantasy. Today's world is linked by media systems and awash in advertising images; political policies are packaged by public relations experts and celebrity gossip is considered news" (p. 5). The Enlightenment tradition's now myopic view initially relied on the visual—anatomical diagrams, documentation, and "transparency" in scientific and legal proceedings—as essential to establishing and supporting valid and reliable epistemological claims.

One Spectacular Strategy: Sociological Warfare

Promoting alternate understandings of the minds of animals—e.g., the perception of sentience—will necessarily involve engaging and influencing the social imaginary to bridge gaps between audience and the subject animals. *Sociological warfare* is any strategic effort to alter the public moral imagination about some aspect of the social imaginary (i.e., what is taken for granted). Sociological warfare is deployed by nonstate actors who advocate ideals and alternative visions of social life (see Lowe 2008). Sociological warfare bears some parallel to psychological or political warfare undertaken by state actors in order to pursue specific policy or geopolitical objectives (see Waller 2007). In the case of animal advocacy, sociological warfare might be applied to opposing the sale of fur garments or to promoting an alternative consciousness regarding the positions of animals within human societies (Gusfield 1981).

The primary impetus of sociological warfare is to communicate complex ideological messages to audiences with the intention of influencing the social imaginary itself. This is accomplished by providing a synthesis of direct informational appeals and emotional and/or aesthetic appeals. These efforts often involve a personalizing and subjective narrative that is widely distributed through cultural artifacts. Harriett Beecher Stowe's 1852 *Uncle Tom's Cabin* personified the suffering created by slavery in the antebellum American South. Roald Dahl and the British Security Coordination generated domestic support in the United States for involvement in World War II and eroded the credibility of isolationists through publishing pro-British articles in popular magazines, including *Colliers, Harper's, Ladies Home Journal,* and *Town and Country* (Conant 2008:41–42). Tom Dooley's 1956 best-selling (and CIA-supported) first-personal account, *Deliver Us From Evil,* of Dooley acting as a devout Roman Catholic doctor selflessly treating many of the victims of the Vietminh became the first perspective on the conflict in French Indochina encountered by many Americans (Wilford 2008). Ingrid Newkirk's *Free the Animals* (1992) presented a dramatic account of "Valerie," a police officer who becomes a member of the Animal

Liberation Front and participates in the "liberation" of animals (Newkirk 1992). In addition to providing compelling and subjective narratives of macro-level societal and political controversies, these cases also interweave data, argument, and emotionally compelling appeals with the goal of altering the public moral imagination.

Case Studies of Persuasion Within the Spectacular

One noteworthy case of data and spectacle in animal rights protests involves cruelty in the name of science. On May 28, 1984, members of the Animal Liberation Front (ALF) penetrated the laboratory of Thomas Gennarelli at the University of Pennsylvania. This raid resulted in the seizure of what were later termed the "Watergate tapes of the animal rights movement": over sixty hours of videotapes of Gennarelli's own research on test subject baboons deliberately subjected to head traumas. These tapes were quite graphic in nature and contained understandably troubling depictions of animal research. These tapes were potentially defensible in that they represented animal-based research that might hold benefits for humans who suffered from head traumas. Such justifications were eroded as the research tapes made in Gennarelli's laboratory documented several violations of the Federal Animal Welfare Act. These violations were damning in that Gennarelli had received roughly one million dollars annually in federal grants to conduct that research since 1971, and neither these tapes nor other documents revealed any scrutiny or sanction by federal government officials regarding violations of the Federal Animal Welfare Act (Finsen and Finsen 1994).

As the Gennarelli research tapes were graphic in nature, derived from an unimpeachable source (Gennarelli's own data), and documented government-supported animal abuse, they were ideal candidates for deployment in efforts to alter the public moral imagination about animal experimentation. The research tapes were compiled by People for the Ethical Treatment of Animals (PETA) into a thirty-minute documentary titled *Unnecessary Fuss* (Newkirk and Pacheco 1984). The title was derived from a 1983 interview in which Gennarelli stated that he did not wish for his research to become publicly known because "it might stir up all sorts of unnecessary fuss among those who are sensitive to these sorts of things (Finsen and Finsen 1994:68)." Copies of *Unnecessary Fuss* were distributed to the *New York Times*, the *Washington Post*, NBC's *Nightly News*, and the Cable News Network (CNN), which broadcast clips from the compilation. Two screenings of *Unnecessary Fuss* also occurred on Capitol Hill. The mediated publicity of these tapes was amplified by a campaign of civil disobedience organized primarily by PETA and the Animal Legal Defense Fund (ALDF) at both the University of Pennsylvania and the National Institute of Health (the primary source of Gennarelli's funding). The synthesis of traditional protests, civil disobedience, and public outcry generated as a result of the dissemination of *Unnecessary Fuss,* congressional pressure, and a lack of a systematic response from animal experimentation advocates led to the closure of Gennarelli's laboratory—the first time in American history that a federally funded research laboratory had been closed as a result of public protests informed by mediated images (Finsen and Finsen 1994; Jasper and Nelkin 1992; Blum 1994; Beers 2006).

Another example of sociological warfare using data and spectacle involves the Sea Shepherd Organization. Paul Watson founded the Sea Shepherd Organization in 1977 after a break with Greenpeace regarding strategies to pursue environmental or animal protection. The core strategy of Watson and the Sea Shepherd Organization was direct confrontation between their ships and whaling vessels at sea in order to interrupt and possibly disrupt commercial whaling. The ships of the Sea Shepherds are named after publicly recognized, quasi-celebrity animal advocates, including Farley Mowat, Steve Irwin, and Cleveland Amory. As noted by Heller in his 2007 account of being aboard the *Farley Mowat,* Watson succinctly summarizes the Sea Shepherd strategy as "sink ships, but don't break laws." The actions of the Sea Shepherds have been the subject of a series aired on Animal Planet titled *Whale Wars* for three consecutive seasons and is currently in its fourth season.

The Sea Shepherd Organization legitimizes its actions through appeals to international law and the scientific community in order to convince audiences that the actions undertaken by the Sea Shepherd organization defend both marine life and legal statutes. Watson justifies the Sea Shepherd's actions by noting that commercial whaling—especially in international oceanic sanctuaries—is illegal, and therefore actions undertaken to disrupt such hunts are legally protected. Watson argues that the primary justifications cited by Japanese whalers—that they are actually gathering data on whales and thus their actions are sanctioned—are false and indefensible. Conversely, Watson argues that the Sea Shepherds are acting on behalf of whales and other marine mammals in keeping with international law: "Our intention is to stop the criminal whaling. We are not a protest organization. We are here to enforce international conservation law. We don't wave banners. We intervene. . . . I don't give a damn what you think. My clients are the whales and the seals. If you can find me one whale that disagrees with what we're doing, we might reconsider" (Watson, quoted in Heller 2006:58). The Sea Shepherd Organization's strategies include placing the *Farley Mowat* between the whalers and their targets, throwing foul-smelling butyric acid onto the decks of the Japanese fleet to make the decks impassible and to contaminate whale meat, deploying "prop foulers" in the hopes of paralyzing or destroying the engines of the Japanese vessels, and then rapidly disseminating evidence of their activities and interactions with the Japanese whaling fleet to news media and supporters. In their disseminations the protesters elevate the status of whales and seals to "clients" worthy of protection. Heller reports that, while the activities of the *Farley Mowat* have limited effectiveness in actually inhibiting the Japanese whaling fleet, the press dispatches transmitted to the Sea Shepherds' supporters and others become tactically significant in this conflict. For example, Heller notes that, due to negative public sentiment generated by communications from the Sea Shepherds and Greenpeace, the Japanese whaling fleet was unable to enter an Australian refueling area, thereby delaying and limiting their hunt. In his dispatches Heller stressed the endangerment of the *Farley Mowat*'s crew members, Australian citizens who were placed in jeopardy while pursuing their goal of ending whaling in the Antarctic sanctuary. He knew that a near collision between the *Farley Mowat* and the much larger *Nisshin Maru* would channel political pressure and public outrage against the Japanese practice of whaling, as he revealed in a conversation between Heller and Watson:

> "If he would've ended it [referring to the *Nisshin Maru* ramming and sinking the *Farley Mowat*] there, that would've probably ended commercial whaling. But I still believe that not sacrificing people for that, in that way, is probably a better choice."
>
> "But personally, you're willing to make that trade off—trade your own life to stop whaling?"
>
> "Absolutely. But I'm not going to engage in some suicide mission. It's gotta be a calculated risk."
>
> The captain said, "If they had sunk us, there'd be such bad PR for them. The Australian navy would be down here in no time. They'd be hauled in for investigations. Australia would have to intervene at that point. We have Australian citizens on board. . . . "

Watson ducked into the radio room. By 0605 he already had his first press release posted. It began: "No whale will be killed this Christmas day . . . "

(Heller 2007:207–208)

This exchange demonstrates Watson's strategy of intermingling direct action, mediated accounts of direct action, and efforts to manipulate governments to intervene on behalf of the Sea Shepherds (and the organization's interpretation of international law).

Trends in Sociological Warfare

These examples suggest four interrelated strategies in sociological warfare: the truncation of time and space, the dissemination of evocative information, the personalization of animals, and the undoing of fetishism.

Time and space truncation occurs when perceptions of distance and duration are diminished because of developments in information technology and transportation (see Giddens 1984 on time and space disembedding). For example, when people engage with media accounts about the Sea Shepherd Organization's documentation of the activities of the Japanese whaling fleet, they do so from thousands of kilometers away. Likewise, they experience the events of Gennarelli's head injury studies conducted at the University of Pennsylvania as here and now, rather than in some distant temporal period or spatial location. In spite of this temporal and geographic distance, animal advocates are able to use these activities to alter contemporary public perceptions of practices involving animals and to involve the public in animal advocacy.

Evocative information refers to presenting factually correct data in an emotionally compelling manner. Rather than presenting purely informative or data-driven argumentation and claims making, evocative information sensationalizes and visualizes. Those who can offer evocative information can more readily traverse the spectacular environment. The *Unnecessary Fuss* compilation released by PETA to the media is an excellent example of evocative information in that it provided data and compelling visual images of animal suffering from an unassailable source: Gennarelli's own research (Beauchamp et al. 2008). Related to the production of evocative information is what Arluke (2006) termed, adopting the imagined perspective of Humane Society officers and animal advocates, the "beautiful case" of cruelty, that is, a case that evokes sympathy from an audience without being too graphic to bear or too mundane to be perceived as unworthy of consideration.

The *personalization of animals* (a subcategory of evocative information) refers to presenting animals as individuals with their own histories, memories, and preferences. Such personalization can be used strategically to erode boundaries of species or societal categories, as when animals widely consumed for food are named and represented as beings with life histories. "Personalization" may occur at a species level (as in the elevation in status of whales as intelligent beings) or in more microsociological scenarios whereby specific animals become subjects of sympathy (as in the case of the primate test subjects of Gennarelli's head injury laboratory, photographs of which were displayed at public protests outside of the National Institutes of Health). Personalization serves the dual functions of emphasizing the suffering of animals while simultaneously challenging the status of those who impose such suffering on these animals and the politics in which certain activities become more or less normative (Gusfield 1986). For example, in an ongoing controversy over potential deer culling by the town of Cayuga Heights, New York, opponents to the proposed culling have protested at town meetings by showing "posters of individual deer in chairs facing the trustees . . . [with captions reading] 'My life matters to me' and 'I am an individual with a family'" (Gashler 2009).

Undoing fetishism concerns making consumers aware of the origins of the products they consume. Karl Marx argued that under conditions of capitalism (which extended the social and geographic distances between workers and consumers), consumers of commodities tended to ignore the fact that commodities are the products of human labor and instead perceived these commodities as though they were endowed with qualities that labor did not provide them with (and hence commodities were like fetishes, which are endowed with qualities that they do not, of themselves, have). Revealing this "fetishism of commodities" was one of the concerns of Marx's analysis of capitalism as unique relative to other relations of production (Tucker 1978). Similarly, many animal advocates are charged with revealing to audiences how many of the products and services they consume are provided through an exploitation of animals. Animal advocates routinely face the difficulty of emphasizing the connections between products (such as food) or activities (such as animal-based research) and the unseen suffering of animals. One method to address the connection is through the utilization of media in order to inject either information or cultural resources (Swidler 1986) into the social imaginary in order to show that the production of certain commodities or activities involve animals (such as animal experimentation). In the case of the Sea Shepherd Organization, media are used to document whales being killed and processed upon whaling ships for the purpose of providing products with no apparent connection to a scientific need for whaling. Undermining facades or simulacra (see Baudrillard 1994) that may dominate the social imaginary regarding how animal-based products are created by revealing the realities of factory farming also serve to create conditions conducive for an alternative consciousness to the practices in question (Gusfield 1981).

These few examples illustrate trends from a "sociological warfare" perspective. As states are increasingly beholden to powerful economic actors that benefit from animal exploitation and are able to successfully increase their entanglements with legislative and judicial bodies (Kenner 2008), and as the capacity for animal advocates to engage in policy formation is weakened (for example, through the passage of the Animal Enterprise Terrorism Act), more direct efforts to shape the social imaginary become an important route for animal advocacy. As the social imaginary about the minds of animals is increasingly informed by popular culture, more struggles to recast the social imaginary will occur there.

Note

The research for this project was funded, in part, by the SUNY Oneonta Faculty/Professional Staff Research Grant Program. The presentation of this project at the Minds of Animals Conference in Toronto, Ontario was made possible by a faculty development grant from the Office of the Provost at SUNY Oneonta.

References

Arluke, A. 2006. *Just a Dog: Understanding Animal Cruelty and Ourselves.* Philadelphia: Temple University Press.

Baudrillard, J. 1994. *Simulacra and Simulation.* Trans. S. F. Glaser. Ann Arbor: University of Michigan Press.

Beauchamp, T. L., E. B. Orlans, R. Dresser, D. B. Morton, and J. P. Gluck. 2008. *The Human Use of Animals: Case Studies in Ethical Choice.* 2d ed. New York: Oxford University Press.

Beers, D. L. 2006. *For the Prevention of Cruelty: The History and Legacy of Animal Rights Activism in the United States.* Athens, OH: Swallow.

Best, J. 2007. *Social Problems.* New York: Norton.

Bob, C. 2005. *The Marketing of Rebellion: Insurgents, Media, and International Activism.* Cambridge: Cambridge University Press.

Blum, D. 1994. *The Monkey Wars.* New York: Oxford University Press.

Conant, J. 2008. *The Irregulars: Roald Dahl and the British Spy Ring in Wartime Washington.* New York: Simon and Schuster.

Debord, G. 1995. *The Society of the Spectacle.* Trans. D. Nicholson-Smith. New York: Zone.

——. 2002. *Comments on the Society of the Spectacle.* Trans. M. Imrie. New York: Verso.

Duncombe, S. 2007. *Dream: Re-imagining Progressive Politics in an Age of Fantasy.* New York: New Press.

Edelman, M. 1988. *Constructing the Political Spectacle.* Chicago: University of Chicago Press.

Finsen, L., and S. Finsen. 1994. *The Animal Rights Movement in America: From Compassion to Respect.* New York: Twayne.

Gashler, K. 2009. "Cayuga Heights Reviews Deer Shooting Plan." *Ithaca Journal,* November 25, p. 3A.

Giddens, A. 1984. *The Constitution of Society: Outline of the Theory of Structuration.* Berkeley: University of California Press.

Gusfield, J. R. 1981. *The Culture of Public Problems: Drinking-Driving and the Symbolic Order.* Chicago: University of Chicago Press.

——. 1986. *Symbolic Crusade: Status Politics and the American Temperance Movement.* 2d ed. Urbana: University of Illinois Press.

Heller, P. 2006. "The Whale Warriors." *National Geographic Adventure* 8, no. 4 (May): 58–100.

——. 2007. *The Whale Warriors: The Battle at the Bottom of the World to Save the Planet's Largest Mammals.* New York: Free Press.

Jasper, J. M., and D. Nelkin. 1992. *The Animal Rights Crusade: The Growth of a Moral Protest.* New York: Free Press.

Kalof, L., and A. Fitzgerald, eds. 2007. *The Animals Reader: The Essential Classic and Contemporary Writings.* New York: Berg.

Kellner, D. 2003. *Media Spectacle.* New York: Routledge.

Kenner, R., dir. and coproducer. 2008. *Food, Inc.* Los Angeles: Magnolia Pictures Participant Media.

Lowe, B. M. 2008. "Animal Rights Struggles to Dominate the Public Moral Imagination Through Sociological Warfare." *Theory in Action* 1, no. 3: 1–24.

Newkirk, I. 1992. *Free the Animals! The Untold Story of the U.S. Animal Liberation Front and Its Founder, "Valerie."* Chicago: Noble.

Newkirk, I., and A. Pacheco, producers. 1984. *Unnecessary Fuss.* Washington, DC: PETA.

Regan, T. 1983. *The Case for Animal Rights*. Berkeley: University of California Press.

Ross, S. 2008. "Chimps Aren't Chumps." *New York Times*, July 21, p. 19.

Singer, P. 1975. *Animal Liberation: A New Ethics for Our Treatment of Animals*. New York: New York Review of Books.

Steger, M. B. 2008. *The Rise of the Global Imaginary: Political Ideologies from the French Revolution to the Global War on Terror*. New York: Oxford University Press.

Swidler, A. 1986. "Culture in Action: Symbols and Strategies." *American Sociological Review* 51:273–286.

Taylor, C. 2004. *Modern Social Imaginaries*. Durham: Duke University Press.

Tucker, R. C., ed. 1978. *The Marx-Engels Reader*. 2d ed. New York: Norton.

Waller, J. M., ed. 2007. *The Public Diplomacy Reader*. Washington, DC: Institute of World Politics.

Wilford, H. 2008. *The Mighty Wurlitzer: How the CIA Played America*. Cambridge: Harvard University Press.

PART VI

Synthesis

FIGURE 21.1. Sarolta Bán's lovely image evokes questions pertinent to the chapters of this book: Is the human view of nonhuman animals always an anthropomorphic projection? In other words, do we always only see ourselves when we look at animals? (Is that the meaning of the man being tied to the mirror?) Do our preconceptions (donkeys are stupid) inhibit us from seeing profound connections between us and them, for example, our shared evolutionary past? Are we too influenced by what other humans think (always looking over our shoulders) to see animals straightforwardly? Does our human self-consciousness render us less good observers of animals than they are of us? Do animals or our relationship to them cause us anxiety, as is suggested by the man's expression? Like this book, Bán's artwork raises these questions and many others. Her image of the grizzly and teddy bears on the cover of this volume also confronts us with contradictions and problems in our interpretations of animal minds. Importantly, however, both book and images leave open the possibility of many answers. Photo courtesy of the artist.

21

Animal Ethics and Animals' Minds

Reflections

Julie A. Smith and Robert W. Mitchell

We wrote in the introduction that we hoped readers would make connections across chapters, would compare and contrast ideas they found provocative. We assumed that authors' points of view derived from their experiences, academic disciplines, personal interests, heartfelt beliefs, and the debates of our culture. We have included authors who have an everyday, caring relationship with particular animals and those whose interests are more dispassionate, as well as combinations in between. Such a medley is reminiscent of the Victorian era, when science writing, poetry, art, narrative, travel literature, personal essays, and other genres were part of public discourse about human and animals' minds . From that era until today, animals' minds have been a topic of public fascination and concern. As a result, what humans think about them is complicated. We felt that only by bringing together many points of view under one cover could we start to understand the complexities of our collective thinking. In this concluding chapter each editor relates the essays to a topic of interest to him or her, comparing and contrasting them within one framework that brings them into relation.

Animal Ethics: Descartes's Theory of Knowledge and Animal Ethics

Readers of the present volume may well be familiar with the long and persistent intellectual history of assessing mental capacity through interpretations of the seventeenth-century philosopher René Descartes. Descartes may be best known for his mind-body dualism, which equates mind with an exclusively human rational soul. But his discussions of the nature of knowledge have also contributed to the persistent view of animals as all but mindless and therefore of negligible moral concern. Many essays in this collection reflect both pro- and anti-Cartesian ideas about the relationship between the nature of knowledge and animals' minds. Some reformulate those positions.

For Descartes, knowing was possible only by humans; it was that special capacity for "reflexive" or "reflective" mental activity, that is, an ability to think about one's thoughts, to know that one was thinking by virtue of being able to have thoughts about those thoughts. This was possible only because of language: through language one might separate one's thoughts from oneself, so to speak, so that one might make them available for inspection as one would investigate external phenomena. Only then could one understand them, or even know that one was having them. Thus, much of what has been posited as making humans mentally different from all other species is the capacity to "off-load" the mind's contents into a representation, to the end that they might be evaluated objectively; only in this way could a person know that he or she had a mind or even that he or she existed. This singular capacity to step back from one's own mental contents has been elaborated over the centuries as the requirement for all so-called higher-order mental activity, for example, the ability to think about the self, about one's death, about the future, and so on, all of which have traditionally been granted only to the reflective/reflexive human mind that can turn back on itself as a function of human language.

The chapters that eloquently express a neo-Cartesian view of knowledge are by Alain Morin and Gary Steiner, although others seem to accept it as a given. After granting certain kinds of awareness to animals, Morin is cautious about granting them what he calls higher-order mental functions because they lack language, specifically a capacity for inner dialogue in language: "Self-directed speech allows us to verbally label our internal experiences and characteristics; as a result these become more salient—more conscious." For Steiner, the ability to have conscious thought is directly related to the ability to have abstract concepts separate from perceptual experience, and those are possible only because of a human linguistic capacity that enables their formulation. Indeed, for Steiner, intentions (beliefs and desires) are structured predicatively, that is, according to the subject-predicate structure of language, and without language one cannot have them. Concepts (and Steiner sees intentions such as beliefs and desires as concepts) make possible higher cognitive functions such as taking inventory of one's thoughts.

The Cartesian view has always had its critics, most recently from biology and posthumanist philosophy, sometimes in tandem. (However, many scientists secularize the Cartesian view by assigning to humans higher-order [linguistic] mental functions as a result of brain activity in place of Descartes's immaterial mind-soul). Scientists, and the postmodern critics who integrate scientific views into their theories, have questioned the long-held assumption that higher-order mental states are inevitably tied to a singular capacity, that is, linguistic representation. One argument is that sophisticated thinking is the result of complex, mechanistic, distributed brain function, not a single capacity such as linguistic ability. Paraphrasing Hauser (2001), Haraway writes, "organisms possess heterogeneous sets of mental tools, complexly and dynamically put together from genetic, developmental, and learning interactions throughout their lives, not unitary interiors that one either has or does not have" (2007:374, n. 47). Another approach has been to challenge, not the Cartesian definition of consciousness as "knowing that you know," but its importance in human life. Indeed, some scientists have drastically downplayed the role of consciousness in comparison to other brain functions that humans obviously share with animals. Neuroscientist David Eagleman (2011) writes for a popular audience that "your consciousness is like a tiny stowaway on a transatlantic steamship, taking credit for the journey. . . . Brains are in the business of gathering information and steering behavior appropriately. It doesn't matter whether consciousness is involved in the decision making. And most of the time, it's not" (pp. 4–5). Thus, "knowing that you know," the hallmark of consciousness for Cartesians, appears be in short supply in human life even if it is an exclusively human capacity, according to this view.

Some posthumanists have argued as well that the very of idea of "knowing that you know" (rather than just "knowing") is an artificial distinction. They have suggested that scant evidence exists that these operate separately most of the time or that the model of separate functions licenses the application of the so-called lower function ("merely knowing") to animals. They take exception to assumptions in neo-Cartesian arguments about what language does. They say that in fact language cannot "re-present" our thinking to us because language does not have the capacity to copy or translate nonlinguistic states of mind—these are different modes of mental activity, not two versions of the same thing. Therefore, language does not provide a mechanism by which we can think about our thoughts. What it produces is something different from our mental states without language (see Wolfe 2010, especially pp. 31–47). We may well talk about ourselves in language in a way nonhuman animals do not, but that does not mean we are more aware of our internal states than they are, some posthumanists maintain. One might even say that language distracts us from our internal mental states. Finally, postmodern commentators have argued that mental states are not entities that exist in a static, stable, originary condition that can be accessed by the rational human mind as it would an external object. Rather, mental states are processes that entail a continuous interaction between themselves and an environment that continually changes both (see Wolfe 2010:31–47 for a discussion of systems theory, a possible model for the mind/environment relationship).

The essays in the collection that fall most clearly into the anti-Cartesian camp are those that refuse the Cartesian mind-body separation that assigns the body the role of providing the mind with records of perceptual states that it then reworks into linguistic thought (the only kind of thought there is and only for humans). Explicitly locating her essay in posthumanist theory, Rohman argues that one form of language, poetry, must be understood not as the expression of a uniquely human reworking of mental contents but as the production, or generation, of intensity and the transfer of sensation or affect to others. And those feelings that lead to art are derived from an excess of energy in which animals fully participate. Thus, for Rohman, poetry, which is for some the highest form of artistic expression in language, is not at all about the exclusively human linguistic mind. Dillard-Wright would also decenter language as the sole route to knowing, arguing that languaged thought is only part of mind and ought not to be privileged as higher order over the biological or vital processes with which the mind is intertwined. Both Rohman and Dillard-Wright implicitly question the distinction between "knowing" and "knowing that you know" in the sense that they discuss powerful mental states that are independent of language yet presumably mentally available to the beings who have them.

Although all of the essays cited thus far may be read as either neo-Cartesian or anti-Cartesian, they may also be interpreted as blurring the boundary between the two. Morin grants to animals an awareness of external events and an ability to experience mental states such as sensations and emotions, even though he does not afford animals what he sees as higher-order awareness, particularly self-awareness. Steiner, both in his chapter here and more explicitly in other writings (2008, 2009), has criticized the Western philosophical tradition linked to Descartes that licenses humans to harm and kill animals on the basis of their lack of higher-order thinking; and he has maintained that perceptual experience is enough to grant animals moral worth. Both Morin and Steiner might well agree that animals have complex and interesting minds in all the ways outlined by the authors writing from a more postmodern bent, even as they exclude those forms of nonlinguistic knowing from what they have designated as higher-order thinking. For their part, those authors friendly to postmodernism would, or actually do, acknowledge the importance of human language; they simply see it as one of several ways to be "minded."

CONTEMPORARY (POST-CARTESIAN) ANIMAL ETHICS

None of the chapters in the collection takes as its primary focus the relationship between animals' minds and human treatment of animals, in spite of a long historical association between these subjects; that is, none is specifically about animal ethics. Nevertheless, many seem to imply something about human obligation toward animals; and, even if they do not, they will most likely be read in that way. One approach to relating the essays thus might be to compare and contrast them in terms of their ethical implications. To do this, I have chosen to look at the essays through one particular ethical framework called the doctrine of perfectionism. This philosophical view proposes a hierarchy in moral status among beings based on their capacities. Compatible with Cartesianism, it holds that the interests of some beings count more than those of others—that some beings deserve more consideration and some less—on the basis of the possession of (or level of or lack of) important capacities, invariably mental ones (see Cavalieri 2008, especially pp. 1–14). Some of the essays seem to work against this view; others provide information that might be mustered for one side or the other; some remain agnostic about animal moral worth as derived from mental capacities.

The doctrine of "animal personhood" might easily be read against the doctrine of perfectionism. It accepts the idea that treatment ought to be determined by mental capacities, but it holds that in fact animals possess many mental abilities similar to human persons; in other words, it maintains that animals have sophisticated mental traits that ought to compel the same moral obligations as those that humans impose on each other. Desmond explicitly notes the capacity of some primates to paint pictures, and she explains that for many people this challenges the supposed impermeable moral boundary between humans and animals because it demonstrates apes' humanlike subjectivity and appreciation of aesthetics. Even though hers is a personal experience essay rather than a philosophical or scientific treatise, Karen Davis's essay fits well into this framework too. It details many moments of interaction between herself and chickens that may be read as affording chickens such capacities as identity (mirror self-recognition), intention (in the sense of having a plan and then executing it), and reciprocity (the ability to act considerately of others). Julia Schlosser's essay maintains that artists photographing moments of touch between humans and their companion animals convey animal and human minds coming together within the touch experience; she concludes that, in this respect at least, the animals and the humans are "like-minded" in that they are able to comprehend each other's intentions. Albert Braz tells of the nineteenth-century naturalist Grey Owl who insisted that beavers had speech, just as humans do, and that they also possessed other high-level human mental capacities, such as reasoning and even wisdom.

As we see from these examples, the discourse of animal personhood reworks the doctrine of perfectionism in order to replace the classic Cartesian higher-order capacities of reflexive/reflective thinking with abilities that proponents claim are equally important as definers of humanity and by extension the personhood of animals—things like emotions or agency or individuality. In other words, personhood discourse often seeks to substitute the classic criteria of consciousness with broader and more affect-based mental states. In spite of these attempts to reconfigure the doctrine of personhood so as to more securely include animals, critics, even among animal advocates, have argued against making a case for animals on the basis of their similar mental abilities to humans. They say that this further entrenches the human model of mind and requires animals to demonstrate levels of competence equal to humans in frameworks not appropriate for them but for another species. Critics of personhood ethics, nevertheless, acknowledge that it has made people aware of the practice of anthropodenial, a term coined by Frans de Waal (1997): the refusal to acknowledge that animals in fact share many important mental capacities with humans.

Supporters of animal personhood often point to scientific research for support of their views that animals have humanlike mental abilities. However, determining personhood with all its moral obligations toward animals through science can be a dicey affair, say those opposed to the personhood approach, not least because little agreement exists within the scientific community on what the evidence says about animals' minds. Paula Droege's chapter on scrub jays challenges the view that the jays are conscious in what she sees as the fully human sense, even as they demonstrate sophisticated memory as part of their food caching behavior. Robert Lurz's chapter outlines an experimental model for moving forward on the question of whether chimpanzees have theory of mind (that is, whether they know that others have mental states). Nevertheless, as Lurz himself explains, his essay is imbricated in a long-standing and highly charged controversy about chimpanzee mental capacities that is freighted with competing claims and contested research protocols. In short, scientists do not agree. In an article on the role of science in finding similarities between animal and human minds, Sandra Mitchell (2005) writes, "the manner and degree to which nonhuman animals are similar to human beings becomes an even more pressing scientific problem in a context in which the very morality of our actions depends on the answer" (p. 102). Nevertheless, by the end of the article, she acknowledges that "the most controversial and consequential claims about the similarity between humans and nonhuman animals are the most difficult to substantiate" (p. 115).

Science does not lend itself to addressing ethical problems for other reasons, say those who are skeptical of the scientific route to animal personhood. Science is not always as impartial as we tend to assume. Sara Waller documents the ways that human agendas have affected the choice of scientific research on primates, as well as research designs and the reporting of research results. Others have noted that scientists map their scientific data onto their own value systems, as if they too emerged from disinterested inquiry. Often accomplished through a comparison between animal and human minds at the conclusion of scientific reports, scientists can take for granted their own evaluations of what it means to be advanced or can unwittingly change the topic to a trait that remains undiscussed and yet naturalized as self-evidently indicative of humans' special status. Were science the thoroughly disinterested and fact-driven enterprise we imagine it to be, scientific fact still does not easily translate into ethical imperatives. The Harts' essay explains how the structure of the cerebral cortex of elephants' brains correlates to their particular kind of memory, empathic behavior, and problem-solving ability. Yet they do not claim that such scientific facts "speak for themselves" as ethical directives. This is not to say that science might not provide some insight into animals' experience, the necessary starting point of an animal ethics. In her chapter on human-animal encounter programs in aquariums, Traci Warkentin is explicitly interested in a methodology to study the subjective experiences of these animals. In the end, however, most scientific discourse elides the animals' experiences, their points of view, and the impact of human behavior on them, with the result that animals are often represented as not having a point of view of their own. The compelling research project outlined in the Sickler, Fraser, and Reiss essay presents information on the various ways the public thinks about dolphins in an attempt to determine the best means to educate it about scientific data on dolphins. At the same time, it leaves unexplored as irrelevant ethical questions about the experiences of dolphins in aquariums, the manner and consequences of the ways in which they were acquired, and the larger issue of whether dolphins should be in captivity at all.

Another contemporary approach to animal ethics was alluded to in the previous discussion of postmodern responses to Cartesianism. As stated earlier, many postmodern thinkers promote the principle of "different mindedness" in a way that takes on ethical implications. They maintain that humans' moral obligations must be founded not on finding in animals mental capacities similar to those of humans but rather on recognizing in animals their own kinds of minds with their own special capacities, different and sometimes superior to those of humans. This view rejects ordering mental abilities in a hierarchy, which its proponents say has in past and contemporary theory been biased toward humans (the theorists). Entitling her chapter "Toward a Privileging of the Nonverbal," Argent reviews the rich literature on nonverbal communication, which elaborates different modes of conveying nonlinguistic thought through the body; she then goes on to explain the ways that horses use synchronicity to convey information about their mental states. Likewise, Smith's discussion of what she calls kinesthetic-visual transfer, or the transfer of the feeling of movement from one being to another through the look of movement, is intended to suggest not a lower-order (unconscious or semiconscious) faculty but one that dogs are fully aware of, not because they ruminate about it in language, but because the state of mind includes such strong mental experience and is so often employed as a mode of communication that dogs must have their own meanings for such perceptions. If animals then have sophisticated mental capacities tied to their particular kinds of bodies, ones that entail knowledge of their own mental contents, as well as their own ways of conveying them, then those ways command as much moral consideration as human ways of knowing, imply theorists of different mindedness. The difference between the way this kind of discourse uses science and the use of science in the service of animal personhood is its greater interest in nonhuman ways of being minded and a strong desire to disarticulate moral worth and humanlike minds.

In addition to different mindedness, another approach to dismantling the doctrine of perfectionism might be called relational ethics as opposed to capacity ethics. This has been variously elaborated as a feminist care ethic (Donovan and Adams 2007) that sees humans as obligated to develop empathic abilities necessary to understand the experiences of other animals; as an "ethic of causal relations" (Palmer 2010) that holds that empathic care ought to be aroused by the distress or harm which human behavior causes to animals, however indirectly and absent from view; as an ethic based on a sense of "felt kinship" that founds human obligations toward animals on a powerful sense of shared sentience and its accompanying vulnerabilities (Steiner 2009); or as an ethic of "corporal compassion" (Acampora 2006), a cross-species conviviality based on shared embodiment within different but overlapping body-worlds with points of contact. All these approaches stress animal and human embodiment rather than minds alone as a route to determining human obligation to animals. They do this in two ways: they claim that at least partial access to animals' inner experience comes through knowledge about how the bodies of individual species operate in the world and they say that this knowledge can be used to create an informed empathy for animals' experiences. Mitchell's essay lends support to this thinking: it proposes that we have just as much access to animals' inner experiences as we have to that of other people. That access is possible because much private experience is similar to perceptual experience; therefore, once we understand that both we and others have interior states (which we do through kinesthetic-visual matching), and once we understand that those states resemble in some respects perceptual experience, then we can achieve at least some level of understanding of others' mental contents.

A profound reworking of a relational or care ethic might be called an acceptance in oneself of a state of "woundedness." It is not an ethic per se, because in postmodern fashion it eschews any moral obligations or behavioral rules based on rationally derived formulae. It is also cautious about a care ethic based on affect or feelings as a guide to treatment of animals, as it maintains that this route can easily lead to a kind of ethnocentrism wherein those animals with whom a person or group feels a close relation are privileged because of those attachments (see Wolfe 2010:49–98). Instead, it posits human openness to the raw reality of what humans do to animals. Wolfe develops this idea by using as his starting point Cora Diamond's essay (2003) on the character of Elizabeth Costello in J. M. Coetzee's *The Lives of Animals* (2001): "The awareness we each have of being a living body, being 'alive to the world,' carries with it exposure to the bodily sense of vulnerability to death, sheer animal vulnerability, the vulnerability we share with them. This vulnerability is capable of panicking us. To be able to acknowledge it at all, let alone as shared, is wounding; but acknowledging it as shared with other animals, in the presence of what we do to them, is capable not only of panicking one but also of isolating one" (cited in Wolfe 2010:72).

Several of the artworks discussed by Ullrich seem intended to bring the viewer into an awareness of his or her own state of inconsolable woundedness for the plight of animals under human control. These works proceed by recreating for humans the circumstances into which they have put animals foreclosed of any and all possibility of escape—the horse in the alien environment of the city, the rat in the laboratory. Nobel prize-winning author J. M. Coetzee (in Cavalieri 2008) hypothesizes that there takes place in those of us who care about animals "something like a conversion experience, which, being educated people who place a premium on rationality, we then proceeded to seek backing for in the writings of thinkers and philosophers" (p. 89). Thus many of the essays that articulate a capacities-based ethic perhaps express something derived from the unconscious channels of distress described by Diamond and Wolfe. Braz emphasizes the way that Grey Owl passionately responded to his own earlier cruelty toward beavers. Was his insistence that beavers could speak in part a metaphor Grey Owl needed as an explanation for himself of the beavers' ability to impact him so profoundly, a way to credit them for his dismay at what he had done to them in the past? How does anyone who identifies with chickens, and who refuses to turn away from the degrading representations of and unspeakable brutality toward them in this culture, live in anything but a perpetual state of woundedness, a state that would terrify most of us? Surely this rawness to animal reality is not one that a single human can impose on another, however. In his chapter on the strategies of animal activists, Brian Lowe describes the ways they use, and must necessarily use, spectacle, that is, dramatic visual media and performance, in order to impact the social imaginary (the collective mind of the culture). One wonders after reading his essay whether the use of spectacle is an attempt to coerce reform through manipulative, shocking strategies or to simply make possible awareness of a personal but disallowed grief for animals' vulnerability to death and harm.

Of course, none of the chapters in the volume can properly be reduced to one ethical position, although I have simplified them in the interest of comparison. I would expect that for most of the contributors, as for most people, animal ethics derives from many entangled strands of thoughts and feelings.

Animals' Minds

As suggested earlier, ethical treatment of animals often, although not invariably, emerges from beliefs about animals' minds. But why do humans think that animals have minds at all? And how do we go about learning more about animals' minds, if we believe they have them?

MINDS

We have focused on views of animals' minds in Western culture, but one could argue that, cross-culturally, we humans commonly experience ourselves and other animals as infused with mind, consciousness, and thought—unless we are trained not to. Of course humans as a class have no monolithic view on animals' minds, but rather engage with diverse ideas about them, from ready acceptance that not only animals but everything else is minded (animism—see Harvey 2005) to complete denial that anybody has a mind (behaviorism—see Mackenzie 1977). The fact that humans and other animals share, for example, vocalizations, mating rituals, bodily processes, perceptual systems, and sociality indicates important mental connections between us and them (amid sometimes striking differences across species) for many authors, much as it did for natural historians (see Richards 1987) and early ethologists like Lorenz, Hediger, the Heinroths, and Uexküll (see Burghardt 1985). Indeed, Davis's descriptions of chickens remind one of these early ethologists' descriptions of other birds, though in a more feminine voice. Contrary to assertions that it is only recently that investigators are acknowledging intelligence in nonmammalian species, in fact the origins of comparative psychology and ethology began with assumptions of intelligence in a diversity of nonmammalian species (see Boakes 1984; Burghardt 1985), a belief also present in late nineteenth- and early twentieth-century naturalists—not only Grey Owl but also Ernest Thompson Seton, Charles Roberts, and William Long (see Lutts 2001).

One reason for belief in animals' minds is their behavior. Animals appear self-motivated and purposive, which seems central to many conceptions of mindedness. When we think of the relation between mind and behavior, we might posit one or several ways to connect them. Several authors assume that internal experience is inherent in the behavior itself; in this view mental states and behavior are not differentiated but entwined. Alternatively, others interpret behavior in terms of psychological causes that produce it but are distinct from it. And some consider mental states as both causing action and inherent in it.

Different assumptions about the locus of mental processes result in different ways of viewing animals' minds. Initial attempts at comparative psychology led, in America, to behaviorism largely because the internal psychological states (e.g., sight of box, smell of food) imputed to animals that were thought to cause their behavior were essentially redescriptions of the external situations they found themselves in (e.g., in a box with food); by contrast, comparative psychology and ethology in Britain and Germany had no need for behaviorism because researchers assumed that mental states inhered in (rather than caused) the behavior of animals, such that they formed a unit (see Mackenzie 1977). In creating behaviorism, scientists ignored aspects of animal experience like seeing, hearing, and otherwise perceiving, and the stimuli that the animals presumably perceived were taken as the only objective data, thereby obviating explicit concern about the animals' point of view.

How we understand animals' minds also depends on each observer's background assumptions (e.g., world hypothesis: Pepper 1942), including his social imaginary. As delineated by Lowe, the social imaginary—the norms, expectations, and practices that are taken for granted within a group—constrains how information will be understood even as it facilitates communication within the group. One group's social imaginary and background assumptions can make it difficult to understand another group's views. Thus social imaginaries, like human language and conceptualizations, can create blinders. Indeed, any representational system delimits how the world is perceived and appropriated for knowledge; perhaps articulating alternative ways of experiencing the world is merely elucidating the categories that facilitate and constrain perception and knowledge appropriation (Jones 2009). On the other hand, social imaginaries may simply direct attention one way rather than another. As Sickler, Fraser, and Reiss discovered, visitors to the aquarium with divergent perspectives on dolphin psychology remembered the same facts but found most engaging those facts that fit with their previously held perspectives. Some may argue that our current social imaginary does not permit recognition of animals' minds because such recognition is viewed as a threat to human uniqueness (Asquith 1997). However, English-speaking popular culture is apparently intrigued by scientific findings about similarities between humans' and animals' minds, as Waller notes, as long as implications of too much similarity are contained through rhetorical strategies. In Western culture cognitive similarities between people and animals are engaging because surprising whereas emotional similarities are assumed to be the norm; in Japanese culture the reverse is true (Asquith 1997).

Using language to understand minds creates difficulties that might not be immediately discernible. Accurately or not, most of us think about the language we use to describe what we experience as being obviously related to things in the world (e.g., "That's a *cat*"). By contrast, psychological language presents complications for this perspective on the relation of language to the world. One is that many psychological terms refer (minimally) to behavior in context, such that actions with similar antecedents or consequences in similar contexts are called the same thing, whether enacted by animals or humans and whether or not enacted with identical understanding of consequences or implications (e.g., a "threat" can be made by a baboon and a president; see Asquith 1997; Beer 1997). Hobhouse (1915) argued that our terms about mental states might capture only the functional qualities of mental states—for example, what they do for the animal to connect its actions and their consequences—and do not indicate a similar consciousness, content, or organization for mental states across individuals (see also James 1890:267). Thus, although many of us are inclined to attribute mental states to animals on the logic that similar behavior in us and them means similar states of mind, this can be problematic. The problem is, as Lurz and Steiner note, that different organisms exhibiting identical responses to a given situation need not indicate that identical mental processes are used to produce those identical responses. Thus exists the standard conundrum for comparative psychology: in relation to our knowledge of animals' mental states, if it looks like a duck, smells like a duck, and acts like a duck, it still might not be a duck! Steiner employs reasoning of this kind in discussing whether a dog believes that his master is at the door, calling the dog's mental state "equivalent" to a human's state of belief, but without the same intentional structure. This slipperiness of psychological terms is evident when you compare uses of *intentionality, consciousness, thinking,* and *intelligence* across chapters. Hence, even if we use similar terms to describe mental states in both humans and animals, this does not mean we are talking about identical phenomena.

Another complication is that when we use language to describe internal experiences, or other mental states that are not publicly available, we use language describing public phenomena to describe them. As Graham Richards (2010) writes, language about psychology is essentially metaphorical (think of "grasping" this concept): "We are in effect saying 'I am like that' where 'that' refers to some public phenomenon or property. Moreover . . . we structure and explain our private psychological experience in terms of how the public world is structured and explained in our culture. Insofar as we can actually communicate about the psychological it is therefore a sort of internal reflection of the outside world" (pp. 8–9). Thus the problem is not that the language we use to describe animal mental states is metaphorically related to the same language used to describe human mental states (one standard objection to anthropomorphism) but rather that using language to describe *anyone's* psychological states is largely metaphorical. New ways of talking about the mental result, in Richards' view, in new psychologies: "When psychologists introduce a new concept or theory about the psychological they are therefore directly engaging in *changing* it. . . . To classify and explain the psychological in a new way is to be involved in changing the psychological itself" (p. 9), because we use these new concepts and theories in thinking about minds. For example, folk psychological terms like *going off the rails* began after railroads appeared and *being on the same wavelength* after radio appeared (p. 9). Consequently, whatever we believe about minds is influenced by the terms we use to describe minds. Hence our understanding of minds is always contingent upon our assumptions, norms and expectations (i.e., our social imaginary); and culture and history are essential components in this understanding of animals' minds and our own.

PERSPECTIVE TAKING

Any attempt to understand another mind employs perspective taking, which can be viewed as having four components: our *aim* is understanding something about another's mind; we use diverse *sources of information* that we interpret via multiple *processes*; and we arrive at a psychological *result* through the confluence of these components (Davis 2005). The chapters of our contributors can be read against these components.

The *aims* of our contributors concern understanding diverse mental phenomena. Authors focus on propositional thinking, skill at phenomenological perspective taking, episodic memory, emotional experience and expression, feelings of connection via tactile experience, similarities and dissimilarities between animals' and humans' experiences, psychological union, communication, embodiment. An author's initial aim to understand animals' minds can be tied to larger aims: Some authors want to describe why they feel connected to animals, others want to know what animals experience or how to treat them ethically, and still others seek to evaluate theories about animals' minds. These larger aims are tied to decisions about which sources of information authors accept as evidence of animals' minds.

For almost all authors the main *source of information* about another organism's mind is its behavior in context, whether experienced directly, experimentally manipulated, recorded on videotape, photographed, or considered after these activities. Sources of information about the mind of an animal also include the shape and physiology of its body, its sensory-perceptual apparatus, its evolutionary history and adaptations, and the general organization of its brain, but these are usually interpreted within the behavioral repertoire of the animal. For example, the Harts argue that the nervous system of animals may provide clues to how animals think, elaborating a model of elephant thought as having some psychological systems in common with and others divergent from human thought. Mitchell and Droege, in different ways, focus on sensory or perception-based representations as a basis for animals' thoughts (see also Smith 2005).

Overwhelmingly, authors take for granted that the mind is embodied. Dillard-Wright presents the body itself as minded, an organized and organizing entity, reminiscent of Gregory Bateson's (1979) view of mind as immanent in any functioning system employing feedback. Even if we think of the mind as a function of the brain, the brain itself is inherently dependent on a functioning body, and the form of an organism's body and its ecological circumstances influence the kinds of experiences, capacities, and skills the organism can have. The sensory and perceptual systems of an organism's acting body bear knowledge for action and interaction: Schlosser describes how touch creates psychological connection; Argent and

Smith present other nonverbal behavior as often immediately understood between organisms, either as a natural outgrowth or through entrainment to be mutually responsive; and Mitchell suggests that inner experiences are transformations from the perceptual systems of organisms. For example, inner speech as described by Morin is usually an audition-based representation, so much so that sometimes we think we've said something aloud when talking to ourselves. Touch and feelings of vibration are viscerally felt, and visceral sensations can be experienced through perceiving another animal's actions, seeing a photograph of touch, or listening to or performing a song. As Rohman notes, the cry of the sexually engaged tortoise made D. H. Lawrence (and us, through Lawrence's poetry) aware of the tortoise's embodied mind.

A spin-off of embodiment might be called *interembodiment.* In interembodiment, animals incorporate other animals' embodied minds into their minds when interacting. Interembodiment is a source of information about another's mind because that mind is felt to be directly experienced between organisms. Animals and people may experience the mental and the behavioral together as a complex, in effect experiencing themselves in a fusion of their own and others' embodied actions, as Smith describes for dogs. As a way for self-conscious people to experience such a lack of boundaries, Dillard-Wright recommends singing or acting, where one loses oneself in the performance, Argent might recommend riding a horse, and Warkentin, imitating a killer whale. Interembodiment is salient in blind people, who describe their guide dogs as inseparable from themselves (Sanders 2000). Schlosser and Ullrich discuss art that attempts to express interembodiment in order to entice the viewer into an understanding of a relationship between a particular human and animal or of a particular animal's experience. Schlosser presents some aspects of interembodiment in photographs of people with domestic animals. Ullrich questions how effective artists are in capturing other aspects of interembodiment with another species. An alternative form of interembodiment is Mitchell's description of knowledge of one's own and another's psychology as inherently interdependent but distinguishable: one's own as known from direct experience, another's from behavior in context.

Multiple *processes* are employed by our authors to understand animals' minds: logical inference, literary criticism, experimental manipulation, historical interpretation, mental state attribution, anthropomorphism, empathy, imagination, stereotyping, projection, behavior reading, imitation. All these methodologies employ background assumptions about a panoply of psychological components. Commonsense observation, using psychological terms to describe what animals are doing, is the standard practice among most authors in this collection, whether allied with a scientific gloss, an artistic one, or an everyday acceptance of psychological terms as appropriate for use across species. Theory, perhaps followed by experimentation, is another method offered to understand animals. In some cases theoretical expectations of embodiment and interembodiment allow another's mind to be understood from behavior or mutual interaction. For example, Warkentin observed bodily imitation between children and a young killer whale; using her own synthesis of embodiment theories, she interprets the mutual imitation as expressive of a desire for reciprocity in both species. Some theorists offer components that they posit as essential to animal thought (for example, awareness of time, representation, embedding, perception-like imagery, kinesthetic-visual transfer or matching, distinct neural structures), but few of these offerings are elaborated. Rather they are suggestions for kinds of mental processes or structures that might allow animals to think and behave and communicate as they do without language (see Lurz 2009). Droege proposes that determining the content of the representations animals use, derivable from their behavior, informs us about their first-person perspective, much as Mitchell argues that we can use animals' behavior in context to determine their attitudes toward things in the world. Steiner hypothesizes that animals have concepts based on associative webs, which he contrasts with human language-based concepts (but note that associative models can be used to explain human concepts: Clark 2003), while Waller remains open to alternative interpretations of animal thinking based on representations without propositionality (see also Beer 1997; Millikan 1997).

Psychological theories can become quite complicated, as psychological phenomena do not occur in isolation but are tied to other psychological phenomena. Steiner notes that having one belief implies having a whole set of related beliefs. Droege observes that you cannot have a concept of "now" without a concept of the past and future, nor can you have episodic memory without a sense of autobiographical time. Mitchell, following Strawson, posits that you cannot have a concept of your own consciousness without a concept of other possible consciousnesses. Lurz and Droege query whether or not animals can embed a point of view (of another animal's conscious experiences or their own) within a point of view. Can chimpanzees, like people, embed their knowledge of another's perceptual experience of the world in their interpretation of the other's actions? Can birds, like people, have conscious experiences in which they embed an experienced memory of a past event, recognized as from the past, within a current experience?

Finding appropriate methodologies by which to evaluate our sources of information can be problematic. Indeed, disagreements about the appropriateness of methodologies abound. Some authors suggest that living with animals allows us to know them, others devise experiments or perform empirical investigations to explore theories, and still others express doubts about what can be claimed from everyday observations or scientific studies. Anecdotes—singular instances of an event described in a narrative—retain their power to provoke thinking about animals' minds for many authors, despite scientific antagonism to anecdotes. Whether it be the coital cry of a tortoise, the mutual imitation between killer whale and child, the preening of a chicken looking in a mirror, the first spontaneous transfer in the use of the word *none* by Alex the parrot, Derrida's noticing his cat looking at him naked, footage of particularly cruel treatment of a monkey in a laboratory, or an elephant helping a conspecific in need, the anecdote unsettles our complacencies. In some cases these singular instances beg for further confirmation of a species' capability; in others they induce thoughtful consideration of what we do and don't know about particular species.

Anthropomorphism is used strategically in several authors' accounts. Indeed, the beautiful artwork of Sarolta Bán prefacing this chapter questions whether we ever use any other strategy! Some authors imagine translations between animals' actions and English, seeming to employ the easy anthropomorphism of children's stories, in which animals think linguistically. Rather than offering these translations as exact descriptions of the meaning or indication of animals' thoughts, or implying that these thoughts are propositional, authors present such translations as an approximation of the intent behind animals' actions. By contrast, Braz describes Grey Owl's belief that beavers actually communicate via language; other observers have also viewed some animals as having vocal language (Garner 1892) or words (Thompson 1968). Grey Owl believed beavers the most humanlike of animals; that descriptor today belongs to chimpanzees or bonobos (de Waal 2002), but one early twentieth-century naturalist viewed bears as the most humanlike animal (Mills 1976 [1919]). Waller shows that anthropomorphism is the basis for a great deal of scientific research (however much scientists deny its usefulness) and rightly notes that standard measures of intelligence in humans (IQ tests) are incommensurable with those used for animals. (Piagetian measures of intelligence may provide an appropriate method for cross-species comparison at early developmental periods [Parker 1997].) Boddice examines the heyday of anthropomorphic and anthropocentric understandings of animals, the Victorian era, showing that recognition of psychological similarity between animals and humans did not necessarily induce empathy toward animals. Anthropocentrism, the flip side of anthropomorphism, implies that always interpreting animals' mentalities as projections of human ones is overly human centered. Indeed, denial of propositional character to animal thought on the basis of animals not having language may be short-sighted. Prairie dogs, for example, appear to name colors via their vocalizations (Slobodchikoff, Paseka, and Verdolin 2009); their vocalizations may be linguistic. Can something like inner speech be far behind if animals have their own forms of vocal representation? In humans, internal speech (whether vocal or expressed in imagined sign language) is presumed to derive from external speech to the self (Levina 1981). Might Davis's muttering rooster be capable of internal (silent) vocalizations with representational (affective) content? Steiner imagines that a vocal language like that learned by Alex the parrot might allow for development of propositional thought. One expects that Washoe, the sign-using chimpanzee, experienced inner speech following her use of signs to talk to herself (Fudge 2002).

Concern to avoid anthropomorphism and anthropocentrism is emphasized in some chapters. Rohman finds art in anything that is creatively both productive and repetitive, while Desmond is reluctant to include as art any production resulting solely from aesthetic impulses rather than emerging from representational systems dependent on culture. The excess energy theory of art described by Rohman reminds one of similar theories about play, from which art was presumed to derive, that were promulgated in the nineteenth century by Herbert Spencer and Karl Groos (see Müller-Schwarze 1978) and remain open today (Burghardt 2005). Rohman's theory extends beyond this nineteenth-century view to include even the evolutionarily repetitive yet variable creation of species members (cf. Bateson 1979), thereby decentering art as human or even as intentional. Desmond also decenters art from the human but retains art for productions by individuals or groups that incorporate aesthetics, intentionality, design, and culture, promoting the idea of animals as agents. Waller asks us to resist the human point of view (or, more specifically, one's own point of view) for understanding animals and instead to look to animals for psychological processes or experiences that do not resemble one's own (cf. Millikan 1997). Essentially, her approach is that taken by researchers who avoid applying their culture's psychology to the people they study (Lillard 1998;

Rosaldo 1980). Waller espouses a form of zoomorphism, asking researchers to apply knowledge of "polygamous primate societies to explain Mormon culture, rules, emotions and mores"; behavioral ecologists have begun this project already (Moorad et al. 2011).

The *results* of our inquiry into taking the animal's perspective can be knowledge, belief, empathy, a feeling of closeness, an ethical concern for animals, theoretical comprehension, a tweaking of a theory, a new way of looking at animals, misunderstanding, understanding, an awareness of one's limitations, or several of these at once. The chapters themselves, many of which present or imagine animals' perspectives, stand as the results of authors' inquiries. We leave it to the reader to determine how successful each chapter is in satisfying its aims, but note that a common theme is awareness of the difficulties of the enterprise of understanding another's perspective. Schlosser plays with perspective in her own photographs as the viewer attempts to figure out what exactly is happening between owner and pet. Ullrich promotes and acknowledges the difficulties of heterophenomenology, the study of consciousness of other organisms from either a first- or third-person perspective (promoting Radner's [1994] take on Dennett's [1991] original solely third-person view). Ullrich points out how human conceptions can both plague artistic attempts at heterophenomenology and create some understanding of animals' experiences via empathy. She also anticipates that recognizing differences in ways of experiencing and viewing the world will itself produce empathy, whereas Warkentin suggests that shared experiences produce empathy and Boddice shows that recognition of animals' psychological similarity or dissimilarity to humans is no guarantee of empathic responding.

However much our thinking about animals' minds is constrained by our social imaginary or by the slippery nature of human language or by limitations of experimental protocols and their reasoned interpretation, it is filled with conjecture, evidential gaps, and, most interestingly, contradictions. As Waller notes, we say that animals make inferences but don't think. Steiner elucidates a propensity among some scientists to posit that animals have rich, humanlike inner lives, yet they acknowledge that animals have nothing like language, which seems to be the basis (as Morin notes) for what we take to be the richness of human inner life. Droege exposes a related conundrum: Many philosophers who accept that animals have consciousness also posit that this consciousness is subjective and not directly knowable to anyone other than individual animals themselves. Specifying such contradictions is the start to resolving them, and resolving them may allow us to understand and treat animals more intelligently than we do now.

References

Acampora, R. R. 2006. *Corporal Compassion: Animal Ethics and Philosophy of Body*. Pittsburgh: University of Pittsburgh Press.

Asquith, P. J. 1997. "Why Anthropomorphism Is Not Metaphor: Crossing Concepts and Cultures in Animal Behavior Studies." In R. W. Mitchell, N. S. Thompson, and H. L. Miles, eds., *Anthropomorphism, Anecdotes, and Animals*, pp. 22–34. Albany: SUNY Press.

Bateson, G. 1979. *Mind and Nature: A Necessary Unity*. New York: Penguin.

Beer, C. 1997. "Expressions of Mind in Animal Behavior." In R. W. Mitchell, N. S. Thompson, and H. L. Miles, eds., *Anthropomorphism, Anecdotes, and Animals*, pp. 198–209. Albany: SUNY Press.

Boakes, R. 1984. *From Darwin to Behaviourism: Psychology and the Minds of Animals*. Cambridge: Cambridge University Press.

Burghardt, G. M,. ed. 1985. *Foundations of Comparative Ethology*. New York: Van Nostrand Reinhold.

——. 2005. *The Genesis of Animal Play: Testing the Limits*. Cambridge: MIT Press.

Cavalieri, P. 2008. "The Death of the Animal: A Dialogue on Perfectionism." In P. Cavalieri, ed., *The Death of the Animal: A Dialogue*, pp. 1–41. New York: Columbia University Press.

Clark, A. 2003. *Associative Engines: Connectionism, Concepts, and Representational Change*. Cambridge: MIT Press.

Coetzee, J. M. 2001. *The Lives of Animals*. Princeton: Princeton University Press.

Davis, M. H. 2005. "'Constituent' Approach to the Study of Perspective Taking: What Are Its Fundamental Elements?" In B. F. Malle and S. D. Hodges, eds., *Other Minds: How Humans Bridge the Divide Between Self and Others*, pp. 44–55. New York: Guilford.

Dennett, D. C. 1991. *Consciousness Explained*. New York: Little Brown.

de Waal, F. B. M. 1997. "Are We in Anthropodenial?" *Discover* 18, no. 1: 50–53.

——, ed. 2002. *Tree of Origin: What Primate Behavior Can Tell Us About Human Social Evolution*. Cambridge: Harvard University Press.

Diamond, C. 2003. "The Difficulty of Reality and the Difficulty of Philosophy." *Partial Answers: Journal of Literature and the History of Ideas* 1, no. 2: 1–26.

Donovan, J., and C. Adams, eds. 2007. *The Feminist Care Tradition in Animal Ethics*. New York: Columbia University Press.

Eagleman, D. 2011. *Incognito: The Secret Lives of the Brain*. New York: Pantheon.

Fudge, E. 2002. *Animal*. London: Reaktion.

——. 2006. "Two Ethics: Killing Animals in the Past and Present." In Animal Studies Group, eds., *Killing Animals,* pp. 99–119. Urbana: University of Illinois Press.

Garner, R. L. 1892. *The Speech of Monkeys.* New York: Charles L. Webster.

Haraway, D. J. 2007. *When Species Meet.* Minneapolis: University of Minnesota Press.

Harvey, G. 2005. *Animism: Respecting the Living World.* New York: Columbia University Press.

Hauser, M. 2001. *Wild Minds: What Animals Really Think.* New York: Owl.

Hobhouse, L. T. 1915. *Mind in Evolution.* 2d ed. London: Macmillan.

James, W. 1890. *The Principles of Psychology.* Vol. 1. New York: Henry Holt.

Jones, R. 2009. "Categories, Borders and Boundaries." *Progress in Human Geography* 33:174–189.

Levina, R. W. 1981. "Vygotsky's Ideas About the Planning Function of Speech in Children." In J. V. Wertsch, ed., *The Concept of Activity in Soviet Psychology,* pp. 279–299. Armonk, NY: Sharpe.

Lillard, A. 1998. "Ethnopsychologies: Cultural Variations in Theories of Mind." *Psychological Bulletin* 123:3–32.

Lurz, R., ed. 2009. *The Philosophy of Animal Minds.* Cambridge: Cambridge University Press.

Lutts, R. H. 2001. *The Nature Fakers: Wildlife, Science, and Sentiment.* Charlottesville: University Press of Virginia.

Mackenzie, B. D. 1977. *Behaviourism and the Limits of Scientific Method.* London: Humanities.

Millikan, R. 1997. "Varieties of Purposive Behavior." In R. W. Mitchell, N. S. Thompson, and H. L. Miles, eds., *Anthropomorphism, Anecdotes, and Animals,* pp. 189–197. Albany: SUNY Press.

Mills, E. 1976 [1919]. *The Grizzly: Our Greatest Wild Animal.* Sausalito: Comstock.

Mitchell, S. D. 2005. "Anthropomorphisms and Cross-species Modeling." In L. Daston and G. Mitman, eds., *Thinking with Animals: New Perspectives on Anthropomorphism,* pp. 100–117. New York: Columbia University Press.

Moorad, J. A., D. E. L. Promislow, K. R. Smith, and M. J. Wade. 2011. "Mating System Change Reduces the Strength of Sexual Selection in an American Frontier Population of the Nineteenth Century." *Evolution and Human Behavior* 32:147–155.

Müller-Schwarze, D., ed. 1978. *Evolution of Play Behavior.* Stroudsburg, PA: Dowden, Hutchinson and Ross.

Palmer, C. 2010. *Animal Ethics in Context.* New York: Columbia University Press.

Parker, S. T. 1997. "Anthropomorphism Is the Null Hypothesis and Recapitulationism Is the Bogeyman in Comparative Developmental Evolutionary Studies." In R. W. Mitchell, N. S. Thompson, and H. L. Miles, eds., *Anthropomorphism, Anecdotes, and Animals,* pp. 348–362. Albany: SUNY Press.

Pepper, S. C. 1942. *World Hypotheses: A Study in Evidence.* Berkeley: University of California Press.

Radner, D. 1994. "Heterophenomenology: Learning About the Birds and the Bees." *Journal of Philosophy* 91:389–403.

Richards, G. 2010. *Putting Psychology in Its Place: Critical Historical Perspectives.* 3d ed. London: Routledge.

Richards, R. J. 1987. *Darwin and the Emergence of Evolutionary Theories of Mind and Behavior.* Chicago: University of Chicago Press.

Rosaldo, M. 1980. *Knowledge and Passion: Ilongot Notions of Self and Social Life.* Cambridge: Cambridge University Press.

Sanders, C. R. 2000. "The Impact of Guide Dogs on the Identity of People with Visual Impairments." *Anthrozoös* 13:131–139.

Slobodchikoff, C. N., A. Paseka, and J. L. Verdolin. 2009. "Prairie Dog Alarm Calls Encode Labels About Predator Colors." *Animal Cognition* 12:435–439.

Smith, J. A. 2005. "Sensory Experience as Consciousness in Literary Representations of Animal Minds." In M. S. Pollock and C. Rainwater, eds., *Figuring Animals: Essays on Animal Images in Art, Literature, Philosophy, and Popular Culture,* pp. 231–246. New York: Palgrave Macmillan.

Steiner, G. 2008. *Animals and the Moral Community.* New York: Columbia University Press.

——. 2009. "Animal, Vegetable, Miserable." *New York Times,* November 22. Retrieved from http://www.nytimes.com/2009/11/22/opinion/22steiner.html?pagewanted=all.

Thompson, N. S. 1968. "Counting and Communication in Crows." *Communications in Behavioral Biology,* part A, 2:223–225.

Wolfe, C. 2010. *What Is Posthumanism?* Minneapolis: University of Minnesota Press.

Contributors

GALA ARGENT is an adjunct assistant professor in anthropology at Sacramento City College and in the animal studies program at Eastern Kentucky University. She is currently editing a multidisciplinary volume that examines, through time and across diverse societies, the manner in which humans and horses might be seen to construct mutually interdependent selves, identities, and cultural realities. She is a lifelong equestrienne.

ROB BODDICE is COFUND Fellow of the Languages of Emotion Excellence Cluster, Freie Universität, Berlin. He is author of *A History of Attitudes and Behaviours Toward Animals in Eighteenth- and Nineteenth-Century Britain: Anthropocentrism and the Emergence of Animals* (Edwin Mellen, 2009) and editor of a collection entitled *Anthropocentrism: Humans, Animals, Environments* (Brill, 2011). He is currently working on a history of compassion in the context of early Darwinism.

ALBERT BRAZ is an associate professor of comparative literature and English and the director of the comparative literature program at the University of Alberta. He is the author of *The False Traitor: Louis Riel in Canadian Culture* (University of Toronto Press, 2003) and is currently finishing a book on Grey Owl as writer and myth.

JANE C. DESMOND is professor of anthropology and affiliated faculty in gender and women's studies and in the unit for criticism and theory at the University of

Illinois at Champaign-Urbana. She is also director of the International Forum for U.S. Studies. A specialist in American studies, performance studies, tourism, public display, and embodiment, she has applied these interests to human-animal relations. Her first book is *Staging Tourism: Bodies on Display from Waikiki to Sea World* (University of Chicago Press, 1999); her current book project, "Displaying Death/Animating Life," explores animal issues in engineering, art, and mourning.

KAREN DAVIS is the founder and president of United Poultry Concerns, a nonprofit organization that promotes the compassionate treatment of domestic fowl and maintains a sanctuary for chickens in Virginia. Her articles have appeared in *Animals and Women: Feminist Theoretical Explorations, Terrorists or Freedom Fighters: Reflections on the Liberation of Animals, Critical Theory and Animal Liberation, Spring: A Journal of Archetype and Culture, Sister Species*, and the *Encyclopedia of Animals and Humans*. Her several books include *Prisoned Chickens, Poisoned Eggs: An Inside Look at the Modern Poultry Industry* (Book Publishing, 2009) and *More Than a Meal: The Turkey in History, Myth, Ritual, and Reality* (Lantern, 2001).

DAVID DILLARD-WRIGHT is assistant professor of philosophy at the University of South Carolina, Aiken. His book *Ark of the Possible: The Animal World in Merleau-Ponty* (Lexington, 2009) explores issues of mind and animal ethics. He has received research fellowships from the Animals and Society Institute and the Institute for Philosophy in Public Life. His articles can be found in *Society and Animals* and *Janus Head*.

PAULA DROEGE is a senior lecturer in philosophy at Pennsylvania State University, University Park, Pennsylvania. Her research on philosophical theories of consciousness proposes an essential role for temporal representation in conscious states. She is the author of *Caging the Beast: A Theory of Sensory Consciousness* (John Benjamins, 2003) and several articles on consciousness theory.

JOHN FRASER is a conservation psychologist, educator, and architect. He is currently president and CEO of New Knowledge Organization, an entrepreneurial think tank devoted to understanding how people develop knowledge and collaborate on solving society's significant challenges. His research has focused on understanding animal cognition, the expansion of the scope of justice to include wildlife, and the impact of social perspectives on degrees of concern afforded different species.

BENJAMIN L. HART is distinguished professor emeritus in the School of Veterinary Medicine, University of California, Davis. His work on animal behavior in species ranging from cats to elephants has led to more than 180 research publications. He has studied cognitive behavior of elephants in India and Africa with colleague Lynette Hart, which has led to scientific publications on the importance of the huge brain of elephants, particularly its relation to their cognition, long-term memory, and social-empathic behavior.

LYNETTE A. HART is professor of human-animal interactions and animal behavior at the University of California, Davis, School of Veterinary Medicine. In addition to research on large mammals with Benjamin Hart, she spearheads studies of psychosocial effects of companion animals for people. Assistance dogs and companion cats currently are a special emphasis, such as for children with autism or persons with other types of disabilities.

BRIAN M. LOWE is associate professor of sociology at the State University of New York, Oneonta. He is the author of *Emerging Moral Vocabularies: The Creation and Establishment of New Forms of Moral and Ethical Meanings* (Lexington, 2006), a contributor to *The Sociology of Morality Handbook* (Sage, 2010), and the author of several articles on topics such as civil religion and qualitative methodology. In 2008–2009 he served as chair of the Animals and Society section of the American Sociological Association. His current research involves the utilization of visual media by contemporary animal advocates.

ROBERT W. LURZ is professor of philosophy at Brooklyn College, CUNY. He has written a number of articles on philosophical issues related to animal consciousness and mindreading. His books include *The Philosophy of Animal Minds* (Cambridge, 2009) and *Mindreading Animals: The Debate Over What Animals Know About Other Minds* (MIT Press, 2011).

ROBERT W. MITCHELL is Foundation Professor in the Department of Psychology and coordinator of the animal studies program at Eastern Kentucky University. He is the editor of *Pretending and Imagination in Animals and Children* (Cambridge University Press, 2002), and the coeditor of *Spatial Cognition, Spatial Perception: Mapping the Self and Space* (Cambridge University Press, 2010), *The Mentalities of Gorillas and Orangutans: Comparative Perspectives* (Cambridge University Press, 1999), *Anthropomorphism, Anecdotes and Animals* (SUNY Press, 1997), and *Self-Awareness in Animals and Humans: Developmental Perspectives* (Cambridge University Press, 1994). His research publications focus on animal cognition and social cognition.

ALAIN MORIN is associate professor in the Department of Psychology at Mount Royal University in Alberta, Canada. His research concerns the cognitive basis of self-awareness, with an emphasis on inner speech. He has published extensively on the topics of self-recognition, the localization of the self in the brain, the split brain phenomenon, neurophilosophy, fame and self-destruction, and the antecedents of self-consciousness. His most recent work includes papers in *Social and Personality Psychology Compass*.

DIANA REISS is a cognitive psychologist and professor in the Department of Psychology at Hunter College and the biopsychology and behavioral neuroscience graduate program of CUNY. As a research scientist, she directs a program on dolphin cognition and communication at the National Aquarium and investigates elephant cognition at Smithsonian's National Zoo. She and her colleagues demonstrated that dolphins and elephants share the ability with humans for mirror self-recognition. Her work has been widely published in international scientific journals as well as popular media outlets. She served as scientific adviser to the 2009 documentary *The Cove.*

CARRIE ROHMAN is assistant professor of English at Lafayette College. Her research interests include animal studies, modernism, posthumanism, ecocriticism, and aesthetics. Her essays have appeared in such journals as *American Literature, Criticism,* and *Mosaic*. Her book, *Stalking the Subject: Modernism and the Animal* (Columbia University Press, 2009), examines the discourse of animality in modernist literature. She coedited *Virginia Woolf and the Natural World: Selected Papers of the Twentieth Annual International Conference on Virginia Woolf* (Clemson University Digital Press, 2011). She is currently writing about animality and aesthetics in twentieth-century literature, dance, and performance art.

JULIA SCHLOSSER is a Los Angeles–based artist, art historian, and educator. Her artwork and writing elucidate the multilayered relationships between people and their pets. She was a presenter at Animals and Aesthetics, Universität der Künste, Berlin, Germany, 2011. Recent photographs and videos were seen at the Seminário Internacional Arte e Natureza (International symposium on art and nature), São Paulo, Brazil, and Tierperspektiven (Animal perspectives), Souterrain Gallery, Berlin, Germany. Currently she is a lecturer at California State University, Northridge and California State University, Los Angeles, where she teaches the practice and history of photography and writing and critical theory in the arts.

JESSICA SICKLER is a senior research associate at the Institute for Learning Innovation. She has a background in museum education, research, and evaluation. Her research interests include positive youth development, public engagement with science, and public understanding of animal cognition. She has served as adjunct faculty at Bank Street College of Education and Seton Hall University and is currently on the board of directors of the Visitor Studies Association. Her publications have appeared in *Leisure Studies, Society and Animals, Curator: The Museum Journal, Visitor Studies, Journal of Interpretation Research, International Zoo Yearbook,* and *State of the Wild, 2008–2009.*

JULIE A. SMITH is associate professor emeritus of the Department of Languages and Literatures at the University of Wisconsin, Whitewater. Her animal-related academic articles focus on representations of animals' minds in literary works and understandings of rabbit minds within the rabbit rescue community. She is active in animal rights and rescue and she founded the Wisconsin chapter of the House Rabbit Society. She regularly writes for popular publications, including the *House Rabbit Journal.*

GARY STEINER is John Howard Harris Professor of Philosophy at Bucknell University. He is the author of *Descartes as a Moral Thinker: Christianity, Technology, Nihilism* (Prometheus/Humanity, 2004), *Anthropocentrism and Its Discontents: The Moral Status of Animals in the History of Western Philosophy* (University of Pittsburgh Press, 2005), *Animals and the Moral Community: Mental Life, Moral Status, and Kinship* (Columbia University Press, 2008), and *Animals and the Limits of Postmodernism* (Columbia University Press, 2012).

JESSICA ULLRICH is a member of the Faculty of Fine Arts in the Department of Art History and Aesthetics at the University of Arts in Berlin, Germany. She has curated fifteen art exhibitions on modern and contemporary art, including five on animals in art, and written the exhibition catalogs. She coedited *Ich, das Tier: Tiere als Persönlichkeiten in der Kulturgeschichte* (I, the animal: Animals as personalities in cultural history) (Reimer, 2008) and is the editor of *Tierstudien,* the first journal on animal studies in Germany. Her current research interest is human-animal relationships in art, and she is working on her second book on live animals and aesthetics.

SARA WALLER is an associate professor of philosophy at Montana State University. She researches consciousness and the intersection of philosophy, neurology, and cognitive science with an emphasis on animal minds. Her empirical research focuses on the conceptual structures and communication of social predators (coyotes and dolphins). She has published in *Synthese* and the *Journal of Cognitive Neuroscience.* Her edited collection is entitled *Serial Killers: Philosophy for Everyone* (Blackwell, 2010). She is currently focusing on the human fear of death and cultural notions of animality.

TRACI WARKENTIN is an assistant professor in the Department of Geography at Hunter College, City University of New York and a member of the advisory board for the CUNY Institute for Sustainable Cities. Her research interests include human-animal relationships, animal ethics, feminist environmental ethics and epistemologies, environmental and geographic education, and animal and cultural geographies.

Index